OIL PROPERTY VALUATION

OIL PROPERTY VALUATION

RICHARD V. HUGHES

Professor Emeritus of Petroleum Engineering
Colorado School of Mines
Golden, Colorado

ROBERT E. KRIEGER PUBLISHING COMPANY
HUNTINGTON, NEW YORK
1978

Original Edition 1967
Second Revised Edition 1978

Printed and Published by
ROBERT E. KRIEGER PUBLISHING CO., INC.
645 NEW YORK AVENUE
HUNTINGTON, NEW YORK 11743

Printed in the United States of America

Library of Congress Cataloging in Publication Data

Hughes, Richard V.
Oil property valuation.

Includes bibliographies and index.
1. Oil fields–Valuation. I. Title.
[TN871.H8 1978] 333.3'39 77-2945
ISBN 0-88275-402-5

PREFACE TO SECOND REVISED EDITION

This revised edition of Oil Property Valuation has the same primary purpose as the original edition, namely, to be used as a textbook in those petroleum engineering schools which have course work dealing with the fundamentals of the subject and related aspects of the economics of petroleum and natural gas production. It also is intended to update the subject as much as possible at the time of preparation, particularly from taxing viewpoints, to serve as a refresher for those who may have studied the subject years ago, and to introduce the basic principles of valuations to interested self-studying individuals.

This revised edition is literally divided into three parts, the same as in the original presentation. Subject matter, order, and method of presentation are essentially unchanged. This plan was an outgrowth originally of personal experience in oil property valuation work and years of teaching the subject. It is repeated herein in part also upon recommendations of former petroleum engineering and geological students and other professional acquaintances dealing with various aspects of the subject. Small prospective investors, legislators, and members of the legal profession also have found all the subject matter informative and useful. Statistical data pertinent to all topics and discussions are still intended to provide in part a basis for studying past, present and future trends within the industry, all of which have a direct bearing upon present and future values of oil properties. Extensive direct and indirect related references are provided again for those wishing to pursue the subject matter further.

Thanks are offered to all former students, associates, and other professionals who said they found the former edition interesting and useful, and to those technical and related petroleum societies, publishing organizations, petroleum and natural gas producers, and oil-loan banks used and referred to in the preparation of the present work. And especial thanks are due W. Waring, P. F. Drucker, and the former Douglass Evans Rankin.

Richard V. Hughes

Americus, Georgia
November 30, 1977

PREFACE

This book is intended primarily for use as a textbook by undergraduate students in petroleum engineering and petroleum geology curriculums whose work includes courses dealing with the valuation of oil properties and related aspects of petroleum economics. It also is intended to serve as a refresher, and perhaps in part as a self-study textbook, for those who may have studied the subject some years ago and now find themselves engaged in some of the many aspects of valuation work. Oil property valuation work is now much broader and more demanding than when Paine's classic book on the subject was published in 1941. A large number of references at the end of each chapter will aid the student and the reader in finding and developing more background knowledge on most every phase of valuation work.

The book is literally divided into three parts. Chapters 1 through 7 are essentially introductory and deal with the such subjects as the uses and needs for valuations, land subdivisions, kinds of oil properties, the many meanings of the world *value,* the various concepts of capital, the laws of interest and amortization, depreciation, depletion, and taxes. These subjects seldom are covered as broadly or from the same viewpoints in economics and other courses open to undergraduate engineers and geologists, yet they are fundamental to the proper understanding of all valuation work. A valuation study usually has for its primary purpose the determination of the *value* of the property. The student and one inexperienced in such work are surprised and sometimes confused to find so many meanings and shades of meanings for that common word. The value of a property depends primarily on gross income, expenses, and a going or desired rate of earnings. Net income after taxes depends in part on the depletion allowance, and considerable discussion is given to it in view of its justification, widespread misunderstanding, and legislative agitation for its reduction or elimination. One's attitude toward the depletion allowance depends in part on what one would include under capital.

Chapters 8 through 10 are intended to be an elementary discussion of that relatively new branch of petroleum engineering called reservoir engineering or reservoir mechanics. These chapters will be neither comprehensive nor particularly instructive to present students and recent graduates in petroleum engineering. They are intended to provide an elementary background to old-timers in the profession, geologists, and others who have thought the subject too difficult for self-study, as well as to indicate the need and value of the subject to everyone dealing with reservoir behavior.

Chapters 11 through 16, the balance of the book, are largely the "how-to" chapters dealing with the classification of properties, estimation of reserves, economics, calculation of present worth, and the valuation report. An actual present-worth calculation of an oil property is given in Chapter 15 using various discount factors. Industry uses many variations to the procedure as given in order to accomplish specific purposes. Except for the estimation of recoverable reserves and anticipated production rates, the present-worth calculation is simple arithmetic. Tables of the present worth of one dollar discounted at various rates and compounded at various frequences are given in the Appendix.

Many of my geological and petroleum engineering friends and colleagues have given advice and assistance in preparing this book, and to each I offer my thanks. Especial thanks go to Giuliano G. Verdina for preparing the originals of the illustrations, to David M. F. Parker for a part of the drafting and for critically reading parts of the manuscript, and to Raymond E. Cassidy for computing the present-worth tables in the Appendix.

Richard V. Hughes

Golden, Colorado
October 1966

CONTENTS

1 INTRODUCTION

The words *valuation, evaluation,* and *appraisal* often are used synonymously as nouns, adjectives, or in such verbal forms as to evaluate and to appraise. The up-to-date unabridged dictionary gives many meanings and many shades of meanings to each word.[1] Yet there would seem to be little difference between a valuation and an appraisal except that the latter under certain circumstances may be limited to valuations made by an authorized person. Valuations of oil properties for other than intracompany use are usually made by authorized persons such as registered petroleum engineers and geologists, yet it is to be questioned whether all such valuations could properly be called appraisals.

Some of the more important objectives of this book are to describe and discuss those subjects and matters that are of major concern in arriving at acceptable valuations of oil properties. The majority of oil property valuations deal almost entirely with properties that have been proven by the drilling of one or more wells to have commercially productive accumulations of oil. Some valuations include properties that have only a promise of commercial accumulations; and very few valuations deal only with nonproved properties. The topics covered are those steps to be followed by finding the value or worth of an oil property. Certain background material dealing with pertinent historical developments, economics, and business law are included in order to provide a complete understanding of the overall subject.

One who prepares oil property valuations usually is spoken of as an oil property valuation engineer, irrespective of his professional training and background. He utilizes principles of oil reservoir engineering to predict future production rates and future ultimate recoveries. He estimates development and exploitation costs, taxes, depreciation, and depletion allowances, and in consideration of economic conditions calculates future incomes, cash flow, and values for a multitude of purposes.

Many oil property valuation engineers prefer to use the words *valuation, evaluation,* and *to evaluate*, rather than *appraisal* and *to appraise*. They evaluate an oil property irrespective of the purpose of the valuation. Despite the similarities of the definitions given in the dictionary, they think of the words *appraisal* and *to appraise* as signifying the actual observation of physical things in order to establish either a value or a loss. The appraiser enumerates the numbers of things, either separately or in related groups, and establishes individual values that may be totaled to obtain the whole. Actually the oil property valuation engineer seldom visits or sees the property under study. It is rare for him to be concerned with the establishment of a loss. He cannot count directly the unit volumes of oil that give value to the property. He usually must rely on data furnished by his clients or associates within his company, and then provide estimates based on experience and mathematical methods of calculation.

THE ENGINEERING VALUATION

The basic principles of the engineering method of oil property valuations originated in the mining industry. Yet the use of engineering methods in mining valuations is scarcely as old as the oil industry. An engineering mine valuation usually consists of two related parts:

1. The examination and sampling of the property in order to establish the the tonnages or ores and their assay values, development costs, and operating costs.
2. The mathematical calculations based on the examination in order to find the dollar value of the property at a specific time.

The principles of the mathematical calculations were called to the attention of the mining industry in 1877 by Hoskold.[2] He stressed the basic concept that the operation of a mine deals with the depletion of a capital asset, presupposed uniform earnings, uniform interest return on the invested capital at high speculative rates, and provision for reinvestment of the balance of the yearly earnings at a safe rate of interest. Important modifications were suggested by O'Donahue in 1906,[3] by Morkill in 1918,[4] and Grimes and Craigue in 1928.[5] O'Donahue objected to the use of speculative rates of interest because they resulted in absurdly low present-worth values when the operating life of the property would be long. Morkill disagreed with charging interest on the whole of the original capital during the entire operating life. Grimes and Craigue suggested a complicated use of three interest rates. Because of the similarity of mining and oil field operations, in that both are dealing with a wasting asset, it is not surprising to find a chapter or two dealing with oil property valuation in present-day texts on mine valuation.

Today's engineering valuation of a producing oil property is based on Hoskold's premises to the extent that we are dealing with a wasting asset, that a fair profit should be made, and that all the capital should be returned by the time the operation is abandoned. Such valuations consist almost entirely of a series of estimates dealing with the following:

1. The total amount of oil, gas, and liquid petroleum gases that can be produced economically.
2. Annual rates of production.
3. Deductible charges for development and operating expenses, depreciation, depletion, (if allowable), taxes, and other costs.
4. Net annual income.
5. Present-worth value of the property on a given date and under certain specified conditions.

The term *oil property* is used here in its broadest sense to include any property with either proven or semiproven underground accumulations of liquid and/or gaseous hydrocarbons that might be produced at a profit. Thus the engineering valuation of a productive gas property would consist of a series of similar estimates. Strictly speaking, properties that have not been proven economically productive should not be classed as oil properties, but values often must be assigned to them. Obviously, the foregoing engineering methods are not applicable and simpler methods must suffice.

Oil property valuations are concerned with dollar values, which basically are measures of prospective profits. Prospective profits come from the development and exploitation of oil and natural gas accumulations. Properties lacking potential earning power cannot be evaluated by engineering methods; there must be a profit motive. Drake's backers undoubtedly had the profit motive in mind when they decided to drill for oil near Titusville, Pennsylvania. Our modern petroleum industry would have been impossible had it not been for the continued plowing back of a large part of the earned profits. And it is the profit incentive that creates today's demand for oil property valuations in this country. Inflation and ever higher exploration, development and production costs, special tax situations, and overall economic conditions create the need for more and more valuations for various purposes and involving properties of ever higher dollar values. For these reasons oil property valuations are almost entirely limited to properties owned and operated under governments with free enterprise systems.

Valuation studies of foreign properties being taken over or confiscated and all "nationally-owned and operated" oil properties are hardly within the scope of this work. Rarely are compensations for such takeovers based upon established valuation principles, even in part. They are based largely upon depreciated values of capital investments in the drilling, well-completion, production, operating, maintenance, marketing, and other surface facilities. In general, all hydrocarbon reserves are considered as belonging to the nation until brought to the surface. It is the reserves of the hydrocarbons still in the ground that give most of, and sometimes all, the exchange value to any oil property under the free enterprise system. But many of the basic principles with respect to the estimation of recoverable reserves, production rates, income trends, and cash flow are everywhere applicable at all times to obtain some idea of the economics of the operation.

EARLY SALES OF OIL PROPERTIES

Hoskold's principles of mine valuations could not be applied successfully to oil property valuations until more than fifty years after the birth of the industry. This was due to a lack of sound methods for estimating oil reserves and future rates of recovery. As a result the oil industry was in a quandary for many years after passage of the first federal income tax law in 1913. The early oil industry had realized that, like the mining industry, it was dealing with a wasting asset, a natural resource that is exhausted in the process of extraction regardless of the prudence and care exercised in producing it. Economists today like to refer to oil, gas, and similar natural resources as "fund resources."

The oil industry has never been able to block out its oil assets before exploitation, as had the mining industry. It was not until the beginning of income taxes that the oil industry paid particular attention to reserves in the ground. Certainly the very early producers did not worry about them for they had no control over underground reserves. If production rates declined too much in one area, operations could be moved to some new flush area. There was no orderly development of fields, no sound production practices, and no sound abandonment practices for unwanted wells.

Many interesting yet true stories of the early operations around Titusville, Pennsylvania, have been recorded. The following examples undoubtedly occurred before the industry even had time to develop "rule-of-thumb" methods of oil property valuation. They clearly demonstrate why crude oil soon became known as "black gold."

J.W. Sherman of Cleveland bought a small oil lease near Titusville in 1861. A well drilled by the spring-pole method was completed in March 1862 with an initial production of 1500 bbl per day. The well averaged 900 bbl per day for two years and netted Sherman an income of $1.7 million.[6]

The 200-acre James Tarr farm near Titusville was sold for $60,000 in 1861. It was located near other producing oil properties. A short time later production was proved on it and the farm was resold for $2 million.[7]

Messrs. Noble and Delemater, merchants in Titusville, had a hole drilled by spring-pole methods to a depth of 134 ft in 1860. It turned out to be dry. It was deepened to 452 ft in the spring of 1863 and was completed with an initial oil production rate of 2500 to 3000 bbl per day, selling at the time for $4 per bbl. The well produced more than 1.5 million bbl, which sold at times for as much as $13 per bbl and netted the owners more then $5 million. This well is said to have been the most profitable well in history to both owners and land proprietors.[8, 9]

Some Pittsburghers, under the name of Ritchie, Hartjio, and Company, bought the 500-acre Storey farm near Titusville in 1859 for $30,000. The Columbia Oil Company of Pittsburgh was organized in May 1861 with a capitalization of $250,000 and with Andrew Carnegie as one of the major stockholders. The company paid Ritchie, Hartjio, and Company $128,000 for the farm and drilled a number of good wells. Dividends were declared as follows:

July 1863	30 per cent
August 1863	25 per cent
September 1863	25 per cent
October 1863	50 per cent
January–June 1864	160 per cent

The company was recapitalized at least once during the next several years. The original stockholders recovered their original investment forty-three times in cash and still had ten times their first stock.[10, 11]

These and other accounts of early-day profits in the oil industry in many parts of the world exemplified the speculative nature of the industry. The oil man simply drilled wells that literally made or lost fortunes for him overnight. He was a gambler, gambling not only in the eyes of the public, but also in the eyes of his associates in the industry. All any oil man needed to be a success was some money to invest, a willingness to take big risks, and luck. Purchasers of properties, drilled or undeveloped, had no sound concepts of the actual values of properties. Some purchasers were naturally lucky and made huge profits; others, perhaps the majority, were happy not to have lost everything.

CONDITIONS BEFORE 1930

The really prolific wells during the early days of the oil industry were located along the Allegheny River, or close by in the flat valley lands of its tributaries in Pennsylvania. Properties lying in the same valleys and close to prolific producing wells were most in demand and brought the highest prices.

Buyers and sellers judged prices more on the size of the wells or, if the property was not drilled or entirely developed, proximity to the streams and other producing wells. A property on a hillside, even if near good producing properties located on the valley floor, had little or no potential oil value until approximately 1865. Later the most valuable prospective properties were located, regardless of topography, on and along extensions of established "oil belts."

Early promoters and operators soon learned to think in terms of daily oil production per well, daily oil production per acre, and recovery per well and per acre. These factors soon became acceptable units for use in arriving at values of oil properties. Probably because of their simplicity, these factors are still used to make sales and trades of nearly exhausted properties in many of the older fields.

Naturally the market price of crude oil in the stock tank always has been a very important factor in determining the value of any oil property. Market price of crude oil always has been very sensitive to the laws of supply and demand. Supplies in general had been adequate, if not excessive, at the wellhead. Shortages had occurred almost entirely at points of consumption and were due primarily to a lack of, or failures in, transportation facilities. This was particularly true during the early days of the industry when all oil had to be shipped down the Allegheny River in barrels. It was particularly true again during 1942-1943 among the Atlantic coastal and New England states when so many tankers were sunk by German submarines. Before 1932 countrywide production rates were governed almost entirely by the amount of exploration effort and the numbers of development wells drilled. Production rates at the wellhead rarely were restricted and there were no restrictions on the spacing of wells.

There was considerable worry during the "twenties" that the Nation might soon run out of oil, primarily because of the very rapid increases in demands for gasoline. Since the very early days of the industry, the greatest increases in demands for domestic production occurred during that period. Crude oil production had to be essentially doubled. Demands for automotive gasoline nearly quadrupled during those years meaning that the percentage of gasoline refined from a barrel of crude oil was nearly doubled also.

A broader and better understanding of subsurface geology, the development and crude application of geophysical prospecting methods, and improvements in rotary drilling equipment and practices were beginning to open up wider and new prospecting areas and deeper prospective horizons. Production rates at the wellhead rarely were curtailed and there were no restrictions on locations and spacing of wells on one's property. A landmark court decision referred to as the *Barnard Case* of 1907 recognized ownership of all production from one's property.[12] It was immediately and heartily accepted and practiced and led to the production abuses of the early 1930's under what was called the "Rule of Capture." Exploratory, drilling, and production costs per barrel of produced oil were quite low. On the whole, natural gas production was of no value. Taxes were low to non-existent except for a short period during and immediately following the Civil War. Needlessly close well-spacing and very inefficient production practices offset to some extent those low production costs. The promise of a quick payout for a well or a property was the chief consideration in making oil property purchases. This was true not only with respect to undeveloped properties and flowing wells, but also with respect to pumping wells. It took the industry many years to learn that pumping wells might be equally profitable in comparison with flowing wells, despite the longer payout time. Industry also learned that few fields ever became totally exhausted. The Bradford field in Pennsylvania is the outstanding example of many of these older fields. It was drilled twice for primary recovery, first during 1871-1880, again during 1890-1900, and beginning in 1922 for what became known as secondary oil recovery. Engineers estimate that little more than 50 percent of the original oil content of the Bradford field will will have recovered at the termination of present-day secondary recovery operations. A field-wide fourth era based upon some method of "tertiary recovery" is anticipated by progressive operators.

The petroleum industry had shown very little interest prior to 1930 in any scientific aspects of oil production. However, the U.S. Bureau of Mines and a few in the industry had published papers on the underground functions of natural gas in oil production operations and its stupendous waste at the time from many viewpoints. A federal board pointed out in a 1929 report titled *The Conservation of Gas* that unitized operation of oil-fields was probably the only solution for controlled waste.[13]

CONDITIONS SINCE 1930

The discovery of the East Texas field in 1930 marked an important turning point in the producing branch of the oil industry. The producing branch has become thoroughly "conservation-minded" since that time. The attitude of the oil industry shortly after discovery of the East Texas field was expressed as follows in a publication of the American Petroleum Institute:[14]

. . . the oil industry has changed completely its original concept of producing oil, which was founded on drill-as-you-please and produce-as-much-as-can methods. Viewed in the light of the limited knowledge and experience originally available to the industry, the early concept was a natural one, and the early adjustments by the courts of the problems of property rights involving ownership of oil and gas were reasonable at the time. The studies of engineers, geologists, and scientists, and the ensuing application of their knowledge and experience in producing oil pools, have caused a marked change in those old concepts. Studies of the native expulsive forces present in all oil-producing reservoirs have led to, and are continuing to lead to, more efficient producing practices based on more specific knowledge of these forces and the part they play in producing oil.

The art of finding and producing oil had progressed to such an extent by 1930, when the great East Texas oil pool was discovered, that the country was confronted with such an avalanche of overproduction of oil that comprehensive regulation clearly became necessary. Such regulation was effected under the conservation statutes of the various oil-producing states, for the facts disclosed that much waste was taking place, aboveground and underground. This was the end of the drill-and-produce-as-you-please period, and inaugurated the era of conservation.

The Board of Directors of the American Petroleum Institute issued a declaration of policy for the oil industry in 1930, which stated in part:[15]

. . . that it endorses, and believes that the industry endorses, the principle that each owner of the surface is entitled only to his equitable and ratable share of the recoverable oil and gas energy in the common pool in the proportion which the recoverable reserves underlying his land bear to the recoverable reserves in the pool.

This new idea has been called the recognition of the principle of correlative rights. It indicates the desire for fair play and the equitable adjustment of property rights that are inherent in the philosophy of a free and democratic people, . . .

Despite the suggested policies of the American Petroleum Institute and early state and federal attempts at regulation, the East Texas oilfield remained inadequately controlled for several years. There had been strong disagreement at all times as to whether abuses should be controlled by the federal government or by the individual producing states. Passage by Congress of the "Connally Hot Oil Act" in 1935 was very helpful in that it prohibited movement in interstate and foreign commerce of oil produced, or withdrawn from storage, in violation of state statutes or regulations. However agreement was reached in 1935 upon formation of the "Interstate Compact to Conserve Oil and Gas" and the setting up of the "Interstate Oil Compact Commission" which was quickly approved by Congressional resolution. The Commission had no regulatory authority, but its influence through widespread cooperative committee work and multitudinous publications and public meetings helped immeasurably in the adoption and continuation of true conservation practices in all the oil-producing states. Only six states originally approved it, namely:

Texas, Oklahoma, Kansas, Colorado, New Mexico, and Illinois. Over the years nearly all oil-producing states have become members or associates and have so influenced in adopting and enforcing wise conservation practices and regulatory statutes.

There is no doubt that this radical change in the philosophy of oil production has worked very successfully for the industry, the government, and the consuming public. On the whole the producing branch of the industry is now highly regulated in every oil-producing state. The industry actually has requested many of the regulations. State permits have to be obtained to drill, complete, recomplete, and abandon wells. Well spacings with respect to acreage, minimum distances between wells, and the distances from property lines are regulated. Production rates of wells have been regulated according to the national demand and to produce maximum recoverable oil with minimum waste of reservoir energy. Up to the mid-seventies production rates in some states were so restricted that many wells were allowed to be produced at only a fraction of their maximum efficient rate. Open-flow conditions are no longer permitted in most states except for short testing periods, or for a time under a "discovery allowable" permit that a portion of the cost of the well may be returned quickly. Wells must be produced without undue waste of natural gas and original reservoir energy. State regulatory bodies have shut in oil fields to prevent the production of dissolved and associated natural gas where it could not be sold or returned to the producing sands for various reasons. The purchase of oils by the refineries has become more selective since it has been realized that crude oils of the same gravity may be widely different in quality and as sources of specific refined products. Today the higher gravity crudes do not necessarily bring the higher prices. Costs of exploration, drilling and development, production operations, transportation, and storage all have increased alarmingly. Many companies have found it to be much cheaper to buy producing properties than to try to explore for and to develop their own. The markedly increased costs in all branches of the industry followed the same pattern as all business in general after World War II. Payout time for producing wells, properties, and fields in most of the states is no longer a matter of three or four years. Periods of ten, twenty, and even thirty years are common. Long periods for payout combined with ever rising costs and narrower profit margins forced even the larger companies to turn to outside sources for capital to develop and to expand operations sufficiently for domestic needs. Some turned to foreign operations where capacities to produce could be developed much cheaper than domestically. As an end result our Nation's financial centers with their large banks, investment houses, and insurance companies have loaned hundreds of millions of dollars to the industry to help it meet its capital needs.

Despite all the regulations since 1930 the oil industry always more than met consumer's demands for both oil and new products at very reasonable prices. The production of oil rapidly developed into a science. Production methods and recovery efficiencies have completely superseded the old "produce-as-much-as-you-can" practices. Prior to 1974 there had been no great fluctuations in crude prices for approximately 40 years and inflationary trends had not caused product prices to be increased to nearly the same extent as with other widely used commodities. The stability resulting from the practice of true conservation principles by the production branch of the domestic industry aided the valuation engineer in his quest for reliability in valuation work. The domestic search for oil has become ever more intensive and competitive despite the productive capacities of the nation's producing fields and the continually narrowing margin of profits. The narrower the margin of profits and the more intensive the search, the more careful and accurate the valuation engineer must be in his work. He must conscientiously and scientifically consider every factor, large and small, that may determine the income and profit to be derived from operation of the property under study.

Estimations of total recoveries and future production rates of oil and gas from a property and prospective market prices at the wellhead were once the only major uncertainties in valuation calculations. Several other uncertainties were added during the mid-1970's, largely as a result of changes in former policies of the Federal Power Commission with respect to pricing of natural gas and Congressional discriminatory tax measures with respect to the major oil companies. Those factors combined with unknown future costs of environmental controls, offshore drilling and leasing policies, increased coal supplies, Congressional threats of forced divestiture within the major companies, and Middle East production and pricing policies were all added uncertainties effecting future income from any property.

Because of all these factors which are subject to sudden change, the specific date of every oil property valuation must be decided upon once the specific purpose has been determined. In fact, no oil property valuation should be started until both the specific date and the specific purpose have been determined. These are largely the responsibilities of the employer or client. It is also advisable to establish at the same time the scope of the valuation and the probable time necessary to prepare an acceptable report. Failure to heed these preliminaries immeasurably weakens the valuation study.

THE VALUATION ENGINEER

The oil property valuation engineer cannot turn out a first-class valuation report unless he has been properly trained, has had wide experience in the industry, and is at all times ethical with both his employer and his profession. Actual educational background is not necessarily the most important qualification, provided other required abilities have been acquired. From practical and theoretical viewpoints the petroleum engineer with a fundamental knowledge of petroleum reservoir engineering, geology, mathemtaics, and economics should make the best oil property valuation engineer. Many primarily trained in geology, however, have acquired the other requirements and have become reputable oil property valuation engineers. Some would limit the most important qualification to an economics-engineering viewpoint. The oil property valuation engineer must be a man of many talents.

Certain members of the petroleum engineering sections in our larger oil companies usually are assigned the valuation work. In the smaller companies and in the consulting field, either geologists or petroleum engineers usually prepare valuations depending on individual experience and availability. In the larger companies, those actually preparing the valuations may be assisted by such others as the accounting, tax, geological, transportation, and legal departments. The oil property valuation engineer, however, is normally responsible for the entire study, the report, and its accuracy.

The experienced oil property valuation engineer knows that the value he determines for a property under study is not necessarily unequivocal. This is true even though the most applicable mathematical methods may have been used in combination with adequate and reliable data. Often an employer or a client finds this difficult to understand, even though he may have held back certain pertinent data and information as irrelevant or too confidential. It behooves the oil property valuation engineer and his employer or client to establish complete confidence in each other at the start, as is expected in lawyer-client and doctor-patient relationships. No so-called "bugger factors" should be used in preparing the engineering valuation. The employer or client should be told, however, of data believed to be in error, and if factual data cannot be supplied, he should be informed of the approximate magnitude of error occasioned by its use or omission.

The oil property valuation consultant will save himself much trouble, and perhaps his reputation, if he determines before accepting a valuation assignment that there will be no conflicts of interest with other commitments. Naturally the oil property valuation engineer can maintain his professional status only by being completely ethical at all times. He would do well to follow the advice given by Taylor[16] to the professional engineering witness. Taylor admonishes him to remember the concluding paragraph of "Faith of the Engineer," which states the following:

> To my fellows I pledge, in the same full measure I ask of them, integrity and fair dealings, tolerance and respect, and devotion to the standards and dignity of our profession: with the consciousness, always, that our special expertness

carries with it the obligation to serve humanity with complete sincerity.

Within a company, and especially within a large company, it may or may not be important that the one responsible for making valuation studies be a licensed engineer. In most states the consulting petroleum engineer is required to be registered as a professional engineer and currently licensed to practice his profession. In some states the same is true with respect to geological engineers. Any valuation report or legal testimony with respect to valuations would carry little or no weight before a court or a regulatory body if the opposing lawyer cared to discredit the one responsible for it because he was not a registered engineer. In all cases in which points of law, state, or federal, are involved all supporting documents and reports usually must be signed by a currently registered engineer.

Thus it behooves every engineer who has any contemplation of going into consulting valuation work to become a registered engineer duly licensed to practice in the state in which his office or most of his work may be located. McCawley[17] has prepared a book and periodic supplements which list the requirements of the various states. All requirements for registration are intended to protect the public and anything concerning the public interest and safety from unscrupulous persons and inferior services. Engineers hope that registration eventually will raise the status of the profession to something comparable with that enjoyed by the legal and medical professions. Registration and licensing imply that the registrant is ethical and reasonably competent.

REFERENCES

1. *Webster's Third New International Dictionary of the English Language*. unabridged, G. and C. Merriam Co., Springfield, Mass., 1961.
2. Roland D. Parks, *Examination and Valuation of Mineral Property*, third ed., Addison-Wesley, Cambridge, Mass., 1949, pp. 155-190.
3. *Ibid.*, pp. 345-346.
4. *Ibid.*, p. 350.
5. *Ibid.*, p. 353.
6. John J. McLaurin, *Sketches in Crude Oil*, publ. by the author, Harrisburg, Pa., 1896, p. 117.
7. *Ibid.*, p. 136.
8. *Ibid.*, pp. 113-114.
9. Paul Giddens, *Pennsylvania Petroleum, 1750-1872*, Pennsylvania Historical and Museum Commission, Titusville, Pa., 1947, pp. 235-239.
10. McLaurin, *op. cit.*, p. 133.
11. Giddens, *op. cit.*, pp. 251-252.
12. *History of Petroleum Engineering*, American Petroleum Institute, New York, 1961, pp. 1118-1123.

13. *Ibid.*, p. 1127.
14. *Progress Report on Standards of Allocation of Oil Production within Pools and among Pools*, American Petroleum Institute, Dallas, Tex., 1942.
15. *Ibid.*
16. R. W. Taylor, "The Petroleum Engineer as a Professional Witness," *J. Petr. Tech.*, August 1960, p. 39.
17. Alfred L. McCawley, *Professional Engineering Registration Laws*, Alfred L. McCawley Trustee Publication Fund, Jefferson City, Mo., 1954, and periodic supplements.

Additional References

History of the Petroleum Industry, American Petroleum Institute, New York, 1961.
Max W. Ball, Douglass Ball, and Dan S. Turner, *This Fascinating Oil Business*, The Bobbs-Merrill Company, Inc., Indianapolis, Ind., 1965, 464 pp.
T. Murray Robinson, "The Practicing Attorney Gives Some Tips on Courtroom Guidance for Young P.E.'s," *Oil and Gas J.*, Jan. 7, 1957, pp. 117-121.
Harold F. Williamson and Arnold R. Daum, *The American Petroleum Industry–The Age of Illumination, 1859-1899,* Northwestern University Press, Evanston, Ill., 1959.

2 PURPOSES OF OIL PROPERTY VALUATIONS

Early-day oil property valuations, such as they were, had only one purpose. That purpose was to determine what a property might sell for. It was the responsibility of the purchaser to think in terms of the actual worth of the property under operating conditions. Present-day oil property valuations had their inception with the passage of the first federal income tax law in 1913. Between 1913 and 1926 income tax calculations stimulated the need for and the development of acceptable methods of determining recoverable reserves and future rates of production of oil properties. Naturally some of the procedures developed during those years were used to determine the worths of properties for sales and trades. Since 1926 oil property valuations have been made for a number of added important purposes. The particular methods to be used and the form and scope of the valuation study depend primarily on the intended purpose. The ethical engineer will understand clearly the purpose of the valuation before starting the study, clearly state the purpose as he understands it in the beginning of his report, and clearly show in his conclusions that the purpose was consistently carried out. Some oil property valuations may consist of little more than estimates of reserves, prospective yearly production rates, gross or net cash flows before or after taxes, and include no data with respect to profits and values. Other more complete studies may include prospective profits, gross or net, payouts, and total cash values at specified times under various development and operating conditions.

From both theoretical and practical viewpoints the value of an oil property should be a measure of the profitability of its operation. The meanings of value have become complex, and as a result some oil property valuations likewise have become complex, not only with respect to methods of preparation but also to purpose. Almost all present-day oil property valuations, however, are prepared with one or more of the following broad purposes in mind:

1. Determining security values for loans.

2. Consummating sales, trades, and mergers of properties.
3. Aiding management in formulating policies and operations.
4. Guiding the investor.
5. Complying with governmental rules and regulations.

DETERMINING SECURITY VALUES

Oil properties very rarely were considered as acceptable security for loans by banks or other lending institutions before the early 1930's. Yet today a high percentage of valuations are prepared for purposes of obtaining a loan of cash or credit. An owner of an oil property may make or have made some form of valuation to be presented to a bank or other financial institution to warrant its value as security.[1, 2, 3, 4] The lending institution may or may not accept the study without considerable revision. Much time may be saved and many doubts eliminated if such studies are prepared by experienced valuation engineers who are well known to all parties concerned. Obviously the studies should be of recent date and complete enough to meet the particular requirements. Often short initial and approximate valuations may serve to create interest of the lending institution because "oil loans" have become "big business." Large banks and insurance companies that cater to oil loans are usually well staffed with competent engineers and geologists who can either check an operator's valuation or prepare an entirely new one. As a general rule a complete valuation may not be needed for loan purposes because the lending institution is not always interested in the present worth of the property. However, it is always concerned with the certainty that the principal and interest payments can be paid when due. Thus, it is primarily concerned with the production rates and the net cash balances that may be generated from that production each month during the period of the loan. It is not interested in having to foreclose on properties and tries to avoid taking any such chances. Money lending is its business, not the oil producing business. The small independents and the major oil companies alike depend on the banks and other lending institutions for much of the cash used for expansion purposes. Expansion may range from drilling of wildcat and development wells to the buying of other properties. So-called "banker's loans" have been profitable to both the lending institutions and the industry.

CONSUMMATING SALES, TRADES, AND MERGERS

Every progressive oil company, regardless of its size, seems increasingly eager to acquire properties that will enhance its competitive position in the industry. Its competitive position may be enhanced by such factors as increased productive capacity, either for sales or to meet the demands of its refineries, higher recoverable reserves, a higher percentage of domestic to foreign income, and increased retail outlets in poorly served and new sales territories. Industry has found trades, purchasers, and mergers to be the quickest, easiest, most certain, and cheapest way to accomplish such benefits.

Some small independent producers are always ready to sell at a profit if they can take advantage of the "capital gains" tax. A quick net profit on a property through a sale is sometimes greater than the present worth of the property when operated under high income tax payments. Others who hold attractive producing properties and reserves of oil may be forced to sell for reasons which may vary from the need to settle an estate to difficulties in maintaining sufficient operating funds. They may have overexpended on borrowed money or drilled too many wildcats and dry holes. Or state allowables may have severely restricted anticipated production income and increased payout time considerably. The primary trend is for the larger companies to acquire the smaller companies, although sales of properties in the 100 to 350 million dollar range are unexpectedly frequent. The largest transaction may involve a multicompany purchase and more trading of stocks than cash may be involved in the payment. So many large sales occurred in 1961 that the editor of one petroleum journal[5] referred to them as a "rash of sell-outs by oil producers." Sales and trades are much easier to agree upon than are mergers. Mergers, especially of larger companies, are delayed and halted by the Justice Department more frequently than are sales and trades. The Justice Department is constantly on the alert to prevent any transaction, be it sale, trade, or merger, that could lead to a company having what the Department considers to be a monopolistic position in industry.

Complete and detailed valuations of oil properties entering into sales, trades, and mergers usually are made by both sellers and buyers. Occasionally an outside consulting firm is called in to prepare an independent and disinterested valuation. These engineering valuations seldom represent the true worth of the properties to all parties concerned in the transaction. Large transactions may include refineries, natural gasoline plants, filling stations, and other assets that require physical appraisals to be made. In all cases the anticipated profitability of the transaction must be integrated with that of the existing business.

Sometimes oil property valuations are made for dissolution purposes. Partnerships are dissolved more often than are corporations. A valuation for this purpose usually is made by a disinterested party and may consist of little more than a check and split of "book values" with little attention being paid to the actual worth that might result through continued operation.

AIDING MANAGEMENT

More or less complete valuation studies of particular properties are needed by management to arrive at certain policy-making and operational decisions.[6, 7] These valuations may be made of properties not for sale at the moment, yet they may be needed on rather short notice. The studies may call attention to the advisability of a sale. At times valuations are made of adjoining properties not being offered for sale. These occurrences may relate to offset operators who, through permissible means, may be producing oil or gas from beyond the boundaries of their properties or may be jeopardizing overall field recoveries. Some producers may make their operations as offensive as possible so that the adjoining owner may be willing to pay more than a fair price. Purchasers of certain properties may be the only way to include acreages held by obstinate owners in areas undergoing unitization proceedings. Various kinds of profitability studies are needed at times to compare income with costs for those operated properties approaching their economic limits.

Companies sometimes make more or less complete valuations as a routine part of operating studies. Every expenditure should pay out at some reasonable time or at least should contribute to the profit picture. For instance, the company may want to find what the added profits might be by following an infill drilling program on some property, by practicing secondary recovery operations on another, by joining some unitization program, by delayed exploitation of a large gas cap, or by making multiple completions in a multizone field.

Company valuations of some properties may result in the decision to farm out certain areas, to abandon certain leases, or to put back on production certain closed-in wells. In these and similar cases a payout study is usually made which follows established valuation practices in whole or in part. Valuations are often made in connection with the pooling of interests in unitization projects. Such studies may or may not be complete valuations depending upon the needs of the participating companies and regulatory agencies. The need sometimes becomes necessary when a small operator holds out because of being pressed to accept what he believes is inadequate participation in an enforced project.

Valuation studies for purposes of bidding on leases is perhaps a misnomer because valuations in the strict sense of the word rarely are made for this specific purpose. Perhaps they should be referred to as economic studies. The leases under bid usually are limited to lands, waterways, and water bottoms under the jurisdiction of certain branches of government. They may be either prospective or producing areas in the continental United States or more often offshore. Prior to the early 1970's many United States companies were operating world-wide in large exploration and production areas under basic terms usually similar to those of domestic operations. By 1975 many of those foreign operations had been relinquished to the host nations long prior to expiration of their concession contracts. Some are still operated under minority-ownership agreements, some under various forms of profit sharing, and others on a fixed fee basis.

When bids, usually called bonus payments, are required as in acquiring federal onshore and offshore leases, a company has to resort to a lot of background information and the combined judgments of its geological, geophysical, engineering, transportation, marketing, and economic advisers in order to arrive at realistic bonus offers. Often such offers are little more than guesses that may be increased or decreased prior to submitting depending upon how much the company wants the acreage and what it thinks some other might bid. A top bid of $1200 per acre offered at the first federal lease sale of tracts offshore from Louisiana was thought to be surprisingly high at the time.[8] Bonuses in the tens – to near forty – thousands of dollars per acre were being offered during the late 1960's and early 1970's for unproven acreages in waters up to 1,000 ft. in depth and 100 miles from shore in the Gulf Coast area.[9] Some of the highest bids offered were for acreages in the Prudhoe Bay of Alaska where drilling costs are extremely high. As prohibitive as these offers seemed at the time on an acreage basis, the bonus cost of a tract was thought to represent only a fraction of the overall investment needed for development should it prove productive and before any income might accure.

The industry was beginning to hedge in its bidding by the mid-seventies on account of suspected inadequacies of future net incomes and potential borrowing limits to maintain a healthy industry. Much greater care was necessary in scrutinizing the probable benefits of all capital expenditures. Federal acceptance of royalty bids in addition to the bonus on some offshore tracts in 1974 was intended to be an inducement to smaller companies, but the policy was hailed by the majors also.

Management's quandries are well illustrated by one company's desire to purchase another's properties which were to be sold to the highest bidder. Acceptable estimates of proven reserves as of a given future date were furnished all interested bidders. The desirous management made 110 "computer-estimates" using various rates of production, production costs, and future prices of crude oil and natural gas. The property was sold to another before the management could decide to eliminate all the lowest calculations.

A small operator sometimes obtains a valuation, or at least some idea of the value of a property that he owns, by pretending that it is for sale. He asks one or more companies to study the property and the records of it and to make him their best offer. If more than one company make offers, he feels assured that the property is worth much more than the offered prices. He may have wanted to apply for a loan originally and needed such information to bolster his confidence in applying for one.

GUIDING THE INVESTOR

The oil property valuation engineer rarely plays more than a small and indirect part in studies intended to guide the investing public. He may contribute a great deal of data pertaining to prospective production rates, recoverable reserves, drilling, development, and operating costs, but he seldom participates in the final preparation of the brochure. This is particularly true if he is a company man, because disinterested engineers are considered to be entirely independent of the objectives of the company. Such brochures usually are prepared by experts who know the kind of information needed and the proper methods of presentation. Certain governmental watchdogs must be satisfied that all information is factual and that no laws are being violated, yet the brochure may be made up in ways that will attract investor interest.

Such brochures may be prepared and used whenever the company wishes to raise additional capital for expenses and/or operations through the sale of preferred or common stocks. Similar abbreviated brochures may be prepared for stockholder's information to be used at meetings called to vote on "stock splits."

Factural information explaining present policies, operations, and profitability, certain major policies to be carried out during the next fixcal year, and prospective per-share earnings are presented periodically by high officials of many of our major oil companies to boards of security analysts. Some of the data presented are based on and may include information prepared by the valuation engineer, although the accounting department plays the major role in presenting the overall profit picture.

COMPLYING WITH GOVERNMENT RULES

Before 1913 the producer of oil and gas had little if any contact with governmental matters and no contact with regulatory bodies. Now the average company finds itself spending a great deal of its time preparing various types of reports and valuations required in connection with tax obligations, suits, and the various rules and regulations under which it operates. During the period 1913 to 1926 the industry had to learn rather suddenly how to calculate reserves of oil and gas, the costs of properties, and production costs for income tax purposes. That part of the studying and reporting was made largely unnecessary in 1926 on passage of the standard depletion allowance which could be used for income tax purposes. Nevertheless, changes in regulations, particularly with respect to accounting procedures, repeal of the depletion allowance late in 1975 for the larger companies and eventual disallowance for all, combined with unrealistic pricing regulations for both crude oil and natural gas have all added tremendous burdens upon the industry. Congress showed an almost complete lack of understanding of the economics of the industry in that its intent was to prevent a reoccurence of the so-called "windfall profits" of 1973-1974. These sudden and unexpected attitudes and Congressional threats of forced divestiture within the major companies have made older valuation studies obsolete and have added many large and puzzling uncertainties to the best of today's valuation.

Valuations made for purposes of settlement of inheritance, estate, and gift taxes are seldom of concern to the larger oil company. These studies usually are prepared by consultants under legal guidance. The specific date of such valuations becomes of great importance because of developments and later information which the valuation engineer must ignore in presenting facts and viewpoints as they existed at the specified time.

The industry is making an ever increasing number of simple to elaborate valuations dealing with production costs of natural gas under advisement by the Federal Power Commission. Most of these needs are an outgrowth of the greatly increased demands for natural gas and requests by the Commission for costs at the wellhead. Interstate marketers must maintain satisfactory ratios of reserves to demand either through outright ownership of gas reserves or through long-time purchase contracts. Prices paid for gas at the wellhead are largely under the control of the Commission, but the market prices paid by the consumer usually are controlled by state or local agencies. For example, prices for intrastate purchasers in Texas during the mid-seventies were approximately fourfold those allowed by the Commission for interstate sales. Producers were allowed to make only modest profits, if any on interstate gas sales, which discouraged exploration for new and added supplies. An immense amount of effort on the part of the producing companies has gone into studies aimed at showing the replacement cost of natural gas being sold. One of the industry's most difficult problems has been to prove the actual cost of gas being produced by oil wells. Valuation and cost studies of nonassociated natural gas are not too difficult, but estimates of costs of producing associated gas can become very controversial. The problem is complicated by the fact that not many years ago most of the gas produced in this country was flared because of a lack of pipelines to transport it to prospective consuming centers.

REFERENCES

1. C. R. Dodson, "Facts the Banker Needs in Making an Oil Loan," *World Oil*, Feb. 1, 1958, pp. 32-34.
2. C. R. Dodson, "Application of the Petroleum Engineer's Report to Financing," *J. Petr. Tech.*, Feb. 1967, pp. 187-192.
3. Lyon F. Terry and Kenneth E. Hill, "Valuation of Producing Properties for Loan Purposes," *J. Petr. Tech.*, July 1953, Sec. 1, pp. 23-26.
4. J. J. Arps and J. L. Arps, "Prudent Risk Taking," *J. Petr. Tech.*, July 1974, pp. 711-716.
5. Gene T. Kinney, "What's Behind the Rash of Sellouts by Oil Producers?" *Oil and Gas J.*, Jan. 22, 1962, pp. 30-32.
6. C. E. Reistle, Jr., "Application of Engineering Valuations in Management," *J. Petr. Tech.*, June 1956, pp. 11-12.
7. John M. Houchin, "Management's Use of Petroleum Engineering Evaluations," *J. Petr. Tech.*, July 1958, pp. 11.
8. "First Offering of Continental Shelf Leases Brings High Bonuses," *World Oil*, November 1954, p. 86.
9. "Eastern Gulf Sale Nears $1.5 Billion," *Oil and Gas J.*, December 24, 1973, p. 19.

Additional References

J. J. Arps, "Profitability of Capital Expenditures for Development Drilling and Producing Property Appraisal," *J. Petr. Tech.*, July, 1958, p. 13.

L. B. Davidson, "Investment Evaluation under Conditions of Inflation," *J. Petr. Tech.*, October 1975, pp. 1183-1189.

Granville Dutton, "The Effects of the 1969 Tax Reform Act on Petroleum Property Values," *J. Petr. Tech.*, December 1970, pp. 1475-1479.

Melvin Kaitz, "Percentage Gain on Investment – An Investment Decision Yardstick," *J. Petr. Tech.*, May 1967, pp. 679-687.

Charles E. Phillips, "The Application of Equity Conecpt and its Relationship to Multiple Rates of Return," *J. Petr. Tech.*, February 1965, pp. 159-163.

Douglass H. Powell, "Evaluation of Oil Properties in After-Tax Dollars," *J. Petr. Tech.*, January 1960, pp. 28-36.

R. S. Wansbrough, "An Approach to the Evaluation of Oil-Production Capital Investment Risks," *J. Petr. Tech.*, September 1960, pp. 25-29.

Wallace, W. Wilson, "Oil and Gas Property Acquisition," *J. Petr. Tech.*, June 1961, pp. 527-530.

3 FORMS OF OIL PROPERTIES

A valuation of an oil company for outright sale, purchase, or merger would include many kinds of assets. A large part of such a complete valuation would be outside the realm of petroleum engineers and geologists. Their valuation work normally has to do only with "oil properties," properties that may or may not be productive on the specific date of the valuation. Occasionally equipment and facilities located on or related to the producing oil properties may have to be included, but these usually will amount to only a minor part of the complete oil property valuation.

Every oil property may be classified in one or more ways depending on the form of title or ownership, its operational status, and its quality. Oil properties take many forms because of the mineral laws of the country in which they are located. On the other hand, a classification with respect to operational status can be rather simple because it is applicable to all forms of oil properties. The same can be said with respect to a classification based on quality.

OIL PROPERTIES ARE LANDS

Property has been defined as "the right and interest which a man has in lands and chattels to the exclusion of others."[1] The terms "oil property" and "mineral property" have been used for decades by both industry and government. The Internal Revenue Service did not define oil and mineral properties until 1954 despite the imposition of taxes on incomes from them continuously since 1913. That definition simply stated:[2]

> . . . the term "property" means each separate interest owned by the taxpayer in each mineral deposit in each separate tract or parcel of land.

The word "land" in the foregoing definition is legally interpreted in a very broad sense with respect to oil properties to include all those continental areas wholly or in part of water bodies such as tide-lands, marshlands, stream and river bottoms, lakes and bayous. Seaward areas under jurisdiction of federal government are usually referred to as offshore tracts, -blocks, or -properties. All must be described in ways that will facilitate their being located exactly geographically and for recording the locations clearly in public archives and company files. Land descriptions are records of surveys and/or descriptions of land boundaries and areas. Boundary surveys and descriptions must start from an actual point or position on the ground which may be unequivocally found or relocated. Each nation has its own systems of land surveys. For many years lands in the United States and Canada, excluding urban areas, were usually measured in rods (poles) and chains of 16½-ft and 66-ft length, respectively. The statutory unit of area is still the acre of 160 sq rods. Today's surveys are usually made in feet using a 100-ft chain (tape). Most other nations in the Western Hemisphere, and particularly the French- and Spanish-speaking nations, use the meter (3.280 ft) as the unit of length in lands surveys and the hectare (2.471 acres) as the unit of area. All measurements and areas with respect to land surveys must be taken in a horizontal plane and so recorded irrespective of topographic relief. Thus an acre of land on a very steep hillside may contain much more than one acre of surface area.

Copies of deeds and transfers of mineral rights with respect to continental properties must be recorded in the county seat or parish in which the property is located. These records normally include a description of the boundaries of the property, the areas involved, and prior ownerships. The public may inspect, use, and copy these records without restraint.

U. S. SYSTEMS OF LAND SURVEYS

Several systems of lands surveys and descriptions are used in the United States. Each stems from the history of the region and the time of the original conveyance to private ownership. The principal systems encountered in oil property valuations are the following:

1. Metes and bounds.
2. U. S. Public Land Surveys.
3. Coordinate systems.

Surveys by Metes and Bounds

The metes-and-bounds method of surveys and lands descriptions is an inheritance from England. Surveys were made with the pole and compass when no simpler method would suffice. Some old properties were described without surveying simply by naming the adjoining land owners. Preferable, however, were natural boundaries and markers such as crests of ridges, banks or center lines of streams, edges of cliffs, and large rocks and boulders. Trees, stumps, edges of fields and forests, fences, and even stakes were used when necessary. Conscientious recorders made an effort to write descriptions of boundaries that would be accurate and unequivocal for long periods of time. Yet a stranger might find it impossible to locate the property or its boundaries at any time without local aid. The 66-ft surveyor's chain with 100 links of equal length and the compass were used to survey, mark, and describe properties more accurately. A large proportion of old properties have been resurveyed in feet and redescribed where older descriptions were inaccurate or particularly vague.

Old descriptions of oil properties in New York, Ohio, Pennsylvania, West Virginia, and Kentucky might range from those difficult to locate and trace on the ground to those based on resurveys with chain and compass or transit. A typical old description of an oil property in Kentucky read as follows:

> Beginning at a large boulder in the northerly line of Stoney Creek road and marking the S. W. corner of lands belonging to Chas. Wilson, and running westerly along said line to the crest of the ridge, thence northerly along said crest to a stone wall bounding the lands of John Houghton, thence easterly along said wall to the center of Stoney Creek, thence following the thread of said creek downstream to said northerly line of said road, and thence along said northerly line to said boulder at the point of beginning, said parcel of land containing a total of 205 acres, more or less.

It would be impossible to plot these boundaries on a blank sheet of paper and to check the acreage unless a survey with chain and compass or transit was made.

"Land grants" are included under the metes-and-bounds method of surveying because the boundaries of most of them were not surveyed originally and had to be so described. Natural physical features were used for landmarks and boundaries when feasible and occasionally a distance with compass bearing might be used if necessary. As a rule the distances and bearings when used were scaled from some crude regional map. As a result the boundary descriptions of lands grants generally are vague and very inadequate.

Three sources of land grants in the United States are of interest in oil property valuations. They are those originally granted by Spain, Mexico, and Texas and to some extent those granted by France. Very few original grants remain intact and industry is interested in them only when property titles must be searched back that far. Most of the colonial grants were given to faithful soldiers and officers of the Revolutionary War by their home states in lieu of back pay. They were located west of the Allegheny Mountains, principally in Kentucky. The size of the grant depended on the individual's rank, the location of the grant, and the amount of the indebtedness. Owners seldom went out to claim or to check their grants. Troubles were started largely by purchasers and remote legal heirs. Resurveys could seldom be made without encountering large overlaps, omissions, and readjustments.

The old Spanish grants of interest to the industry are limited to Texas, New Mexico, Arizona, and California. Most of them were very large originally and had to be described rather vaguely. Natural features served as boundaries where possible. Sometimes straight line boundaries were needed which were given a vague direction and a distance in "leagues" which never may have been measured. Old Mexican grants were probably as numerous as old Spanish grants in Texas, but old Mexican and Texas land sales were far more numerous. The abstracting of titles of oil properties in parts of Texas has been severely handicapped by original overlaps of grants and land sales of which the majority of the latter were fraudulent. While an independent nation Texas was noted for its "private land grabbers."

Old French lands grants were numerous in Coastal Louisiana but on the whole have been subdivided into small tracts. They are not as troublesome to trace as those in Texas.

U. S. Public Land Surveys

Lands owned by either the state or federal government are known as *public lands*. Lands owned by the federal government are of two types: *public domain lands and acquired lands*. Public domain lands are those obtained through treaties or purchases coincident with the expansion of our national boundaries. Acquired lands are those obtained by purchase, condemnation, or gift from previous owners. Indian lands are not federal lands as such but are administered the same as federal lands with respect to mineral deposits.

"The United States System of Surveying the Public Lands" was inaugurated in 1784. Although approved by all the thirteen states, it did not affect them because each had retained title to all its public lands, which in many cases extended far beyond the present-day boundaries of those states. The basic principles of the surveys system are still followed. They provide that federal public lands[3]

> shall be divided by north and south lines run according to the true meridian, and by others crossing them at right angles so as to form townships six miles square, . . . Also, that the townships shall be subdivided into thirty-six sections, each of which shall contain six hundred forty acres, as nearly as may be, by a system of two sets of parallel lines, one governed by true meridians and the other by parallels of latitude, the latter intersecting the former at right angles, at intervals of a mile.

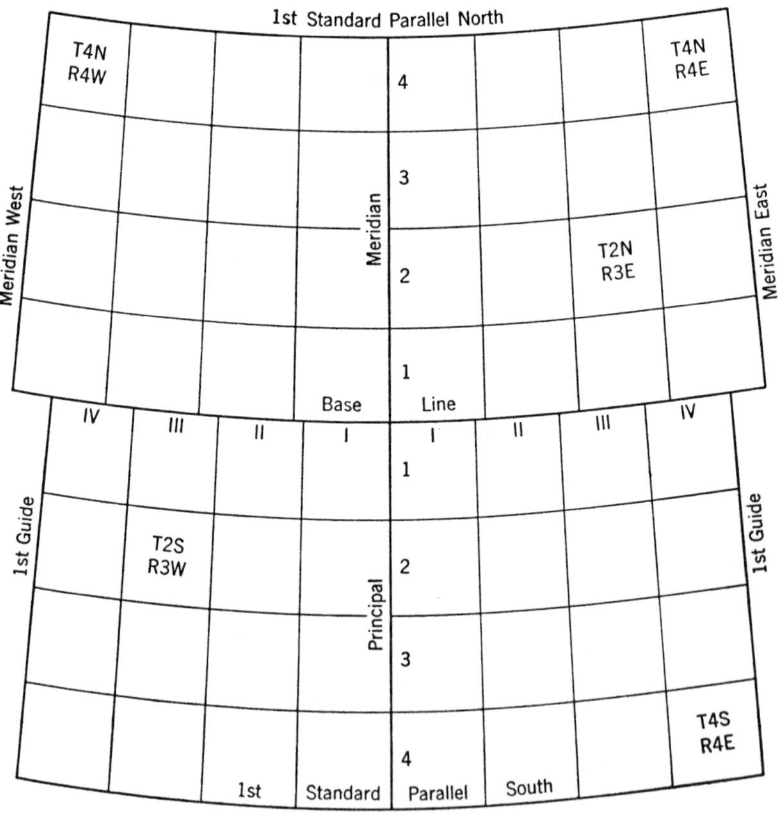

Figure 3-1. Initial division of United States lands by standard parallels and guide meridians into 24-mile blocks and townships.

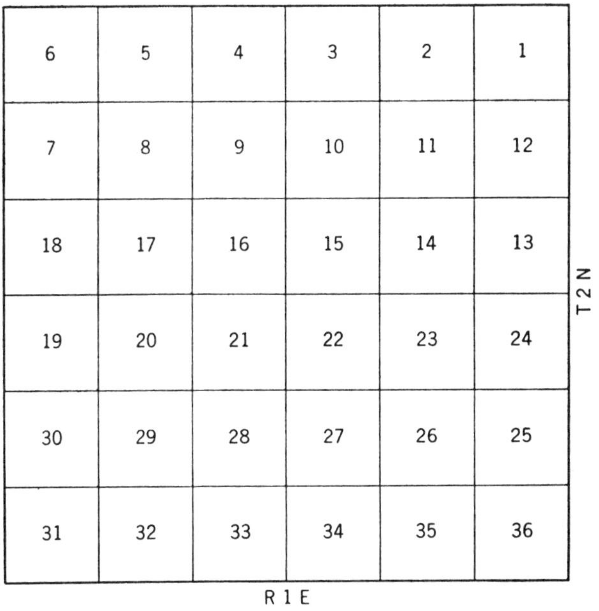

6	5	4	3	2	1
7	8	9	10	11	12
18	17	16	15	14	13
19	20	21	22	23	24
30	29	28	27	26	25
31	32	33	34	35	36

T 2 N

R 1 E

Figure 3-2. The township and method of numbering the sections.

In practice, principal base lines and meridians are established astronomically. Standard parallels and guide meridians are run every 24 miles from the principal base lines and meridians making squares of 16 townships each. The convergence of the meridians necessitates corrections along each standard parallel. The standard parallels are often referred to as correction lines for this reason. Townships are numbered east or west from the principal meridian and north or south from the base lines. The division of the 24-mile square blocks into townships, their designation, and corrections along the standard parallels are shown in Figure 3-1.

The township can never be a full six miles wide at its north end. Thus all 36 sections in a township cannot be exactly 640 acres. The subdivision of the township must start at its southeast corner with section 36 and proceed from east to west, a north-south tier of sections at a time. The sections are numbered from the northeast corner of the township as shown in Figure 3-2.

The system provides for throwing all error due to convergence into the northwesterly half-mile of the township and errors due to the curvature of the parallels into the most northerly half-mile strip. As a result the northwest quarter of the northwest section of a township can be considerably undersize, particularly where the guide meridians and the standard parallels cross each other. The division of a section and the designation of its parts are shown in Figure 3-3. Additional original surveying errors occurred in most areas because of topographic difficulties, methods actually employed, carelessness, and incompetence. The description, location, acreage, and ownership of every tract of land included in an oil property valuation should be checked.

Departures from the original system of subdividing public lands have been authorized many times. For instance, an Act of Congress in 1811 permitted surveyors in Louisiana to divide public lands bounding on rivers, lakes, creeks, bayous, and other water courses into tracts 58 poles in front along the water boundary and 465 poles in depth.[4] This was only a modification of an earlier system established by the French to give every property owner a waterfront for transportation and communication. Subsequent acts permitted smaller

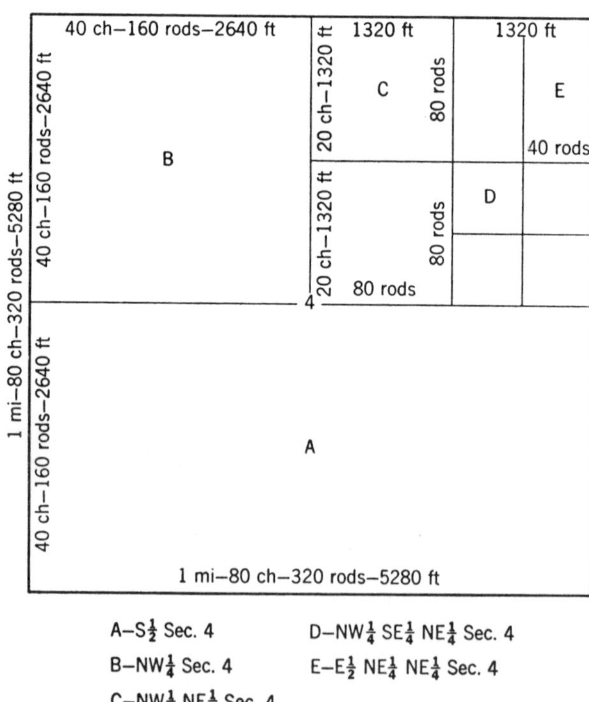

A–S½ Sec. 4
B–NW¼ Sec. 4
C–NW¼ NE¼ Sec. 4
D–NW¼ SE¼ NE¼ Sec. 4
E–E½ NE¼ NE¼ Sec. 4

Figure 3-3. The subdivisions of the section.

and narrower tracts to be laid out. Most of those early tracts have been subdivided many times through inheritances, with the result that today the original tracts may be scarcely more than slivers of land ownership radiating back from present and past water boundaries. They present a picturesque sight along the lower Mississippi River and adjoining coastal region when viewed from an airplane, which belies the many problems created for the exploratory and producing branches of the industry.

Spain granted titles to an estimated 10 to 20 million acreas of lands between 1731 and 1819 in what is now the state of Texas.[5] On gaining its independence from Spain, Mexico began to colonize Texas. Its colonization law of 1825 granted one square league of land of 4251 acreas to each settler who was the head of a family and a stock raiser by occupation. He received only one labor (177 acres) of land if farming was the only occupation. If he both farmed and raised livestock, he received 4428 acres. A single person received only one fourth of these amounts.[6] Another law in 1834 provided for the cancellation of the law of 1825 and called for the survey of vacant lands into lots of one labor each, which were to sold to the highest bidder. Land grants made by Spain and Mexico, and recognized by the independent government of Texas, amounted to approximately 26.4 million acres.[7]

Upon gaining independence from Mexico in 1836, Texas in turn began to colonize its national lands. A system of public land surveys modeled after that of the United States was introduced. The system of townships of 36 sections, the 640-acre section, and its subdivisions was to be used. Most of the early Spanish and Mexican land grants had been laid out either to include the larger streams and rivers or to provide extensive boundaries along and parallel to them for irrigation, stock watering, and transportation purposes. Shapes of tracts back from the streams were determined to considerable extent by topography, purpose, and usefulness of the land. For a time very little attention was paid to the official system of surveying. It was very difficult to join it to older land boundaries, and it made no provision for land use, land values, or topography. Very little land east of the high plains area was surveyed in 640-acre section.

Texas retained jurisdiction and ownership to all its vacant and unappropriated lands when it became a state in 1846. No U. S. General Land Office was established in the state with the result that for many years surveying practices with respect to sizes of sections, numbering, orientation, and subdivision was official from locality to locality yet quite diverse. Extensive areas were surveyed in the high plains areas to be sold to provide revenue, and to provide lavish bonuses to the newly building railroads and other internal improvement companies. The Texas Congress in 1839 allotted three leagues of land in each county for the endowment of public schools. The allotment was increased to four leagues in 1840. In 1850 four leagues of land were allotted to schools in each county that had been created during the previous eleven years. In all, approximately 4.2 million acres of land were so allotted to the public school system, nearly all of which had been sold by 1928. Also, in 1839 approximately 221,400 acres of public lands were set aside for two state colleges or universities. Later grants raised the total to over 2 million acres, most of which was in the western part of the state. A large proportion of these lands have become valuable oil properties.

Checking of boundaries, areas, and titles of oil properties in Texas by local qualified surveyors and title attorneys is very important with respect to all valuation studies.

Official modifications of the U. S. system have been authorized from time to time for mineral areas and mountainous terrain in California, Nevada, and other western states. Oil properties are not common in these areas.

Coordinate Systems

The coordinate system has been defined as the method that[8]

> . . . designates the determination of the location of points on the surface of the earth by surveying methods which take account of the size of the earth and the curvature of its surface, and the description of points thus located by stating their positions with reference to some system of lines of known geographic position.

The position of a point determined by this method and so described is known as a "coordinate position," and the data that show its relationship to the system of lines are called the coordinates of the point. Coordinates may be given in either geographic or rectangular units. Geographic coordinates are designated in degrees, minutes, and seconds of latitude and longitude. The latitude is measured north of the equator and the longitude is measured west of Greenwich, England. Rectangular coordinates are given in some common unit of measurement, such as feet, either east or west and north or south from some well-known point whose latitude and longitude are usually known or established.

The United States Coast and Geodetic Survey has determined the geographic coordinates of hundreds of points throughout the United States and along our state and national boundaries. These points usually are determined accurately by triangulation survey methods and serve as starting and/or check points for boundary and other land surveys. The method is particularly adaptable for locating the position of inaccessible points accurately in high, rough, and mountainous areas, over inland waters, and long irregular coastlines.

Points so located along the coastlines of the United States have been especially useful to the oil industry in conducting seismographic surveys, in locating lease corners of offshore tracts and wells being drilled within a tract. Several electronic triangulation survey systems such as Loran, Shoran, and Lorac have been utilized for these purposes.

California, Texas, and Louisiana started to lease offshore areas to the industry long before the federal government questioned ownerships. These early properties were laid out to fit the indicated geophysical structures by prospective interested bidders. Sometimes a number of small tracts were provided to cover a structure, and at other times a very large tract covering one or more structures might be laid out when only one company was particularly interested in an area. Literally several thousands of offshore blocks, tracts, or other somewhat regularly-shaped areas have been established worldwide on all continental shelf areas of all nations where prospective sedimentary deposits extend seaward. Some areas may cover waters 1,000 ft. or more in depth, particularly off the Gulf Coast states. Only some system of geographical coordinates can be used to locate the boundaries of such tracts. For instance, all blocks or tracts off the coast of Louisiana[9] are based on the Louisian (Lambert) coordinate system, whereby geodetic coordinates have been changed to rectangular coordinates. Offshore boundary lines between parishes run geodetic south by legislative act. All blocks or tracts are bounded then by north-south lines and by east-west lines. No state lease shall contain more than 5000 acres by legislative act. Inasmuch as no boundary lines could be surveyed as such, the problem in establishing the corners was to calculate the rectangular coordinate position of the corners by taking into account the convergence of the north-south lines and the curvature of the east-west lines. The Lambert system was particularly suited for this purpose. A reference point was chosen in the Gulf of Mexico at 91° 20′ longitude and 28° 40′ latitude. This would mean that all foreseeable coorindate positions of corners would have positive X and Y values. Once such a grid is laid out on a map showing all the discovered salt domes, it is relatively easy for a company to decide on blocks it wishes to lease when put up for bidding. The location of any well on a block can be established accurately by one of the electronic means of triangulation.

Texas has laid out a similar system along its coast except that the boundaries are more or less at right angles and parallel to the shorelines. Boundaries of individual tracts or blocks also must be located by the coordinate system.

Corners of leases offshore from California and wells drilled thereon must be located and described by similar systems. All offshore California wells within the foreseeable future will be drilled within sight of the coastline and can be located by direct triangulation. In such cases coordinates do not have to be figured unless desired because one point can be located with respect to a known point by means of a bearing distance.

FORMS OF OIL PROPERTIES IN THE UNITED STATES

Valuations of oil properties deal with specific lands yet rarely include lands that are actually owned by any oil company. Valuations of oil properties are most often merely valuations of rights and interest in rights—not valuations of actual land ownerships.

The exploratory efforts and production operations of the industry extend into almost every state of the union and cover approximately 400 million acres of land. The industry could not afford to search for needed supplies of today or of the future without some convenient and economical way to establish rights to prospective productive properties.

The mineral rights on these millions of acres of properties must be held under strict control for many years for exploratory and productive purposes. Some properties may be relinquished after a time with no exploratory drilling. The economic life of a producing property may last for several decades; for instance, oil properties in Pennsylvania have been producing in commercial quantities since the early 1860's. The industry holds its properties in or as an assignment derived from one or more of the following three basic forms.

1. Ownership in fee.
2. Ownership of surface rights.
3. Ownership of mineral rights.

The earliest instrument granting anyone rights to develop a source of crude oil for commercial purposes is believed to have been a profit-sharing agreement covering land on which Drake later drilled the first oil well. It is purported to have stated the following:[10]

> Agreed this fourth day of July, 1853, with J. D. Angier, of Cherrytree township, in the county of Venango, Pa., that he shall repair up and keep in order, the old oil spring on land in said Cherrytree township, or dig and make new springs, and the expenses to be deducted out of the proceeds of the oil, and the balance, if any, to be equally divided, the one-half to J. D. Angier and the other half to Brewer, Watson & Co., for the full term of five years from this date. If profitable.
>
> (Signed) Brewer, Watson & Co.
> J. D. Angier

Angier's venture into the oil business was short-lived. The property was sold later with a transfer of mineral rights by the new owners to the company that financed the drilling of the Drake well.

Ownership in Fee

Ownership in fee means that the oil company or operator owns both the mineral rights and the surface in a tract of land through one or more deeded conveyances. The farm on which Angier was to work the oil springs was sold to George H. Bissell and Jonathan G. Eveleth in November 1854 for purposes of organizing the Pennsylvania Rock Oil Company of New York. Presumably this was the first ownership in fee of an oil property where purchase was made primarily for oil production purposes. There undoubtedly were many properties purchased in fee immediately following the drilling of the Drake well, but most purchasers of properties carried the promise of a share of the production to the seller. Eveleth and Bissell spent approximately $200,000 in a relatively short time after the discovery of oil in leasing activities and land purchases. Ownerships in fee are still prevalent in the older Appalachian areas, and especially in the fields of Pennsylvania. Ownerships in fee also are met with in most of the older productive areas of Indiana, Kentucky, Ohio, and Oklahoma. Large ownerships in fee are found in parts of California and in some of the other western oil-producing states. These properties consist largely of early acquisitions through or from the pioneer railroads, which had received from our federal government large grants of land on each side of their rights-of-way.

Colonel Thomas A. Scott, an associate of Andrew Carnegie, upon learning of the hugh profits made on the Storey farm in Pennsylvania, bought 277,000 acres of lands in 1863-1864 in what is now Ventura County, California, and some 12,000 acres in Los Angeles and Humbolt Counties.[11] Ever fewer purchases in fee of prospective oil lands were made after the first hectic years of the infant industry. Fee purchases tied up large sums of money at times unnecessarily. Ever wider spacing of wells mean proportionately less occupation and damage of the surface, thus agriculatural or other uses of land could be continued with less interference. All land owners soon learned that they too might share in the riches of oil by retaining ownership of the minerals, and if no oil was found they still had the "old home place." Today an oil company rarely desires or needs to purchase a prospective oil property in fee.

Ownership of Surface Rights

An oil company has little concern with the surface ownership of oil properties except to the extent required for drilling, building roads, and the installation of surface production equipment. A company desiring surface rights at the time of procuring a property would obtain the ownership in fee. The industry is not in a position to practice agricultural pursuits profitably.

Neither can it afford to hold the surface for speculation. As a result surface rights are purchased more often after production has been established than before. These purchased surface rights are usually for lands on which the company has the mineral rights. Surface rights may be purchased to eliminate a nuisance, such as continued unjust claims for surface damages, to obtain sites for tank batteries, storage of materials, field camps, compressor stations, pumping plants, water disposal plants and wells, and extraction plants. The total acreages involved are small in comparison with the producing acreages. Obviously a company could neither drill wells nor legally produce oil from lands to which it held surface rights only.

In general, the costs of surface rights to an oil company are at above-normal prices. Nevertheless, the value of surface rights is ordinarily of little significance in the complete valuation. It is usually included in the appraised value of the facility constructed on the land.

Ownership of Mineral Rights

Originally, there was only one kind of mineral ownership, namely sovereign ownership. Now the term "national ownership" or "federal ownership" is usually used to describe mineral ownerships belonging to the nation. A few nations, the United States and Canada being the outstanding ones, have ceded some original mineral ownerships to lesser political subdivisions within the nation. Canada and the United States, in particular, ceded all mineral rights to the first private owners of former public lands. Private owners could cede their mineral rights to others. Consequently, there are now two basic kinds of mineral ownerships held by the individual. One is the outright ownership whereby title to one or minerals, originally and still in place, is conveyed through a deed. The other is merely ownership of the right to extract one or more minerals from the land. The extractor, i.e., the producer of crude oil and natural gas in the petroleum industry becomes owner only of that share of the production initially agreed upon. This concept is adhered to worldwide and has been resisted by foreign operators upon nationalization of their producing properties.

Any part or all the mineral rights may be sold, traded, deeded, and held separately from the surface ownership in a tract of land in all states except Louisiana. Civil and common law in that state had its origins under both Spanish and French rule and follows much of the principles and text of the Code Napoleon. Only a "servitude," which is the right to procure the minerals and not a ownership, can be assigned in Louisiana and it must be exercised within ten years.

The fact that ownership of minerals, and particularly rights to extract minerals, could be separated from surface ownership of the land, either in whole or in part, has been a major factor in the growth and success of the industry. Transfer of mineral rights, exclusive of conveyance by deed, is usually accomplished through the *lease*, a contractual arrangement between an owner of the minerals, the "lessor" who may or may not own the surface of the land, and a second party, the "lessee," who will extract the minerals with mutual benefits to both. Either an owner of the minerals or of rights to extract the minerals (the lessee) has the right of ingress and egress on the property and cannot be prevented from carrying out all necessary drilling and production operations, subject of course to payments for crop or other damages and in some cases rentals for use of lands beyond the well.

The Lease

Permission must be obtained from the landowner prior to conducting seismograph surveys and at times to do geological field work on a property. Permission usually is granted for seismograph work through a shooting permit or "shooting lease" upon payment of an agreed fee. The shooting lease may include an option to lease the mineral rights on all or part of the property, depending somewhat on whether the landowner is the owner of the minerals at the time, and the amount of the payment. The shooting lease is not to be confused with the "oil and gas lease," usually referred to in industry simply as the lease. The Drake well was drilled on land held by the Seneca Oil Company under lease from the Pennsylvania Rock Oil Company. Dissension within the Pennsylvania Rock Oil Company, which had purchased the property from Brewer, Watson and Company in 1854, led the majority stockholders to organize the Seneca Oil Company in 1858. The Pennsylvania Rock Oil Company had leased the property to Drake and Bowditch for a period of fifteen years. The Seneca Oil Company took over the lease from Drake and Bowditch, hired Drake as agent, and financed the drilling of the well.

The lease as understood and used in the industry is very different from the customary lease drawn between landlord and tenant. The lease must be a written document, must have the signatures of all parties notarized, and must be officially recorded.

The form of the present-day lease has been the result of a slow process of improvement. The earliest leases were short and simple. They developed into longer and longer and more complex documents where every imaginable contingency was included. Both the lessor and the lessee realized that the binding inclusions were no better that the intent of the signatories. As a rule today both lessor and lessee prefer to use one of the standard printed short lease forms knowing that such provides the same treatment for all and that the courts have ruled on the interpretation of all its parts.

In all oil property valuation work it is very important that the lease be examined to see if any out-of-the-ordinary clauses occur that would be detrimental to the value of the property. At the same time the valuation engineer must make sure that the lessee has not assigned or lost any part of the rights called for in the lease.

The lease serves as both a contract and a conveyance. There is an offer on the part of the lessee and an acceptance by the owner of the minerals or of the mineral rights, the lessor, in order that the agreement be legal, and a conveyance of an undivided interest in the mineral estate. According to most of the decisions of the courts, the lessor retains actual ownership of the minerals in place at all times and theoretically relinquishes ownership of the lessee's share only upon extraction and delivery on the surface. The lessee is giventhe exclusive right to explore for, to drill for, to develop, to produce, and to remove his share of the production from the property. In the early days of the industry oil was the sole salable product; today the product includes oil, natural gas, natural gas liquids, condensates, and if included in the lease agreement such associated minerals as salt, sulfur and potash. The product usually must be delivered at the surface free of all costs and expense to the lessor. For these reasons the lessee is said to hold the *operating or working interest* in the property.

As of December 31, 1975, as estimated total of 800×10^3 private, municipal, and state oil and gas leases covering 400×10^6 acres were in force in the United States.[12] In addition 127,504 Federal, Indian, and Outer Continental Shelf leases covering over 102×10^6 acres were in effect.[13]

Most leases permit the lessee to assign all or part of his rights to another for a valuable consideration, in which case the assignee legally assumes all the obligations of the original lessee. Occasionally the right to assign leasehold interests is withheld by the lessor or made subject to his approval. Or at times too many assignees may become involved in a property. In either case the value of the property is impaired to a prospective purchaser or to a bank when reviewing the property as collateral for a loan.

The lessee, who may not necessarily become the operator of a property, usually agrees to pay to the lessor an *initial bonus, rental*, and a *royalty*. These three obligations became standard in leasing operations in the early days of the industry.

The Bonus. The bonus is a cash payment by the lessee to the lessor as an inducement to grant the lease. The bonus is as old as the industry. Its amount depends largely on the location of the property, the prospects of developing commercial production, and the bargaining ability of both lessor and lessee when dealing with privately owned properties. Publicly owned properties and those supervised or controlled by governmental agencies are usually put up at auction on a sealed bid basis.

Bonus payments are considered by the Internal Revenue Service for tax purposes as a capital cost to the lessee and as income to the lessor. Onshore bonuses of $1 per acre for private lands in newer regions have been prevalent for many years, particularly where blocking-in large prospective areas prior to wildcatting. Bonuses on acreages not far from producing wells may range from $10 per acre up. Acreages in or adjoining good producing areas may command hundreds to thousands of dollars per acre. Offshore bonuses usually are granted the highest bidder through periodic sales on a sealed-bid basis. Bonuses offered on offshore blocks in the Gulf of Mexico always have been considered to be quite high on an acreage basis in comparison with private sales. A total of 2644 oil and gas leases on the Outer Continental Shelf of the mainland United States were sold over the period 1954-1975, inclusive.[14] Total bonuses paid amounted to $15.882 billion for 12.46 million acres, an average of $1274 per care. The 1.009 million acres of public lands acquired during the same period averaged only $30.55 per acre. One of the highest bids for a tract offshore from the mainland was paid in 1973.[15] It was for a tract off the Fort Walton Beach area of Florida which brought $36,803 per acre. Bonuses paid for federal oil-shale leases in 1974 averaged $26,328 per acre as shown in Table 7-2, Chapter 7.

The Rental. The concept of the rental has changed since the early days of the industry. At that time it was a rental for the use of land to drill oil wells and was sometimes the only payment made to the landowner. It evolved into what is sometimes called today the *delay rental* so that it may not be confused with any payment for use of the surface of the land. Rental has come to mean an advance penalty payment for the privilege of postponing or delaying the drilling of wells on a leased property. The lease usually stipulates that a well shall be started within twelve months or an annual rental of a specified amount per acre becomes payable in advance. Such rentals must be paid on the entire property. These payments are an expense to the lessee and serve as an inducement for him to start drilling operations as soon as possible or surrender the lease. The rentals eliminate the need for a surrender clause in the lease because if not paid when due it amounts to a notice of intent to surrender the lease which then becomes self-annulled.

Obviously the rentals must be reasonable on a per-acre basis or the industry could not afford to lease so much land far in advance of drilling operations. The lessor likes to obtain as high a bonus and rental as possible because those payments may be the only ones made to him if exploratory wells do not prove-up production. The lessee naturally prefers to keep bonuses and rentals as low as possible to keep his investment at a minimum during the period of no income. Lease rentals seldom exceed $1 per acre per year unless the lease is close to producing wells or the lessor is the federal or a state government. The annual rental, including the minimum royalty, required for leases offered at general Outer Continental Shelf lease sales is $3 per acre for unproven areas. First year rentals for leases offered at drainage sales in proven areas have been $10 per acre. Rentals on all Outer Continental Shelf oil and gas leases from 1953 to 1975, inclusive, amounted to approximately $157 million.[16]

The Royalty. The royalty is that portion of the production retained by the lessor. It is usually referred to as the *landowner's royalty*. The landowner has been the lessor in the vast majority of leases granted since the beginning of the industry.

The word *royalty* comes down from the old English laws when the crown owned all minerals and exacted a payment to the royal family for the privilege of working a mine. Today the term simply infers a share of the production or profits reserved by the owner of the mineral rights when granting another the right to produce from the property. It is actually an income-sharing operation in the oil industry because the royalty normally must be set aside whether the operator or lessee makes a profit or not. On the whole the royalty products are delivered at the surface and into the tanks or pipelines ready for delivery to the purchasing agency at no cost to the lessor except perhaps for his share of the costs of dehydration, treating, water disposal, and related surface expenses.

The land leased for the drilling of the Drake well originally called for a rental (royalty) of one-eighth of the oil. It was changed before the well was started to a fixed cash payment of 12 cents per gallon of production. Royalty payments ranged as high as 50 per cent of the production on some of the very early Pennsylvania oil leases. A one-eighth royalty seemed to be common on unproved lands and became somewhat firmly established by the late 1860's. Until quite recently this customary royalty share of one-eighth of the production has prevailed in most areas of the United States exclusive of Ohio, California, parts of Louisiana, and on public-owned lands where a one-sixth royalty has prevailed. Most of the lands leased from and through our federal government, including offshore acreages, and the university lands of Texas stimpulated a one-sixth royalty prior to 1975. The Texas School Land Board required a one-fifth royalty at an onshore-offshore lease sale in 1975. The first federal lease sales with fixed modest bonus for tracts off the coast of Louisiana attracted 57 bids for eight of the ten tracts offered in the Fall of 1974. Royalties offered ranged from 51.798 to 82.165 percent.[17]

Royalties on production from the federal oil-shale leases in northwest

Colorado and adjoining Wyoming and Utah are 15 percent of all oil and gas produced.

The royalty share of an oil lease may be sold in whole or in part at any time even before the discovery of production. The lessor may do this to reduce his risk in case of failure of the exploratory effort. He would collect something more than just the bonus and rentals in that case. It is relatively easy to dispose of good royalty interests as well as leases through brokerages established for that purpose. Any fractional interest in either may be so purchased by anyone with the finances.

Various forms of royalty interests in oil properties are found in the portfolios of investment groups, charitable organizations, and endowed institutions. Oil property valuation engineers may be called on to make periodic valuations of these interests and particularly to estimate the probable future income to be derived from them.

The prevalence of the one-eighth royalty interest in the early days of the industry led to the acceptance of a unit of royalty interest for sales purposes called the *royalty acre*. It was merely a one-eighth royalty interest on one acre of land, productive or not. Thus a one-eighth royalty on eighty acres of land would amount to 80 royalty acres which could be sold or traded singly or in any multiples. Another unit of royalty interest called an *acre-percent* came into use in California during the early days of the industry and was widely used for a time in that State. It was equal to 1 per cent of the production from an acre of land. Thus the customary one-eighth royalty interest on eighty acres of land would amount to 1000 acre-percents. Both units of royalty interests make it easy for the small investor to purchase a part of the production from what may become a very lucrative property.

Royalty payments can be very profitable to a fortunate royalty owner, particularly when he is the original landowner. He probably came into possession of the land before recognition of any prospects of finding oil or gas. He risked nothing and had everything to gain in leasing his land. Bonus, rental, royalty, and other oil and gas lease payments to farmers, ranchers, and other private owners of prospective and productive properties are estimated to be more than $2.5 billion annually. Royalty payments to the federal government for oil and gas production from the Outer Continental Shelf areas on the West Coast and the Gulf of Mexico have totalled $3.85 billion to the end of 1975.[18] The royalty value of all oil and gas operations on federal and Indian lands for 1920 to 1975, inclusive, amounted to $6.561 billion.[19]

Other Forms of Oil Property Interests. The working interest in a lease is said to be carved out of the mineral interest at the time the lease is granted. The lessee may keep the working interest intact throughout the productive life of the lease or he in turn may carve it into various other property interests.

Many times he is required to do this in order to obtain a lease through a broker. Other times he may find it desirable or necessary to carve out interests in order to raise funds to develop that or other properties, and occasionally it is done simply to spread the inherent risk of developing any property. Many different forms of royalty obligations have developed through a desire to eliminate a fixed royalty either in money or barrels of oil. All these other forms of royalties are to be distinguished from the landowners royalty.

These other forms of royalty commonly used by the industry are:

- Overriding royalty
- Participating royalty
- Term royalty
- Additional royalty
- Offset or compensatory royalty
- Barrel royalty

An *overriding royalty* is that share of the production from a property retained by a broker for handling a leasing transaction. The usual amount obligated by the lessee for the life of the lease amounts to the difference between one-sixth and one-eighth of the production or 4.1667 percent. This would leave the lessee with a net working interest of 83.333 percent of the gross production.

Overriding royalties are also an integral part of what is called a *farmout*. There are serveral kinds of farmouts but most of them apply either to the drilling of wildcat wells, the development of producing areas, or occasionally the operation of producing wells. In all cases a farmout is an assignment of the working interest to another with some of the future production being retained as an overriding royalty. A farmout may include all or a part of the original lease. The assignee of the farmout takes over the responsibility of paying for all drilling, development, and operating costs. The assignor shares in any production free of all expense the same as the original lessor. Most of the farmouts are given to small, independent operators by the larger companies. The overriding royalty seldom amounts to more than one seventh of the original working interest, thus the assignee has an opportunity to share in six eighths of the gross production without having made any leasing investments. The practice of farmouts has been a boon to both the large operator, the small independent, and drilling contractors. The larger operations have been able to save most of valuable leases nearing expiration date by farming out one or more drilling units. Many small independent operators have been able to become soundly established in the production branch of the industry by taking over farmouts, and many drilling contractors have been able to become established in the production end of the business by drilling wells on farmed-out acreages during slack periods.

The term *participating royalty* is somewhat a misnomer because it more nearly represents a partial ownership of the working interest. It differs only in that the liabilities of the assignee are perhaps limited both in amount and time. This form of oil property is most prevalent in promotion wells where interests are sold to a large number of small investors.

The term *minimum royalty* is not as prevalent as it used to be in early-day leases. A minimum royalty is sometimes specified in order to guarantee to the lessor a minimum payment in dollars or in kind where the rate depends on the quality, quantity, and perhaps the price of oil. The term is also used in connection with some leases on public lands where a sliding scale of royalty rate is applicable, depending on rates of production and perhaps drilling depths.

Sometimes a lessor sells all or a part of his royalty to another for a specified period of time. The assignee in this case holds what is called *term royalty*. If the assignment was made before drilling a well or wells and no production is obtained, it reverts back to the assignor.

The term *additional royalty* has many applications. Normally it refers to those leases in which sliding royalty scales have been included calling for higher royalty rates in case of highly productive wells, wells that may be productive from depths shallower than were anticipated, and where production from the lease may be exceptionally high due to thick sands, numerous productive horizons, or other reasons.

The term *offset royalty* is a misnomer. A more applicable and all-inclusive term is *compensatory royalty*. It refers to that monthly payment agreed upon between the operator and the lessor in lieu of drilling a well on the lease or relinquishing the acreage of that drilling unit. The undrilled well may be an inside location or one offsetting a producing well on another's adjoining lease. In both cases the operator thinks that the payment of compensatory royalty is less costly to him than drilling a well which might never pay out. The rate usually is based on the productive capacity of the offset or adjoining wells. If the questionable well is drilled at a later date, the payment of compensatory royalty is discontinued and regular royalty is paid based on its actual productivity.

The term *barrel royalty* is not often included in present-day leases, although it was common in some of the earliest oil leases. It refers to a royalty payment to the lessor based on a certain number of barrels (or Mscf of gas) per day instead of a percentage of the production.

A person is said to have a *participating interest* in a property whenever he is charged with his proportional share or some other fraction of the operating expenses in return for a share of the profits. The owner of an overriding royalty often is said to hold a *nonoperating interest* because he is not charged with any of the operational responsibility or expense.

A *carried interest* comes about through an agreement between two or more partners in the working interest in a property whereby the *carrying party* advances some valuable consideration toward the development and/or operation of the property and the *carried party* does not share in the revenues until the carrying party or interest has recovered a specified sum of money or its equivalent in value. Some farmout agreements and occasionally an overriding royalty will include such provisions. Carried interest may apply to single wells or an entire property.

At times one operator will agree to pay a certain sum of money called *dry hole money* toward the cost of drilling a well offsetting either productive or wildcat acreage owned by the contributor. Usually the agreement stipulates that the well is to be drilled to a certain depth and that geological and other test information is to be provided to the contributor. The contribution agreed upon is paid only if the hole proves to be dry.

Other times one operator will agree to contribute *bottom hole money* toward the drilling of a well in return for geological and test information regardless of whether the well is dry or producing. The contribution of dry hole money or bottom hole money is often made a part of a farmout agreement with mutual benefits to both parties concerned.

Previous owners of properties, lending institutions and purchasers of production may retain or have an interest in properties through what are called *production payments*. A production payment usually specifies that the payee is to receive periodically a portion of the production for such time as is required to retire the principal of the obligation plus interest charges. Production payments have been widely negotiated, sold, and traded since the early days of the industry. For years it was thought to be the same as any other financial interest in an oil property and later was construed as sort of a partnership by the payee for income tax purposes. However, it lost most of its popularity in 1969 when it was defined as a straight loan under the "Tax Reform Act" and was then considered to be the equivalent of a first mortgage against the property until retired. It was coming into use again by 1975, particularly by the major companies. All were literally strapped for capital funds despite the so-called "excessive profits" of 1973-74. Estimates of over $1 billion had been so pledged in 1976.

The seller of a property usually needs or wants cash in full at the time of a sale. A form of transaction which became known as the *ABC Deal* gradually developed to accomplish cash settlements and was widely used during the 1960's. It permitted the seller of a property to get his selling price in cash at the time of the sale, the buyer to get the property with a minimum down payment, and a third party acting as a middleman to make a small profit for handling the transaction. Use of the *depletion allowance* and differences in interest rates made such transactions doubly attractive. Its use was essentially eliminated by the Tax Reform Act of 1969. On the whole the "Act" was a blow to the smaller companies within the industry in that it severely reduced the availability of funds needed by many for exploration and development operations.

Leasing of Public Lands

The general policy of the federal government before passage of the Preemption Act of 1841 was to sell public lands outright to individuals for purposes of raising revenue. Although several subsequent acts were passed, particularly with respect to mineral lands as a result of the discovery of gold in California and its admittance as a state two years later, no provisions were made for oil producing properties until passage of the *Placer Law* of 1872. It provided for the preemption of a tract of not more than twenty acres by an individual and the obtaining of an absolute title after having spent $500 for development on the tract.

The *Alaska Coal Lands Act* of 1914 was the first important leasing measure. This Act was followed in 1920 with passage of the *Mineral Land Leasing Law*, which included special provisions with respect to petroleum, natural gas, and oil shales. It provided for the granting of an exclusive right to explore for oil and gas on an area up to 2560 acres in size for a period of two years. Upon completion of a commercial well, the prospector could obtain a lease on one-fourth of the area with a royalty payment of only 5 per cent. The law of 1920 was amended in 1935 to permit prospecting for oil and gas under a lease rather than under a prospecting permit. The amended law provided for two types of oil and gas leases, depending on whether the area under consideration was over or outside the known geological structure of a producing oil or gas field. Lands lying within the geological boundary of a producing field were to be leased only by competitive bidding. Lands outside the geological boundary of a producing structure were noncompetitive and could be leased to the first applicant. Noncompetitive leases were to be granted for a primary term of five years and so long thereafter as commercially productive on not to exceed 2560 acres on any single geologic structure. Annual rentals were to vary from 25 cents to $1 per acre until royalties became due. The royalty rate was to be 12½ per cent on all production. Competitive leases limited to 640 acres were to be auctioned on a sealed bid basis and granted to the one offering the highest bonus. The primary term was to be for twenty years and as long thereafter as commercially productive. A $1.00 per acre rental was specified, which would be credited against any subsequent royalty payments. Royalty payments were not fixed and might range from 12½ to 25 per cent for oil and from 12½ to 16⅔ per cent for gas.

Four large naval oil reserves in Wyoming, California, and Alaska and the oil shale reserve in Colorado were set aside in 1909 when the government temporarily withdrew from entry all mineral lands on the public domain. Naval oil reserves in Wyoming and California leased to oil companies during the Harding Administration were soon repossessed because of suspected fraudulency. None of the reserves has since been put for lease.

The President of the United States declared on September 28, 1945, that the United States considered[20]

> . . . the natural resources of the subsoil and seabed of the Continental Shelf beneath the high seas but contiguous to the coast of the United States as appertaining to the United States, subject to its jurisdiction and control.

The boundary was to be determined by arbitration where the shelf extended to the shores of another country or where it was shared with an adjacent country such as between the state of Texas and Mexico.

Our Supreme Court ruled in 1947 that the federal government held title to all waters lying seaward of the low-water mark along the coast of any state. This ruling was changed in 1953 by the *Submerged Lands Act* after much wrangling between the federal government and the states. Congress declared it to be in the public interest that title and ownership to all submerged lands beneath the navigable waters of the respective States and . . .[21]

> the natural resources within such lands and waters and the right to manage, administer and develop such areas be vested in the individual States subject to the provisions of this Act.

The Act also confirmed the boundary of each coastal state as *three geographic miles* from its coastline in the Pacific and Atlantic oceans, and up to 3 marine leagues into the Gulf of Mexico for Texas and Florida. The coastline was defined as the ordinary low-water mark along that portion of the coastline in direct contact with the open sea and the line marking the shoreward limit of inland waters. Agreements of the coastlines of Louisiana and southern California have never been settled for some sections.

The related *Outer Continental Shelf Act* passed the same year[22]

> . . . declares it to be the public policy of the U. S. that the subsoil and seabed of the Outer Continental Shelf are subject to the jurisdiction, control and power of disposition of the U. S. The Outer Continental Shelf is defined as all submerged lands lying seaward and outside of the area of the lands beneath navigable waters vested in the individual States by the "Submerged Lands Act."

This act gave authority to the Secretary of the Interior to set up rules and regulations with respect to leasing the oil and gas or other minerals on the Outer Continental Shelf. Lease blocks of not over 5760 acres each are to be awarded on a sealed bid basis. The Secretary has the right to establish a fixed bonus for a tract and allow a competitive bidding on the amount of the royalty at not less than 12½ per cent.

Leases on state-owned lands and on Indian lands and public lands under the jurisdiction of various departments within the federal government also are granted on a sealed bid basis, the bid usually covering only the amount of the bonus, the royalty being fixed.

LAND SURVEY SYSTEMS OUTSIDE THE UNITED STATES

Land survey systems outside the United States of interest to the oil property valuation engineer may be divided into the "Canadian system" and the generalized "metes-and-bounds systems" used throughout most of the rest of the world.

The Canadian System

The Dominion Land Surveys Act of Canada as amended in 1908 provided for a public land survey system for all dominion lands in Manitoba, Saskatchewan, Alberta, British Columbia, and the Northwest Territories similar to the public land survey system of the United States.[23] The system provides for lands being laid out in quadrilateral townships six miles square consisting of thirty-six section each. The numbering, however, starts in the southeast corner of a township and ends in the northeast corner, just the reverse of that in the United States. The Canadian system is also different in that certain road allowances are provided which are exclusive of the 640-acre areas of the sections. The width of the roads is normally 100 links (4 rods). Provisions call for a road allowance along all range lines, along the south boundary of the township, and at the two-, and four-, and six-mile points.

Townships in the provinces are numbered consecutively north from the International Boundary with the United States. The ranges are numbered east to west between meridians established thirty townships apart. The First Meridian begins just west of the northeast corner of North Dakota and passes about 15 miles west of Winnepeg. The Fifth Meridian passes through Calgary.

The sections are divided into four 160-acre quarters and each quarter is divided into four forties, all as in the United States system. The Canadian system uses a convenient method of designating the quarter-quarter sections. They are numbered from 1 to 16, beginning in the southeast corner of the section and ending in the northeast corner. These forties are "Legal Subdivisions" and are designated "Lsd." in a description. In the United States a 40-acre tract of land might be described as the NW/4, SE/4, Sec. 21, T. 8 N., R. 5 W., and abbreviated to NW4-SE4-21-8N-5W. Neither of these descriptions is entirely complete unless the geographical area or the principal meridian and base line are mentioned. Under the Canadian system a similar tract can be described accurately and completely as Lsd. 7, Sec. 21, T. 8 N., R. 5 W., 4, and may be abbreviated to Lsd. 7-21-8-5W-4. The "4" means that it is located west of the Fourth Meridian. As an added convenience the four 10-acre tracts making up a quarter-quarter section may be designated as A, B, C, and D, with A being the southeast quadrant, B the southwest quadrant, and D the northeast quadrant.

Certain properties in eastern Canada where industry periodically explores for new production have been surveyed and described under a generalized metes-and-bounds system.

Metes-and-Bounds Systems

Oil property interests outside the United States with the exception of those surveyed under the Dominion Land Surveys Act in Canada generally are surveyed and described under what might be called a generalized metes-and-bounds method. Rectangular and geographic coordinates may be calculated for corners or other markers along the boundaries. Most foreign oil property interests are granted by the national governments. These property interests may vary in size from a few acres to millions of acres; the smaller ones may be somewhat regular and rectangular in shape whereas the larger ones may be bounded by rivers, mountain chains, or other natural boundaries. Areas wholly or partially offshore, however, are usually more regular to partially or wholly rectilinear in shape. Some of the largest may correspond in area and politically to a whole state in the United States.

The foreign oil company usually is required to survey and map such property interests. The starting point, which may be any well-known landmark, must be located accurately by latitude and longitude. Any accurate system of surveying, including triangulation, may be used to run in and to describe the boundaries. As a general rule what might be called a modified metes-and-bounds method of surveys and descriptions normally is employed. Inasmuch as many foreign oil properties cover land areas that are very inadequately mapped and described, the foreign governments may require generalized topographic and contour maps to be made and in some cases aerial photographs of the entire area.

FORMS OF OIL PROPERTIES OUTSIDE THE UNITED STATES

Since World War I and particularly since the late 1930's almost every American-owned oil interest outside the United States and exclusive of Canada has been held in the form of a "contract" or a "concession." Before World War II only a few of the largest American and British oil companies held concessions abroad. Since World War II an ever increasing number of smaller American companies went into such operations. This increased interest in operations abroad was due primarily to the steadily increasing difficulty and cost of finding and developing new fields in the United States and the rapidly growing demand for petroleum products at home and abroad. The American industry was interested not only in developing sources of petroleum supplies abroad for importation into the United States but also in supplying the foreign markets.

Ownership of minerals in place outside the United States is vested almost entirely in national government. The only important exceptions are Canada, and in Colombia and Australia to a very minor extent.

The laws and regulations governing petroleum exploration and production, as well as related petroleum industry activities, in foreign countries where operations have not been nationalized are said to fall into the following three general types:[24]

1. No petroleum law exists; all operations connected with petroleum are governed by terms of a contract negotiated with the head of state.
2. Federal petroleum law is the basis for all petroleum operations. Such laws often vest important discretionary powers in Federal administrative or legislative bodies.
3. Subsoil rights are governed by laws of individual states, provinces, or other political sub-divisions, with varying control by laws of Federal government.

The first type was usually in the form of a contract negotiated with head of state, or the sheikdoms of the Middle East such as Saudi Arabia and Kuwait.

The second type applies to those nations having a national petroleum law applicable to all foreign operations. Venezuela, Mexico, Bolivia, Peru, and Colombia were early examples. Most of the early contracts of this type had been modified by the early 1970's to give the host nation majority ownership and control. And as of 1976 most of the operations of this type had been taken over completely by the host governments upon payments of depreciated values in cash and/or future oil production payments.

The third type is the least common. Canada and Australia are examples—where laws and regulations of state or provincial governments control the industry with partial to complete control of regional areas of the nation by Federal or Commonwealth Law.

Petroleum laws and requirements vary widely from nation to nation. The most common provisions and regulations have to do with the following:

1. Type of permit, contract, or concession.
2. Size, shape, and geographic limits of area.
3. Initial or primary term and extensions.
4. Fees, bonuses, and rentals.
5. Relinquishment or surrender.
6. Selection and convertibility of acreage.
7. Assignment or transfer of acreage, lease, or concession.
8. Royalty payments and/or sharing of profits.
9. Tax obligations.
10. Official exchange rate.
11. Obligation to supply domestic markets first.
12. Building of local refineries.
13. Employment of nationals.
14. Labor laws and regulations.
15. Equity participation by government.
16. Repatriation of capital,

Canada

Private lands in Canada carrying ownership of petroleum and natural gas rights are called *freehold properties*. All other lands in Canada under which mineral rights are owned by either the federal government or the provincial governments are called *Crown lands*. Each has its own rules and regulations pertaining to exploration, drilling, production rights, fees, rentals, and royalties.

Petroleum and natural gas rights on freehold properties owned by individuals and/or companies and corporations always have been leased, sold, traded, and operated though private agreements similar to those of the United States.

The reservation of mineral ownerships by the Crown excluded all freehold lands. The practice was started in 1887 with respect to lands west of the Third Meridian, which passes north and south through the middle of Saskatchewan, and was extended to include lands east of the Third Meridian in 1890.

The respective provinces own approximately 80 per cent of the petroleum and natural gas rights in Alberta and Saskatchewan, 20 per cent in Manitoba, and 100 per cent in British Columbia. Petroleum and natural gas rights owned by the federal government include all rights in the Yukon and Northwest Territories, the Arctic Islands, Canadian offshore areas, Indian reserves, National Parks, and isolated areas owned by various federal agencies. Rights have been granted by the Federal government to large areas under procedures similar to those of the provinces. Small tracts often have been granted to the highest bidder at sealed-tender lease auctions.

Each province requires that the operator be incorporated or registered in the province in which any kind of oil or gas operations are to be undertaken. In addition an operator's license must be obtained in each province before acquisition of a permit, reservation, license, or lease.

Procedures in obtaining interests in petroleum and natural gas rights in Alberta are of particular interest because of its higher percentage of Crown lands and its longer and greater activity in such matters in comparison with the other provinces. Its early laws and regulations with respect to outside operators combined with particularly favorable geological and geographical essentials made the province especially attractive to the petroleum interests of the United States. Petroleum and natural gas rights of all provincial Crown Lands generally could be obtained through four ways:

1. Exploration rights or reservations; (a) by geological and/or geophysical methods; (b) through drilling.
2. Drilling reservations.
3. Actual oil and/or gas leases.
4. Natural gas licenses.

As of June 1, 1962, a Mines and Minerals Act became effective. It somewhat complicated the title structure of petroleum and natural gas rights in its Crown Lands for a time, but prior leases were largely unaffected. An area termed Block A was established comprising all Crown Lands in Alberta in Townships 1 to 64, inclusive, west of the Fourth Meridian. New petroleum and natural gas reservations and permits were to be divided into three general categories, as follows:[25]

(a) Exploration reservations (lands outside of Block A);
(b) Exploration permits (land inside Block A);
(c) Drilling reservations (all Crown Reserve lands).

Each type of reservation or permit is, in effect, an exclusive option to acquire petroleum and natural gas leases from the Crown and gives to the holder the right to drill wells on the designated area for petroleum and natural gas production.

An interest in provincial petroleum and natural gas rights may be obtained in five ways:

A. Reservations of petroleum and natural gas rights in lands outside of Block A;
B. Permits of petroleum and natural gas rights in lands within Block A;
C. Drilling reservations;
D. Natural gas licenses and natural gas leases;
E. Petroleum and natural gas leases.

Regulations with respect to explorations, drilling, leasing, and production are somewhat similar in all the provinces. Federal rules and regulations are less restrictive for the Arctic Islands and offshore areas and the Yukon and Northwest Territories areas due to the very costly exploration expenses and uncertainties of early deliverability of discovered potential production.

Colombia and Australia

Only a very small proportion of privately owned lands in Colombia carry private ownership of oil and gas in place. Titles to those that do were conveyed before October 1873. Mineral rights on these lands could be leased or purchased from the owners somewhat the same as in the United States prior to the mid-1970's. The lessee or purchaser had to be a natural or judicial person with proven financial capacity. Foreign governments and companies owned by them were excluded.[26]

All six states making up the Commonwealth of Australia have control of their respective mineral rights. The ownership of petroleum and natural gas in place is not everywhere clearly defined because lands grants in fee simple prior to 1886 conveyed all minerals other than gold and silver.[27] All freehold land titles since 1888 have reserved all minerals to the Crown. As in Canada most of the states have established procedures for granting permits to explore, licenses to prospect, and leases for development and exploitation.

Foreign Operations in General

Most foreign operations are carried on through subsidiaries set up within the nation concerned by our international oil companies. Only the large companies can afford the many risks and the large capital requirements.

In addition they are the only ones having established markets for foreign production. Some of our smaller companies lately have joined the majors or pooled their capital to get a foothold in foreign operations.

Some kind of an initial permit or contract usually is required for foreign exploratory work of any kind, including exploratory wells. Before or on expiration of the exploratory agreement, an application must be made selecting that acreage to be retained for development and exploitation. The size, shape, geographic distribution, and per cent of total original area that can be selected by the grantee and converted into an operating concession varies among nations. Many times these restrictions force the grantee to relinquish very promising acreages. These acreages may be put up by the national government at a later date on a sealed bid basis. Such tracts are much smaller than a concession and provide a means whereby a small company can enter foreign operations with a minimum of capital. A tract so obtained often is spoken of as a lease. Several small American independent oil companies entered foreign operations in this way in 1956 when Venezuela offered sales of returned acreages to the highest bidder.

Royalty rates for foreign operations at one time were comparable with those in the United States. Prior to 1973 all oil produced abroad was spoken of as cheap oil. Today this is hardly true because almost every foreign government in time demanded modifications of old contracts and inclusions in new contracts that permit the nation to receive at least 50 per cent of the net income from the oil-producing operation.

The more important remaining American oil property interests outside the United States as of the mid-1970's are in the North Sea, Irish seas, Lybia, Egypt, Nigeria, Angola, Canada, Australia, and the general East Indies area. National control to some extent predominates in all these nations. Majority ownerships or partnerships were pending to nearly accomplished by many of these nations. Working interests can usually be assigned in whole or in part subject to certain restrictions. Naturally there are no assignments of what might be called the "landowner's royalty." Valuation studies of oil interests abroad are rare except for purposes of providing management with information useful in making operational decisions.

INTEGRATION OF OIL PROPERTY OWNERSHIPS IN THE UNITED STATES AND CANADA

Mineral laws of the United States and Canada permit multitudinous and diverse ownerships of oil properties and oil property interests in a single producing area. In general, regulatory bodies have upheld the right of the owner of the working interest of an oil property, irrespective of size, to locate one or more wells on it in order to obtain his fair share of the production. Due to the fugacious nature of oil as established by the famous "Law of Capture" decision, maximum recovery is not necessarily obtained through exercise of this privilege, and property rights may be affected either favorably or adversely. Such conditions also make it difficult, if not impossible, to carry out effective programs for increasing recovery.

None of these conditions exists in the vast majority of foreign operations because of the large areas of individual concessions.

Henry L. Doherty made himself unpopular with the industry when he suggested in 1924 that the federal government should take over control of all oil development and production operations in the United States in order to increase recovery efficiency.[28] He recognized the many wasteful and inefficient practices resulting from diverse ownerhip and operations and felt that the individual would do little or nothing voluntarily to eliminate them. Little was done to correct these conditions until industry recognized the tremendous waste taking place in the East Texas field during the early 1930's. Since that time two general methods of integration of individual oil property ownerships developed: *pooling* and/or *unitization* of properties.

Pooling

Pooling of oil properties has been described in the Doherty memorial volume of *Petroleum Conservation* as relating[29]

> . . . to the combining of small or irregularly shaped tracts for the purpose of providing a well or a production unit. For example, the owners of a number of small contiguous tracts may pool their properties for the purpose of providing a 40-acre drilling unit for an oil well or a 640-acre drilling unit for a gas well, or a small tract may be pooled with an adjacent larger tract in order to become a part of a drilling unit.

Unitization

The terms *unitization* and *unit operation* are often used synonymously. They are defined in the Doherty memorial volume:[30]

> The practice of unifying the ownership and control of a large area or an entire oil or gas pool by the issuance or assignment of units or undivided interests in the entire area with provision for development and operation by an agent, trustee, or committee representing all holders of undivided interests therein is usually called "unitization."

. . . "unitization" involves cross assignments of interests, whereas "unit operation" refers to joint development and operation of the unit which is formed, with or without cross assignments.

The specific intent is to combine separately owned tracts in such manner that they may be developed and operated as if they were a single property, instead of separately. The procedures employed to accomplish this result vary. Whatever the legal procedures employed, the result is that the owner of an individual tract trades his right to develop and produce that tract for a share in the production in the entire unit.

. . . Under unit operation there is no conflict of interests, but, instead, a joint and a common interest in carrying out the operation as efficiently and economically as possible for the maximum recovery.

Normally pooling of interests may be accomplished in a relatively short time because of the small number of parties concerned. On the other hand unitization agreements may take months or years to accomplish. Committees representing the various interests must be appointed to develop the basic engineering information required and to formulate an acceptable basis of participation. Reaching agreement on the basis of participation is perhaps the largest stumbling block to unitization agreements. The following two steps are required:

1. Contact all operating interests and through them or their representatives agree upon a basis or method of calculating the participation of each property interest with respect to costs and production income.

2. After agreement is reached on the basis of participation, each property interest must be specifically evaluated on that basis in order to calculate its percentage of the whole.

Such factors as productive acreage, sand thicknesses, porosity, number of wells, net pore volume of sands, and saturations generally are used to arrive at the estimated remaining amount of recoverable production for each operating interest. Royalty interests also have to agree on the unitization project as an operation before it becomes effective. Hundreds of such interests may be involved in larger fields. State and federally owned or controlled lands as well as Crown lands may be included in such agreements upon approval of the plan by the proper governmental agency.

Each operating participant theoretically pays for his fair share of all costs of the unit operation and theoretically receives his fair share of the production. A committee representing the participants advises and checks all operations. Both pooling and unit operations have proven so successful that some states have passed or have under consideration regulations either requiring or encouraging pooling and/or unit operations under certain conditions.

REFERENCES

1. Francis Rawle, *Bouvier's Law Dictionary*, Vol. II, Boston Book Co., 1897, p. 780.
2. Clark W. Breeding and A. Gordon Burton, *Taxation of Oil and Gas Income*, Prentice-Hall, Englewood Cliffs, N. J., 1954, p. 195.
3. Charles B. Breed and George L. Hosmer, *The Principles and Practice of Surveying*, 7th ed., Vol. I, Wiley, New York, 1940, p. 140 (9th ed., 1958).
4. W. M. Gillespie, *A Treatise on Land Surveying*, Appleton, New York, 1870, p. 370.
5. Aldon Socrates Lang, "Financial History of the Public Lands in Texas," *Baylor Bull.*, Vol. 35, No. 3, July 1932, p. 27.
6. *Ibid.*, pp. 28-29.
7. *Ibid.*, pp. 31-32.
8. John S. Grimes, *Clark on Surveying and Boundaries* (A Treatise on the Law of Surveying and Boundaries), Bobbs-Merrill, Indianapolis, 1959, p. 899.
9. H. M. Hayne, State of Louisiana Dept. of Public Works, personal communication, 1963.
10. E. DeGolyer (Ed.), *Elements of the Petroleum Industry*, American Institute of Mining and Metallurgical Engineers, New York, 1940, p. 91.
11. Geologic Formations and Economic Development of the Oil and Gas Fields of California, *Calif. Div. of Mines, Dept. Nat. Resources, Bull. 118, Pt. 1,* 1940, p. 75.
12. Personal communications, (average of three estimates).
13. *Federal and Indian Lands Oil and Gas Production, Royalty Income, and Related Statistics*, 1920-1975, Calender Year 1975, United States Department of Interior, p. 15.
14. *Outer Continental Shelf Statistics*, 1953-1975, Calender Year 1975, United States Department of Interior, p. 18.
15. "Eastern Gulf Sale Nears $1.5 Billion," *Oil and Gas J.*, December 24, 1973, p. 19.
16. *Outer Continental* . . . . . op. cit., p. 8.
17. *Eastern Gulf Sale* . . . . . op. cit., p. 19.
18. *Outer Continental* . . . . . op. cit., p. 85.
19. *Federal and Indian Lands* . . . . . op. cit., pp. 75-76.
20. Northcutt Ely, "Summary of Mining and Petroleum Laws of the World," USBM *Information Circular* 8017, 1961, pp. 213-215.
21. John Ben Shepperd, "New 'Tidelands' Suits–Texas Will Defend Her Domain," *Oil Forum*, November 1953, pp. 377-378.
22. *Ibid.*, pp. 378-379.
23. Grimes, *op. cit.*, pp. 956-967.
24. John C. Dunlap, "Petroleum Exploration in Foreign Countries," in *Economics of Petroleum Exploration, Development, and Property Evaluation*, Prentice-Hall, Englewood Cliffs, N. J., 1961, p. 103.
25. Oil and Gas – A Guide for Oil and Gas Operators in Canada, Bank of Montreal, Calgary, Alberta, Canada, 1974, p. 16.
26. Ely, *op. cit.*, p. 51.
27. *Ibid.*, p. 190.
28. Robert E. Hardwicke, *Antitrust Laws et al., v. Unit Operation of Oil or Gas Pools*, AIME, New York, 1948, pp. 1-13.
29. Stuart E. Buckley (Ed.), *Petroleum Conservation*, AIME, New York, 1951, p. 280.
30. *Ibid.*, p. 281.

Additional References

F. V. Garcia Amador, *The Exploitation and Conservation of the Resources of the Sea*, 2nd ed., A. W. Sythoff, Leyden, esp. Pt. II, 1959, pp. 86-134.

Earl A. Brown, *The Law of Oil and Gas Leases*, Bender, New York, 1958, and yearly supplements.

Samuel H. Glassmire, *Law of Oil and Gas Leases and Royalties*, Thomas, St. Louis, 1935.

Thomas G. Johnson, "Rule of Capture vs. Common Sense," *World Oil*, July 1952, pp. 142-156.

William Woods Porter II, "Rule of Capture vs. Rule of Thumb," *World Oil*, November 1962, p. 112.

Edward H. Robie (Ed.), *Economics of the Mineral Industries*, Chap. 3, "Mineral Titles and Tenure," AIME, New York, 1959, pp. 81-130.

4 THE MEANING OF VALUE

The accounting department of an oil company keeps current records of the assets, liabilities, expenses, depreciation, taxes, and other items that determine the yearly profits or losses of the company. It has the responsibility of keeping management informed at all times of all financial aspects of the operations of the company. It also must maintain records in such ways as may be required under various statutes dealing with local, state, and federal tax obligations and such regulatory bodies as the Securities and Exchange Commission, Department of Justice, Federal Power Commission, and state oil and gas commissions. It probably will carry a *book value* for the company as a whole and for certain of its component parts.

The book value alone of a company or a property cannot be used as a convenient substitute for an engineering valuation. It does serve to a degree with respect to sales and purchases of securities and as a basis of comparison with engineering or other valuations of properties. The book values of physical things within the company are used primarily in connection with depreciation records. Oil property valuation studies made according to engineering methods for purposes of obtaining loans or in making sales, purchases, trades, and mergers are of major interest to the oil industry. These studies give more meaning to the arrived-at value of the property. Regardless of the specific purpose of the engineering valuation of an oil property, the value to any would-be owner must be not only a measure of the utility or usefulness of the property in the production of income and net profit, but also the rate at which each will accrue. Despite these basic principles the word *value* has many shades of meaning. Everyone has some concept of the value of things with which he is familiar. It comes about through our dealings with others in obtaining and dispensing the necessities and luxuries of everyday living. We see that the value of anything is the integration over the years of each of our willing offerings, if acceptable to willing owners, in obtaining our daily needs and desires. This understanding often underlies the settlement of certain cases in court. And although we may not realize it, we are contributing to the standardization of individual values every time we are a party to a sale or trade.

The oil property valuation engineer will be benefited by a knowledge of the meaning of value from various viewpoints. The very brief review given in this chapter is intended only to indicate the breadth and interest of the subject to him.

EARLY CONCEPTS OF VALUE

The meaning of value has been a favorite subject for philosophers beginning with Aristotle who introducted the concept that the value of economic goods resulted from their individual usefulness, scarcity, and costs, and that it might decrease to the point of becoming negative if availability becomes excessive. He also analyzed the need for money as a measure of value and as a medium of exchange. These concepts were hardly improved on until the seventeenth and eighteenth centuries. Bernoulli introduced higher mathematics into economic thinking in 1738 with his theories dealing with marginal value and the value of money. Adam Smith is the accepted founder of modern political economy. Yet it has been said that he missed an opportunity to establish a new knowledge of value principles instead of "making waste and rubbish out of the thinking of 2000 years."[1] Smith's *Wealth of Nations* is probably best known for the famous paradox concerning the relative values of water and diamonds despite the fact that water has great utility and diamonds do not. He continued with the thought that those things having maximum usefulness very often have little or no value in exchange, whereas those of maximum exchange value very often have little to value in use. Locke who preceded him also knew that value, if measured by price, was not related necessarily to utility, as stated in the following excerpt from his writings:[2]

> What more use or necessary things are there to the being, or well-being of men, than air and water? and yet these have generally no price at all, nor yield any money: because their quantity is immensely greater than their vent, in most places of the world. But, as soon as ever water (for air offers itself everywhere, without restraint, or inclosure, and therefore is nowhere of any price) comes anywhere to be reduced into any proportion to its consumption, it begins presently to have a price, and is sometimes sold dearer than wine. Hence it is, that the best, and most useful things are commonly the cheapest; because though their consumption be great, yet the bounty of providence has made their production large, and suitable to it.

Malthus is much better known for his *Essay on Population* than for his contributions to the foundations of classical English political economy. He wrote at length on the definition of value and related subjects, and like Adam Smith and predecessors believed in more than one kind of value. He used the terms *value in use* and *value in exchange* and added the new descriptive terms *natural value* and *market* or *actual value*. The natural value of commodity was limited essentially to the elementary costs of its production or conditions of supply. The market or actual value at a given time and place was thought of as a measure of the amount of labor or its equivalent required to complete an exchange. Thus Malthus was a firm believer in what we might call today the "law of supply and demand." John Stuart Mill disagreed with both Smith and Malthus; he limited value to *value in exchange*. Karl Marx did not place much importance on exchange value because he considered it to be "accidental" and an "intrinsic" value. He believed that the value of any commodity was no more than that amount of "socially necessary labor congealed in it."

MODERN CONCEPTS OF VALUE

The writings of three well-known American philosophers, Santayana, Dewey, and Perry, are of interest with respect to their philosophies of value. Santayana wrote that man's reason began with impulses. These impulses were modified by reflection and judgment. Values then were "imputed upon reflection which gathered experiences together and perceived their relative worth." Dewey and Perry dealt with value chiefly from moral and social-behavior viewpoints. Dewey never exactly defined value. His thesis appears to have been built around the concept that value is connected inherently with liking. Liking was not considered to be accidental. It was based on judgments which should determine our desires, affections, and enjoyments. Perry stated his view in part as follows:[3]

> According to the definition of value here proposed, a thing–anything–has value, or is valuable, in the original and generic sense when it is the object of an interest–any interest. Or, whatever is an object of interest is ipso facto valuable.

From a philosophical viewpoint one must agree that something can be of value to its possessor for many reasons, but basically there must be an "interest." An oil property may have value to its owner. Conceivably it could have a negative value to him. But it could have no value to another without his having an interest in it. That interest might range from a negative to any degree of positive interest. Oil property valuation engineers try to determine or measure the interest due a property from an economic viewpoint. Haney[4] has stated that "value being the heart of Economics, the economist's philosophy is bound to shape his value theory." Economic principles, however, are subject to change and interpretation. They cannot be expressed as concrete formulae as can the fundamental principles of physics. Gray summarized conditions nicely when he wrote that economic history in times past has been chiefly[5]

> . . . an attempt to explain, within the existing frame-work and assumptions of society, how and on what theory contemporary society is operating.
>
> Thus we have no science of economics because the conditions upon which economic doctrines rest are never everywhere the same at any one time, nor the same in any place over long periods of time.

Carver, on the other hand, called economics a science. He summarized our modern-day concepts of value in part as follows:[6]

> Value is the most important word in the whole science of economics. Fundamentally it means the esteem in which a thing is held, but under ordinary commerical conditions it means power in exchange. . .
>
> Value depends upon two things and two things alone: namely desirability and scarcity. Any material thing which can be transferred will have value if it possesses both of these qualities and will have no value if it lacks either of them. . .

Marx, Marshall, and Keyes also have contributed much to present-day theories of value. However, they are of greater interest to those in the petroleum industry for their views on capitalism, imperfect competition, and monetary theories, respectively. Under a free enterprise system their viewpoints have impacts under capital supply, taxable costs, and "profits", all of which are inter-related with prices. And except during times of government controls, prices, supply, and demand determine each other.

Price is the most common method of stating the value of anything being sold or purchased. Price is readily and conveniently expressed in terms of money which is understood by both buyer and seller. Money, then, also becomes an exchangeable commodity, and when so used is said to have purchasing power. Price expressed in terms of money represents exchange value. Money, therefore, can have no price; it can have only value—value in exchange.

Other economists have stressed the concept that the exchange of goods between those who perform the various steps of production results from the division of labor. Value, money, and prices, then, are the end results of exchange. Taussig stressed the concept that value depends on *marginal vendability*—the price at which the last article and the entire supply can be sold.

VALUE—A WORD OF MANY MEANINGS

Economists, and particularly the theoretical economists, have repeatedly decried the use of the monetary unit as the only reliable measure of value. Money as a measure of value can be an illusion. Money is not a good measure because its value when measured in purchasing power may fluctuate rapidly over relatively short periods of time. Everyone is aware of the decrease in value of the dollar since World War II. There seems to be no hope that inflation or indirect "clipping of the coin" will end within the foreseeable future, and valuation engineers are sometimes asked to make allowances for it in their calculations of present worth. Normally inflation is not considered directly. The monetary unit must be used as an objective standard of measurement because the valuation report goes to a client or to management and they, like the accountants, can deal only in acceptable terms. Even though one oil property is to be traded for another, the monetary unit is the only practical basis for comparison of values.

The oil property valuation engineer arrives at a present-worth value of a property expressed in monetary units through sound reasoning and acceptable methods of calculation. Yet he may be asked to state the real value, fair value, market value, cash value, or some other kind of value of the property. These requests may be entirely proper because he learns through experience that the value he has arrived at may represent only the value to a specific individual or group of individuals who have interest in the property. The value to a prospective buyer may be something entirely different, and the calculated value becomes nothing more than a dollar figure that may be used as a basis for bargaining. The lending institution is interested in the calculated value only to the extent that the property might sell for that much money if foreclosure became necessary. Management may be interested in a calculated value only to see if a proposed expenditure or property will pay out with a profit. Or courts may introduce other and very confusing concepts of value. Some of the most common kinds of values are the following:

Actual value	Income value
Asset value	Inflated value
Book value	Intangible value
Capital value	Intrinsic value
Depreciated value	Investment value

Discounted value	Junk value
Dollar value	Loan value
Enterprise value	Market value
Exchange value	Natural value
Fair value	Nuisance value
Fair market value	Present-worth value
Forced-sale value	Replacement value
Going-concern value	Tangible value
Goodwill value	Tax value
Imputed value	True value
In-the-ground value	

Thus the word *value*, even from the economic viewpoint, has become a word of many meanings. The meanings of many of the terms would appear to be obvious. Some terms appear to be synonymous. Nevertheless, each term has been introduced at sometime to convey a particular meaning with respect to a particular situation. Although these particular inflections of meaning are not absolute, the oil property valuation engineer should become acquainted with them. He will find the many shades of meanings arising most often in court cases, rate-making cases, and in tax matters. He is primarily concerned though with the calculated present-worth value which he might think of as a basis for the exchange value—the dollars that would be given or received for the property if sold at a particular time. Present-worth values are based on certain assumed conditions and buyers and sellers may evaluate the variables differently. Lending institutions use viewpoints that are still different.

At sometime every oil property valuation engineer will insist that the present-worth value of a property calculated by the capitalization-of-income method represents a fair price, fair value, or fair market value of the property. It is when he finds himself in court supporting one of his valuations that he is enlightened. These, and particularly fair market value, are legal terms and should be reserved for such purposes.

FAIR MARKET VALUE

The expression *fair market value* is probably as old as trade and barter in the public market place. Yet today considerable misunderstanding exists with respect to its intended meaning. The expression in connection with oil properties arises most often regarding tax matters. It occasionally arises in connection with condemnation suits, dissolution of partnerships, and bankruptcy. Fair market value also may arise in connection with gas-producing properties under investigation by the Federal Power Commission. It seldom arises in connection with insurance and other claims for damages, investigations of sales of stocks and securities, and under the law of eminent domain. Black defined fair market value:[7]

> The market value of an article or piece of property is the price which it might be expected to bring if offered for sale in a fair market; not the price which might be obtained on a sale at public auction, or a sale forced by the necessities of the owner, but such a price as would be fixed by negotiation and mutual agreement, after ample time to find a purchaser, as between a vendor who is willing (but not compelled) to sell and a purchaser who desires to buy but is not compelled to take the particular article or piece or property.

The discussion of fair market value with respect to oil properties will be limited to tax purposes only because the concept of the willing buyer and willing seller is accepted by all courts as fundamental.

FAIR MARKET VALUE FOR TAX PURPOSES

It often becomes necessary in connection with federal taxation, and occasionally in connection with state taxation, to determine the fair market value of oil properties. Some of these occasions, according to Fiske,[8] are

> . . . to determine the value of the estate of a decedent owning oil and gas properties for computing the Federal estate tax, and to establish the basis for subsequent gain or loss and depletion; to determine the value of a gift of oil and gas properties for gift tax purposes; to determine the amount of a dividend declared in oil and gas properties by a corporation, either as an ordinary dividend, or in liquidation, and establish the basis for subsequent depletion and gain or loss; and to furnish a basis for the allocation of the cost or other basis to individual properties where oil properties have been acquired in a lump sum purchase or by an exchange, or an oil property has been sold and an oil payment retained.

The expression *fair market value* appeared in our first income tax law in 1913 although it was not defined for such purposes until 1919 upon publication of the first manual for the industry. It states:[9]

> The value sought should be that established assuming a transfer between a willing seller and a willing buyer as of that particular date.
>
> No rule or method of determining the fair market value of mineral property is prescribed, but the Commissioner will lend due weight and consideration to

any or all factors and evidence having a bearing on the market value, such as (*a*) cost, (*b*) actual sales and transfers of similar properties, (*c*) market value of stocks or shares, (*d*) royalties and rentals, (*e*) value fixed by the owner for the purposes of the capital-stock-tax, (*f*) valuation for local or State taxation, (*g*) partnership accountings, (*h*) records of litigation in which the value of the property was in question, (*i*) the amount at which the property may have been inventoried in probate court, (*j*) disinterested appraisal by approved methods, and (*k*) other factors.

The attitude of the tax courts with respect to fair market value has not changed since 1919. The Internal Revenue Service still broadly defines it as:[10]

. . . the price at which the property would change hands between a willing buyer and a willing seller, neither being under any compulsion to buy or sell and both having reasonable knowledge of the relevant facts. Assignments of value to the property by parties with adverse interests in an arm's length transaction and strong evidence of fair market value.

Fiske has amplified the attitudes of the Internal Revenue Service with respect to valuations of producing oil properties, as follows:[11]

The income tax regulations provide that the value of oil and gas properties must be established by the owner to the satisfaction of the District Director. These values must be established by evidence. In verifying the value submitted, the District Director will give due weight to all of the evidence presented. Some of the factors which will be considered, listed in the approximate order of preferential weight which will given to each factor, are as follows:

1. An actual sale of the property.
2. A bona fide offer to sell or purchase the property.
3. Actual sales of similar properties in the same or nearby oil and gas fields.
4. Valuations made for purposes other than Federal taxation.
5. Engineer's appraisals.
6. Opinions of qualified oil and gas operators.

Fiske also pointed out:[12]

The regulations state that an engineer's appraisal will not be used if the value can be reasonably determined by another method. In practice the Commissioner will consider engineers' appraisals along with all other evidence of value for what they are worth, which is the rule followed by the courts. The courts have considered appraisals in setting the value of oil and gas properties when supported by expert testimony.

High income tax rates since World War II have brought about new funds for capital investment known as "tax money," money that if not so spent would be collected as income tax. Losses can be written off as losses, and gains are taxed only at the lesser capital-gains rate. The investor in oil properties naturally views the income tax as a cost to be borne by the property. Davis and Stephenson stated with respect to gas properties:[13]

> The impact of the graduated income tax at high rates is so violent that it distorts market forces and relationships which formerly were considered normal. The common definition of fair value should be modified somewhat as follows: "The fair market value of a property is the price at which a willing and informed owner would sell to a willing and informed buyer, with full consideration given to their relative tax positions."

Vance has proposed a more restricted definition:[14]

> "Fair market value" of an oil property is the price agreed upon between a willing buyer and a willing seller both being producers of oil and gas, exclusively, each in possession of all known facts and both in the same tax bracket, the property being purchased without the tax advantage of the oil payment method.

Bonbright some years earlier criticized the legal viewpoints of fair market value as follows:[15]

> A critic of this traditional jargon may raise the point that willingness to buy and sell is a matter of degree and depends in large measure on the price at which the sale shall take place. The difficulty seems to have been recognized by . . . [one judge] for he defined willing buyer as a buyer who is willing to pay the *value* of the property. But this attempt to give precision to the phrase simply impales the court on the other horn of the dilemma, for it runs into a vicious circle. It makes market value depend on a hypothetical sale, and it makes the price of this sale depend on an assumption of the very figure which is to be found, namely the value of the property.

An oil property might have considerable value to its owner but according to the law might not have any value simply because the willing buyer cannot be found. Sherrod[16] has discussed many interplaying and overlapping factors entering into the fair market value of an oil property that make the attempt at an engineering valuation almost seem presumptuous. Valuation engineers and bankers are said to have observed as of the early 1960's that the market value of oil properties as represented by actual sales is approximately 60 per cent of the calculated present worth using a 6 per cent discount factor.

The rule is still of considerable interest despite radical changes within the industry and governmental attitudes and related greatly increased prices for crude oil and natural gas.

In summing up the meaning of fair market value, Bonbright wrote[17]

. . . market value will not qualify as a basis of legal appraisal save in a rather limited number of cases. But no alternative definition of value will serve as a jack-of-all-trades. The use of "market value" as the verbal basis for settling all varieties of legal disputes represents a uniformity of mere words rather than one of principle. The multiformity of value standards is only concealed, not avoided, by the legal definition of market value as the price at which the property would be exchanged between a "willing buyer" and a "willing seller."

REFERENCES

1. Joseph J. Spengler and William R. Allen, *Essays in Economic Thought*, Rand McNally, Chicago, 1960, p. 281.
2. John Locke, "Essay on Interest and Value of Money," *Principles of Political Economy*, Ward, Lock, and Co., London, 1825, p. 246.
3. Ralph Barton Perry, *Realms of Value*, Harvard University Press, Cambridge, Mass., 1954, p. 3.
4. Lewis H. Haney, *History of Economic Thought*, 3rd ed., Macmillan, New York, 1936, p. 16.
5. Alexander Gray, *The Development of Economic Doctrine*, Longmans, Green, New York, 1931, p. 12.
6. Thomas Carver, *The Encyclopaedia Americana*, Vol. 27, Americana Corp., New York, 1959, ed., p. 653.
7. Henry Campbell Black, *A Law Dictionary*, West, St. Paul, Minn., 1910, p. 761.
8. Leland E. Fiske, *Federal Taxation of Oil and Gas Transactions*, Banks, Albany, N. Y., 1975, pp. 157-158.
9. *Manual for the Oil and Gas Industry under the Revenue Act of 1918*, Treasury Department, U. S. Internal Revenue, 1919, pp. 25-26.
10. Tax Guide for Small Business, U. S. Internal Revenue Service, Publication 334, 1976, p. 132.
11. Fiske, *op. cit.*, p. 159.
12. *Ibid.*, p. 165.
13. Ralph E. Davis and Dr. Eugene A. Stephenson, "Valuation of Natural Gas Property," *J. Petr. Tech.*, July 1953, Sec. 1, p. 9.
14. Harold Vance, "Evaluation," Chap. 16 in *History of Petroleum Engineering*, American Petroleum Institute, New York, 1961, p. 1049.
15. James C. Bonbright, *The Valuation of Property*, Vol. 1, Michie, Charlottesville, Va., 1965, p. 60.
16. Gerald E. Sherrod, "Factors beyond the Engineering Evaluation That Affect Fair Market Values," *J. Petr. Tech.*, December 1962, pp. 1307-1310.
17. Bonbright, *op. cit.*, p. 65.

Additional References

J. Oliver Buswell, Jr., *The Philosophies of F. R. Tennant and John Dewey*, Philosophical Library, New York, 1950.

Irwin Edman (Ed.), *The Philosophy of Santayana*, Scribner, New York, 1953.

Robert L. Heilbroner, *The Worldly Philosophers*, Simon and Schuster, New York, 1953.

C. I. Lewis, *An Analysis of Knowledge and Valuation*, Open Court Publishing Co., La Salle, Ind., 1947.

T. R. Malthus, *Definitions in Political Economy*, Reprints of Economic Classics, Kelly and Millman, New York, 1954.

D. H. Parker, *Philosophy of Value*, University of Michigan Press, Ann Arbor, Mich., 1957.

Ralph Barton Perry, *General Theory of Value*, Harvard University Press, Cambridge, Mass., 1950.

Eric Roll, *A History of Economic Thought*, 3rd ed., Prentice-Hall, Englewood Cliffs, N. J., 1956.

Adam Smith, *An Inquiry into the Nature and Causes of the Wealth of Nations*, Great Books of the Western World, Vol. 39, p. 12, Encyclopaedia Britannica, Chicago, 1952.

Edmond Whittaker, *Schools and Streams of Economic Thought*, Rand McNally, Chicago, 1960.

5 CAPITAL—ITS MEANING AND SOURCES

The word *capital* like *value* has been a prolific source of inspiration to a multitude of writers in many walks of life. Capital and the economic system based upon it have been condemned as the roots of all economic evils and praised as the cornerstone of the remarkable growth and prosperity of our nation. Capital and wealth are synonymous in the public mind, particularly if wealth is understood to include all those possessions readily converted into money. Capital often is defined simply as monetary wealth or its equivalent in purchasing power. Billions of dollars are spent annually by the industry to find and develop new producing oil properties. If oil properties are lands, minerals are a part of lands, and oil accumulations are minerals, the question sometimes asked is: "Are oil accumulations capital?" The question is of more than academic interest to the industry. A review of the economist's concepts of capital will show why such a question arises.

EARLY CONCEPTS OF CAPITAL

According to Karl Marx capital had its beginning in the first circulations of commodities, and its modern history began in the sixteenth century with worldwide markets and commerce. Starting approximately at that time and continuing into the latter part of the eighteenth century, many "doctors," most of whom were "clerics," wrote profusely on numerous subjects which introduced many new concepts having to do with social justice and the general welfare of man. Concurrently two widely different schools of economic thought developed; one commonly called the Mercantilistic School evolved in England on account of its developing economic and industrial revolution. The other evolved in France under the name of the Physiocratic School. The Mercantilists promoted manufacture and trade, particularly export trade as a permanent cure for all the economic problems of England. Contrarily the Physiocrats did not believe that the gold and silver gained through trade would have a lasting and real value to England. They believed that land was the only thing of permanent value. Land was productive in itself. It gave to man his food, materials to be used for clothing and shelter, and the metals needed for making all implements and tools. Land continually brought new things into being. Mobile wealth such as money and precious metals used for trade and as business capital was not productive. Only agriculture, fishing, and mining were said to yield a true product. Manufacture and commerce did not produce anything actually new; they just changed the form of things or caused the form to be changed.

Both the Mercantilists and the Physiocrats had large followings at the time of Adam Smith. He was very familiar with the teachings of both schools and was influenced for a time by the teachings of the Physiocrats. Smith amalgamated and added to what he thought were the soundest concepts of all students of economics and social behavior up to the time of writing his *Wealth of Nations*. Smith introduced his concept of capital in this classic by pointing out that one rarely expected to derive any income from his "stock" of wealth or possessions until such time as he had sufficient surplus to last for periods of months to years. Only then would he try to derive any income from a large part of it. He would naturally do this by reserving only that amount needed to tide him over until his revenue started to come in. Smith stated:[1]

> His whole stock, therefore is distinguished into two parts. That part which, he expects, is to furnish him this revenue is called his capital.

Smith also pointed out that capital may yield a revenue or a profit in two ways: first when employed to raise, to manufacture, or to purchase goods to be sold at a profit; and second, when employed to improve land and to purchase useful machines and instruments of trade. In brief, capital without human effort could produce neither flow nor increase of material things. Smith though that the two basic processes of creating wealth are production and exchange. Land, labor, and capital were the basic factors needed for the production processes. Land included everything existing in nature such as soil, minerals, climate, rainfall, flora, and fauna. Labor consisted of the human energy or effort expended in bringing new and more material things into existence. Smith did not limit capital to money alone; it included tools, machines, and buildings and all other useful things that man has created in past from land.

Smith, like Mill and others, divided capital into *fixed capital* and *circulating capital*. Improvements to and on land such as clearings and drainage or buildings and factories were very important forms of fixed capital; money and stocks of goods intended for sale were examples of circulating capital.

Naturally the capital of a community or of a nation included far more than its store of monetary wealth. Not all money could be called capital. Basically money was a necessary medium of exchange and a means to an end.

MODERN CONCEPTS OF CAPITAL

Capital as defined by Smith gradually came to mean the manmade wealth that could be augmented by production. As labor-saving devices were introduced into industry and agriculture it became evident that manmade improvements on land wore out when used for production purposes. Capital in these forms depreciated, that is, it became less serviceable and less valuable with continued use. Hence capital gradually came to mean the manmade wealth that could be augmented by production or diminished by use.

A modern classical economist might define capital as any physical agent of production other than man and natural resources. He would perhaps add that all capital is produced by man, that all capital costs something to produce, and that all capital represents savings or the retention of previous surpluses. He undoubtedly would caution that savings alone are not necessarily capital and savings alone do not create capital; savings become capital only when used to produce something of value. Marx believed that capital as opposed to landed property is always in the form of money first. This money has to be transformed into capital by definite processes. He thought that a difference in the form of circulation is the first observable distinction between money that is money only and money that is capital. Money in the form of capital was not to be considered a factor in production. It was only the means of exploitation of the propertyless through their employment on the "machines" of production.

Marshall, one of the more authoritative late British economists, is often quoted with respect to capital and income.[2] In 1922 he wrote

> . . . the language of the market place commonly regards Man's capital as that part of his wealth which he devotes to acquiring an income in the form of money; or, more generally to acquisition . . . by means of trade.

According to this definition land would appear to be capital and a part of one's wealth if used to produce a money income. Marshall had trouble getting away from the older English concepts, and like most classical economists excluded land as a part of capital from a social point of view; land in these cases being all free gifts of nature, such as mines, fisheries, forests, etc., which yield income. Marshall conceded, however, that land is but a particular form of capital from the *point of view of the individual producer*, and that economists have no choice but to follow well-established customs regarding the use of the term capital in ordinary business; this is, trade capital.

Catlin[3] stated that John Bates Clark, a contemporary of Marshall and often called the dean of American economists, quite consistently discounted any distinction between land and other capital goods. He also added that Herbert Davenport, who is perhaps our leading exponent of commonsense "price-economics," says the only tenable view is that which "conceives capital as including all durable and objective sources of valuable private income."

Land and Minerals Contrasted

The classical economist contrasts land and capital in many ways. To him capital implies productivity for profit. A characteristic of capital and capital goods is that both must be used directly or indirectly to create or produce marketable products. Anything that lacks the ability to create a money return is not capital. The quantity of capital in any form can be increased by leaps and bounds by the work of man according to the demand for it. Capital tends to increase with population growth and the supply tends to keep up with the demand, all with very little change in the costs of using it.

Land is not manmade. It is fixed in quantity from a practical viewpoint. Nature, of course, is constantly creating new land, but at a very slow rate. Man may create some land through the construction of dikes as has been done in Holland; nevertheless, the net increase during lifespan is negligible. The same is true with respect to mineral deposits. They definitely are fixed in quantity in nature.

Land and minerals in place are essentially indestructible except through natural processes. Capital and capital goods can be destroyed more quickly than they can be created (e.g., during wartime). Should a society or government stop payment for the use of capital, the rate of its accumulation would decline rapidly and the available supply would soon be eliminated. Contrarily should a society or a government stop payment to the individual landowners for the use of land, there would be no decrease in the net amount of land. There would be a decrease only in its availability to nonowners.

Land and Minerals Are One

What is true with respect to land also must be true with respect to our mineral resources because they are inseparable from a legal viewpoint. The laws of the United States are unique with respect to ownership of mineral wealth in that ownership of land includes both surface and subsurface rights in a clearly defined tract. Our own courts in the early days upheld the principles of absolute ownership of all the minerals on and in a tract of land. Following the birth of the oil industry they also established the principle that oil and natural gas are minerals and included in any mineral ownership as well as in any conveyance or transfer of mineral rights and ownership unless specifically excluded. Since that time two theories have developed with respect to ownership of minerals; one is referred to as the *ownership in place theory*, which is an outgrowth of the well-known "Rule of Capture" decision of a judge in Pennsylvania, who in 1889 stated in part:[4]

> Water and oil and still more strongly gas, may be classed by themselves, if the analogy be not too fanciful, as minerals ferae naturae. In common with animals, and unlike other animals, they have the power and tendency to escape without the volition of the owner . . . They belong to the owner of the land, and are a part of it, so long as they are on or in it, and are subject to his control; but where they escape, and go into another land, or come under another's control, the title to the former owner is gone.

Thus ownership to oil and gas is either lost or gained when it moves across property lines. It is gained when it migrates into a property and lost when it migrates from a property.

The second theory is known as the *nonownership in place theory*. It too recognizes the fugacious property of oil and gas and for that reason would concede only to the exclusive right to drill on a tract of land and the right to ownership of only that oil and gas produced at the surface.

Sullivan has summarized the working of the two theories in practice as follows:[5]

> In spite of the difference in theory of oil and gas in place, all jurisdictions have reached substantially the same result. Thus, the courts are in agreement as to the following: that when brought to the surface and reduced to actual possession they are personal property; that prior to production that landowner has certain rights, namely, to withdraw the oil and gas through wells drilled on his property and without liability for consequent drainage from neighboring lands, to transfer this privilege to others, to be protected from interference by third persons who have no rights in the reservoir; and under comprehensive conservation statues, not to waste the oil and gas.

Capital from Business and Tax Viewpoints

Despite the theoretical viewpoint of classical economists that there is a marked contrast between capital and our natural resources or the free gifts of nature, one has to be objective, resourceful, and practical to utilize capital profitably for productive purposes. Oil and other natural mineral resources may have existed for thousands or millions of years before discovery or utilization by man. They became of real value to him only when he found them useful and that they could be exploited through production and disposed of at a profit.

In discussing treasure trove Smith stated that it belonged to the sovereigns, not to the finders.[6] Treasure trove was placed in the same classification as gold and silver which were never included in general land grants except through special charter. Mines for lead, copper, tin, and coal were excluded because of their much lesser value. Oil accumulations may be treasure trove or gifts of nature but they are hardly free gifts of nature because industry spends billions of dollars annually in search of them, in reducing them to possession and introducing a marketable product for profit. The oil industry often is said to be an extractive industry and that its very existence is founded on the exploitation of a wasting asset. Actually only the production branch of the industry, that part of particular interest to the oil property valuation engineer, is extractive. The production branch of the industry does deal in the search for and the extraction of a natural resource which is one of the fossil fuels. Fossil fuels are not renewable by man during his life or by any natural process as are some of the other natural resources such as forests and the fertility of soils.

Herfindahl, a noted economist and authority on conservation of natural resources for the future, has included mineral deposits along with other capital goods in discussing the need for the same rate of return on their value with respect to the future of capital accumulation.[7]

On the other hand, Garnsey, also a well-known economist and student of mineral taxation, upheld the classical concept of land and minerals as indicated by the following statements:[8]

> The role of private enterprise, . . . is to discover and process the resource in return for a suitable reward covering costs and normal profits as discovered by competition. From this point of view, discovered oil or oil in the ground, is not capital at all. Rather it is a consumer goods. The analogy between a barrel of oil in the ground and a side of beef in my deep freeze is precise. They are both consumer goods being held for future use.

Coons, like Herfindahl, also took a broad viewpoint in defining capital and income when he stated in part:[9]

> Broadly speaking, capital consists of all useful things which, at any time, have been inherited by society from the past. So defined capital includes

land, buildings, machines, technology, and every other aid which, when combined with human effort, creates the flow of goods and services . . . identified as real income.

Today production from land is overwhelmingly for the market whether it be forest products, mineral products, or agricultural products. It is therefore difficult from legal, practical, and business viewpoints to exclude natural resources in use for production purposes from land and capital. The businessman considers the total amount of investment in a production enterprise as capital. The oil business is primarily a production enterprise. The value of that fraction of the volume of the net working interest of oil in the ground which a company thinks it can produce during the period of its production rights is not included in the annual public corporate report. It is thought of, however, by the company as a capital asset in determining the value of a property if for sale or trade.

Some economists try to get around the problem by distinguishing between what they call real or social capital and business capital. Business capital is construed as a sum of money applied in whatever form to yield a revenue; or what is the same as the present value of a stream of future earnings. Business capital may consist of tangible and intangible things. Business invests in many intangible things. This is certainly true in oil production where it is difficult to pronounce one expenditure an intangible expense and another as an intangible capital investment.

Congress made clear in writing the original income tax law in 1913 that only income and not capital was to be so taxed. Some states tax mineral deposits as well as a stockpile of produced minerals as a capital asset the same as any other personal property. With respect to federal and state income tax laws, however, only the income from the sale of produced minerals should be subject to income tax. Sales of minerals in place and mineral rights are subject to capital gains tax. In all cases provisions were made until late in 1975 whereby a portion of income from production of *all natural mineral resources* were not subject to income tax in order that the tax savings might be used to replenish the wasting mineral asset, that is, to conserve one's capital. The courts and the Congress have tried repeatedly to clarify the distinction between capital and income with respect to income tax purposes. The problem as to whether a mineral resource in the ground, and particularly an oil accumulation, is capital still confuses many legislators when they are faced with the problem of raising additional revenue.

Smith's threefold factor classification of productive agents–land, labor, and capital–may not be obsolete, but from the viewpoint of the modern industrial world, is not wholly acceptable to some modern economists. The courts and legislators have often tried to resolve the differences between capital and income for taxing purposes. Yet the distinctions are often overlooked, particularly with respect to capital. The problem as to whether mineral resources in the ground, and especially discovered oil and gas accumulations, are capital and to be so treated taxwise, has never been resolved satisfactory from the industry's viewpoint.

The United States petroleum industry is one of the largest and most important segments of the nation's economy. Its gross and net investments in fixed assets as of December 31, 1975, were estimated to be $135.2 billion and $71.63 billion, respectively,[10] representing increases of approximately 90 per cent and 100 per cent, respectively, in comparison with December 31, 1964. These net fixed assets represent the gross investment less retirements in wells and in production, pipeline, marine, refinery, chemical plants, marketing, and other related domestic facilities. Accordingly, at the close of 1975 the net investment per daily barrel of domestic productive capacity was approximately $8,500, and $84,500 for each of the industry's employees.

Total domestic capital and exploration expenditures for 1975 totalled $18.92 billion, of which exploration expenditures alone amounted to $1.195 billion or approximately 6.3 per cent of the total.[11] During the period 1965 to 1975, inclusive, the domestic industry spent approximately $115.8 billion for capital improvements and exploration expense as shown in Table 5-1.

Table 5-1

Capital and Exploration Expenditures–United States[12]

	Millions of Dollars		
Year	Total Capital Expenditures	Geological, Geophysical Expense, and Lease Rentals	Total
1965	$ 6,375	$ 610	$ 6,985
1966	7,125	650	7,775
1967	7,630	615	8,265
1968	8,350	715	9,065
1969	8,175	725	8,900
1970	8,225	665	8,890
1971	7,250	715	7,965
1972	9,050	740	9,790
1973	10,640	850	11,490
1974	16,625	1,130	17,755
1975	17,725	1,195	18,920
Totals	$107,190	$8,610	$115,800

Natural gas and petroleum supplied approximately three-fourths of the nation's energy needs from 1965 to 1975. Increasing costs and domestic scarcities of supplies will progressively reduce that percentage in the future. A representative study by the U. S. Bureau of Mines shows that petroleum, natural gas, and shale oils will meet only 64 per cent and 47 per cent of the nation's inputs in the years 1985 and 2000, respectively.[13] Nuclear energy and coal, and hydropower and geothermal energy to minor extents, will have to make up most of the differences.

The question naturally arises as to how the industry can meet future demands for production and consumer's products in view of the rapidly increasing need for capital investment at a time when industry's annual rate of return is abnormally low and capital expenditures are directly related to rate of return.

Capital Formation

Capital formation is the foundation of the capitalistic system of production. The term in a broad sense merely means the creation of a surplus of production over consumption. The surplus of production may be either produced goods that may be converted into money or money itself in the form of savings. There can be no expansion in production operations without capital formation and its utilization.

Historically the capital needs of a newcomer into the industry have been provided through savings, stock sales, and personal credit through borrowing. Savings may have been individual savings or surpluses from another going company or corporation. Stock sales would probably be out of the question except for a successful corporation wishing to branch out into the oil industry. Personal credit through borrowing might be utilized if the newcomer has other property to be put up for security.

Once the newcomer's company is established the process of capital formation depends almost entirely on his receipts from sales of products exceeding operating costs and taxes. This excess, commonly thought of as net profit, must be sufficient together with outside savings which might be attracted to the enterprise to provide for the replacement, growth, and output desired by the owners and needed by the market. In addition there should be something left over which might be distributed periodically to the owner or owners in the form of dividends, rent, or interest on the capital investment.

The capital needs of an established company, whether large or small, are said to be generated primarily from internal financing and secondarily through outside financing. Internal financing consists largely of retained profits, depreciation and depletion credits, and occasionally amortization accounts. Retained net incomes are often thought of as self-generated savings. The term net income is literally synonymous with the term profits, but the latter can be used to arouse denouncement of certain business incomes. Retained net incomes within the petroleum industry since the mid-sixties have had to be augmented by ever greater outside financing, consisting largely of the following:

1. Loans obtained from a bank, insurance company, or other lending institution.
2. If incorporated, sales of capital stock, bonds, or debentures.
3. If unincorporated, the sale or assignment of a part interest in an oil property or properties.
4. Whether incorporated or not, sale or assignment of an oil payment, sale of fixed and other assets, and leaseback of facilities.

Banks usually are somewhat reluctant to loan money for more than two or three years. Insurance companies make long-term petroleum industry loans, but usually do not wish to bother with small accounts. An operator may need money for development and expansion purposes yet his allowable rate of production may be too low to pay off a loan in customary time.

Oil's Financing Changing

Before stabilization of the Texas oil industry in 1933-1934, which followed the hectic days of boom conditions in the East Texas oil field, it was practically impossible to obtain a loan on an oil property from a bank or other commerical lending institution. Since that time banks, insurance companies, and other lending institutions have played an important part in providing the capital funds needed by the industry. Financial analyses of thirty representative oil companies during the period 1934-50[14] show proportionately very little need to borrow for capital expenditures in comparison with two and three decades later. During the depression years of 1934-36 they retired $131 million more than had been obtained from external sources. Conditions began to change in 1937. For the entire period 1934-50, they had borrowed $3,039 million more than had been repaid. Net funds so obtained from external sources represented 15.6 per cent of their capital expenditures. The percentage scarcely increased during the next five years.

It has been more and more difficult since World War II to separate the available financial data of the American oil industry into foreign and domestic operations. However, a study of the financial statistics for 1975 of 29 representative companies, some of whom had operations abroad, showed ratios of long term debt to total capital employed of 40 percent in the United States and 24 per cent for worldwide operations.[15]

The industry broadened its operations horizontally more rapidly during the early 1960's by investing more heavily in chemical plants. More investment was made in such plants during 1966 and 1967 than for refinery construction. Expenditures for chemical plants are estimated to have been approximately 11 per cent of all capital expenditures for each of those two years. Dollarwise these expenditures were only half as much in 1973 and amounted to only 4 per cent of the total. They were nearly doubled in 1974 and 1975, yet represented only a one percent increase of capital expenditures in comparison with 1973.

Some of the majors also have expanded into what they call the overall energy supply field which includes nuclear energy and uranium ore and coal mining operations.

The percentage distribution of the net investment in fixed assets of the United States petroleum industry as of 1965, 1975,[16] and of estimated requirements for the next ten years are shown in Table 5-2.

Table 5-2
Percentage Distribution of Investments of U.S. Petroleum Industry

	Net Investment in Fixed Assets		Future Capital Requirements
	1965	1975[16]	1976 – 1985[17]
Production	58%	51%	63%
Transportation	10	10	7
Manufacturing	17	23	20
Marketing and other	15	16	10
	100%	100%	100%

The major companies have spent hundreds of millions of dollars on research aimed at developing economical methods of producing large volumes of syncrude and syngas from coals and oil shales. Capital requirements for syncrude and syngas for coal are estimated to be approximately the same as for shale-oil plants in 1985 and 85 per cent more in 1990.

Estimated Future Financial Capital Requirements

Projections of the estimated capital requirements of the United States petroleum industry vary widely depending largely upon estimates of domestic demands for products. Theoretically, demands should be a function of prices in the future. Contrary to price trends in the Spring of 1976, increased prices within the near future were expected to reduce former annual rates of increased product demand. Prices would be dependent largely upon anticipated rates of inflation, prices of imported crude oils and products, wildcat well successes, development of oil shales, tar sands, and syncrudes, and future governmental regulations and taxation policies.

Projections made in 1965 of future capital requirements of the domestic industry indicated an average annual need of only $12.5 billion over the decade of 1970-1980. Early nationalization of many of the nations foreign operations and drastic increases in the prices of imported crude oils and products were not anticipated at that time. Annual net earnings were expected to supply approximately 85 per cent of all capital needs. No difficulties were to be expected in obtaining the added 15 per cent or $1.875 billion from outside sources.

Estimates of future financial needs of the industry were suddenly found to be much too low upon passage of the 1969 Tax Reform Act. It not only reduced the amount of self-generated capital, but it also reduced borrowing ability and other capital sources. The institution of price controls in 1971 also hurt. Relinquishment of complete managerial controls and ownerships in major foreign operations and Middle East high crude oil pricing policies quickly followed. All crude oil reserves in the United States and the rest of the world suddenly increased substantially in value. Higher domestic crude and product prices necessarily followed. All resulted in suddenly increased "profits" which triggered the elimination by Congress of the depletion allowance for all the larger companies. All present and future expenditures were subject to continued inflation. And as of the mid-seventies the threats of divestiture, vertical and horizontal, were very upsetting, and especially so to the major companies. Each of these developments greatly increased the estimated future capital needs of the industry while reducing needed self-generated funds. As a result subsequent estimates of future capital needs had to be preceded with certain "ifs" such as future rates of inflation and Congressional action with respect to prices for crude oil and natural gas, divestiture, and environmental requirements.

One estimate of capital expenditures by the petroleum industry for the period 1976-1985, inclusive, totalled $430 billion for the United States and $900 billion for the non-Communist world as a whole. The breakdown is shown in Table 5-3. In addition, other large amounts were needed for debt repayment, added working capital, dividends, etc., totalling $445 billion as shown in Table 5-4. An inflation rate of 10 per cent per annum was assumed throughout the study. Another independent study included three estimates based upon no restrictions to imports, to domestic energy independence by 1985, and by

1990. All estimates were based upon the 1974 dollar with programs starting in 1975 and no inflation. Assuming no import restrictions, $194.8 billion would be needed to meet crude oil and natural gas requirements by 1985 and $298.6 billion by 1990. In addition, $12.8 billion would be needed for facilities to manufacture syngas from petroleum and coal and syncrude from

Table 5-3
Capital Expenditures in Non-Communist World—
10 Years ending 1985[18]

	Billions of Dollars*	
	United States	World
Oil and Gas Production (including exploration expense)	265	435
Natural Gasoline Plants	5	35
Refining and Chemical Plants	85	240
Transportation	30	85
Marketing and Other	45	105
Total	430	900

*Assuming 10 per cent per year inflation.

Table 5-4
Uses of Funds-Non-Communist World
1976-1985[19]

	Billions of Dollars*
Capital Expenditures	900
Investments outside the Petroleum Industry	70
Dividends	200
Repayment of Debt	140
Working Capital Additions	35
Total	1,345

*Assuming 10 per cent per year inflation.

oil shales and coal by 1985 and $35.0 billion by 1990. An estimated $22.5 billion also would be needed to provide for sufficient coal production for synfuels by 1985 and $39.0 billion by 1990. Accordingly, total capital investment in fossil fuels requirements would amount to $230.0 billion by 1985 and $373.0 billion by 1990. Elimination of all imports by 1985 and 1990 would require additional capital expenditures of $123.8 billion and $103.5

billion, respectively.[20] As of the mid-seventies, greater future dependence on imports appeared much more likely than any reductions. Other widely quoted estimates were $250 billion in 1975 dollars over the period 1975-1984 for oil and gas plus $35 billion for coal and synfuels[21] and $380 billion in 1974 dollars for the petroleum industry over the period 1975-1985.[22] Funds needed for debt retirement, working capital, and dividends were not included.

An increased percentage of the industry's capital requirements in the future will need be met through percentage increases in net incomes as compared with the decade ending in 1975. Future returns of 15 to 20 per cent are often quoted, as much or more than the "obscene" returns in 1973. Otherwise, the customary lending institutions may hestitate and perhaps refuse to provide all the extra capital needed.

CAPITAL ASSETS

Anything that has value of salability and is partially or wholly owned or possessed by a business enterprise is a *capital asset*. The oil property valuation engineer evaluates capital assets. Capital assets may be fixed, current, or deferred. They also may be tangible or intangible.

Fixed assets include all assets of a lasting or permanent nature that have been acquired in any manner for conducting a business enterprise. An oil company, and especially an integrated oil company, would have many kinds of fixed assets. An oil property whether owned in fee or held under lease, farmout, etc., and any wells and production equipment on it are fixed assets. *Current assets* are cash, salable things on hand, and other items readily convertible into cash without serious loss of values. An oil company would have many kinds of current assets. The individual oil property might have certain production facilities not needed for future operation and/or some oil in storage. These would be current assets. An asset already paid for, or for which a business is liable, yet not chargeable during the current accounting period, is known as a *deferred asset*. Improvements and production facilities on an oil property may be deferred assets for a time for accounting and income tax purposes.

An oil company may have both *tangible* and *intangible assets*. Its oil properties, pipelines, refineries, service stations, office buildings, office equipment, etc., are tangible assets. Its goodwill, whether originally purchased or not, patents, trademarks, etc., are intangible assets. An oil property also may have certain intangible assets such as location near a pipeline or highway, large size if desired, and low drilling-site costs. Such assets may or may not be included in the evaluation computations, but may be given value in actually consummating a sale or trade.

Capital Gains and Losses

A *capital gain* is the excess of the amount realized from a sale or exchange of a capital asset over the adjusted basis of the asset transferred.

A *capital loss* is the excess of the adjusted basis of the capital asset over the amount realized from the sale or exchange.

The *adjusted basis* from an income tax viewpoint is the original cost or other original basis adjusted for such things as casualty losses, improvements, and depreciation when appropriate. The amount realized from a sale or exchange of a capital asset is everything received for the asset disposed of, reduced by any costs incurred in effecting the transfer such as commissions, costs of preparing deeds, and other legal fees.

REFERENCES

1. Adam Smith, *An Inquiry into the Nature and Causes of the Wealth of Nations*, Great Books of the Western World, Vol. 39, p. 118, Encyclopaedia Britannica, Chicago, 1952.
2. Alfred Marshall, *Principles of Economics*, 8th ed., Macmillan, London, 1922, pp. 71-72.
3. Warner B. Catlin, *The Progress of Economics, A History of Economic Thought*, Bookman Associates, New York, 1962, p. 299.
4. Robert E. Sullivan, *Handbook of Oil and Gas Law*, Prentice-Hall, Englewood Cliffs, N. J., 1955, p. 45.
5. *Ibid.*, p. 44.
6. Smith, *op. cit.*, p. 121.
7. Orris D. Herfindahl, "Goals and Standards of Performance for the Conservation of Minerals," Pt. I, "Minerals and Energy," *Quart. Colorado School of Mines*, Vol. 57, No. 4, October 1962, p. 162.
8. Morris Garnsey (member on panel discussion), Pt. II, "Minerals and Energy," *Quart. Colorado School of Mines*, Vol. 58, No. 1, January 1963, p. 262.
9. Alvin E. Coons, *The Income of Nations and Persons*, Rand McNally, Chicago, 1959, p. 52.
10. *Capital Investments of the World Petroleum Industry 1975*, Energy Economics Division, Chase Manhattan Bank, New York, Schedules 2-3, pp. 12-13.
11. *Ibid.*, Schedule 4, p. 15.
12. *Ibid.*, Schedule 4, pp. 14-15.
13. Walter G. Dupree, Jr., and John S. Corsentino, *United States Energy through the Year 2,000 (Revised)*, U. S. Bureau of Mines, December 1975, adapted from Table 5, p. 10.
14. Frederick G. Coqueron and Joseph E. Pogue, *Capital Formation in the Petroleum Industry*, Petroleum Department, Chase Manhattan Bank, New York, February 1952.
15. *Financial Analysis of a Group of Petroleum Companies*, 1975, Chase Manhattan Bank, New York, N. Y., extrapolated from data, pp. 24 and 31.
16. *Ibid.*, extrapolated from data, Table 13, p. 32.
17. *Capital Investments . . . op. cit.*, extrapolated from data, p. 5.

18. *Capital Investments . . . op. cit.*, p. 5.
19. *Ibid.*, p. 6.
20. *Capital Resources for Energy through the Year 1990.* Energy Group, Bankers Trust Company, New York, N. Y., pp. 16-17, 31, 34.
21. *National Energy Outlook, 1976*, Federal Energy Administration, 1976, Washington, D. C.
22. Wallace W. Wilson, Financing the Oil Industry, *The Future of American Oil – The Experts Testify*, American Petroleum Institute, 1976, Washington, D. C., Chapter 10, p. 222.

Additional References

D. Clayton Arnold, "How to Get an Oil Loan," *Petr. Engr.*, Vol. 24, No. 7, July 1952, pp. A37-A39.

Joel Dean, "Measuring the Productivity of Capital," *Harvard Bus. Rev.*, January-February 1954, pp. 120-130.

Charles F. Parker, "Economics and Finance–The Roots of Corporate Enterprise," *J. Petr. Tech.*, February 1959, pp. 31-34.

6 THE LAWS OF INTEREST

Generally speaking, *interest* may be defined as the income earned on invested capital. Interest usually is restricted to that sum of money paid for the use of money on loan. The amount of the loan or investment on which interest is paid is called the *principal*. A debt is equivalent to a loan and interest may be charged on it also. The amount of the original debt is the same as the principal of a loan. The *rate of interest* is that fraction of the principal that is charged or paid per unit of time. This fraction is expressed as a percentage, that is, in hundredths of the principal per unit of time which is usually a year. Upper limits to rates of interest on loans and debts are set by some states to protect both the borrower and the debtor. The amount of interest can be calculated by many methods, however, some of which provide much more carrying charge than others.

The oil property valuation engineer is concerned only with *simple interest* and *compound interest*. He is particularly concerned with the application of the laws of compound interest. Simple interest is calculated in direct proportion to time and represents a minimum of either charge or income. Compound interest provides larger charges or incomes at the same rate of interest. It is basically interest on both the original principal and on accrued interest that in effect results in periodic increases in the principal amount upon which interest is charged or paid. These periodic increases in the principal may occur one or more times a year and continue indefinitely or until the total amount is paid. It is of advantage to the lender, because in effect his income from interest is automatically reinvested or reloaned at interest. The average citizen can take advantage of compound interest earnings in many forms of "savings accounts."

A great deal has been written in the past concerning the justification for interest charges and the prevailing rates at the time. Like capital it has been condemned and yet has come to play a very important part in our economy. Our present capitalistic system could not have developed without it. The teachings of the Judiac, Christian, and Islamic religions have forbidden the taking of interest. Any charges for a loan were called usury. Roman law permitted it in various forms much of the time and usury came to be regarded as a penalty charged when a loan was not paid on time. The beginnings of the industrial revolution in England and later in Europe undoubtedly led to the change in viewpoint. Interest was made legal in England under Henry VIII in 1545, repealed in 1552, and legalized again for good in 1571. The official interest rate was 10 per cent per year. Fortunately for capitalism, rates declined over the centuries to an average of approximately half that amount. They reached and followed what has been termed a "normal" or "natural" rate—that rate at which the available supply of capital is just absorbed by those needing funds for profitable investment. They were artificially increased during the early half of the 1970's and artificially reduced again in 1975, largely through governmental controls.

Interest rates are now determined by many factors, the chief of which are as follows:

1. The same forces of demand and supply that are applicable to the value of anything desired on the assumption that no risk is taken. Demand depends in part upon general optimism with respect to the future.

2. The Treasury Department of the federal government tries to control demand, optimism, and money supply through changes required in reserves of lending institutions, amount of monies it loans them and rates they must pay, and rates set by it to be paid on government bonds, certificates of deposit, and savings accounts.

3. Federal and state income tax rates and regulations with respect to bonds, certificates of deposit, and other securities.

4. Administration and collection expenses, and reserves to cover risks of reimbursement.

When calculating present worths in valuation work, discount rates (which are handled the same as interest rates) are based principally on prevailing interest rates to which some risk factor may be applied. Discount rates used by a company considering the purchase of a property usually will be determined by the earning power of its capital if used for other available and equally attractive purposes. There is a school of thought that insists, however, that loan rates at banks should never be exceeded. This policy supposedly will give the fairest present-worth values and will permit better comparisons of values when more than one property is under consideration.

SIMPLE INTEREST

Simple interest is the interest paid on the amount of a loan only. It may be prepaid to the lender periodically or on final settlement. The instrument of a loan is a *note*. Usually a note specifies the period of the loan or the specific date on which it must be paid plus any interest charges. It may be a *demand note*, one whose principal amount and interest must be paid at any time upon demand by the lender. Demand notes are usually limited to short periods of less than one year. A note may be signed to cover a debt exactly as for a loan. A debt that draws simple interest is nothing more than the postponement of payment of the original amount of the debt plus interest charges. The principal amount of a note bearing simple interest remains the same throughout the term of the note. The method of figuring the interest is as follows, where

P = the principal amount of a loan or debt, that is, of the note,
r = the annual interest rate,
t = term or period of the note in years,
S = the sum of the principal and the interest at any time t.

$$S_1 = P + Pr = P(1 + r)$$
$$S_2 = P + Pr + Pr = P(1 + 2r)$$
$$S_3 = P + Pr + Pr + Pr = P(1 + 3r)$$
$$S_t = P(1 + tr) \tag{6-1}$$

If r is the interest rate on an annual basis, the interest rate on a monthly basis would be $r/12$ and on a daily basis, $r/360$. The sum of the principal and interest on a monthly basis, where n is the period or term in months, would be the following:

$$S_{n(\text{mo})} = P\left(1 + \frac{nr}{12}\right) \tag{6-2}$$

where n is the period or term in days,

$$S_{n(\text{days})} = P\left(1 + \frac{nr}{360}\right) \tag{6-3}$$

The interest only would be $P(nr/12)$ and $P(nr/360)$, respectively.

Simple interest on long-term loans normally is calculated and payable at the end of each year of the loan. It would amount to Pr dollars annually. The interest for the last year and the principal are paid at the same time and would amount to $P + Pr$ dollars. Interest may be deducted from the principal amount paid to the borrower by the lender at the time of making a short-term note. This practice is called discounting the note and permits the lender to loan the advanced interest at interest to another borrower. Notes are negotiable as money to a limited extent.

The question might be asked what one dollar would amount to in one hundred years at 8 per cent simple interest. If the interest was paid either at the end of the period or on time each year and accumulated by the lender as a simple saving:

$$S_{t=100yr} = \$1.00\,[1 + (100 \cdot 0.08)] = \$9$$

Thus capital, in the form of money, neither grows nor accumulates very fast when loaned at simple interest rates. This is particularly true with respect to long-time loans when the interest is not collected until the loan is paid off or upon maturity.

COMPOUND INTEREST

A lender who collects the interest on a note at the end of each yearly period, and/or upon maturity of the note, has lost the opportunity of re-loaning the interest for periods of up to perhaps one year or more at a time. Compound interest provides a method whereby interest can be earned or charged on unpaid interest by the lender. Compound interest payments permit capital in the form of monetary wealth to grow just as living organisms may grow or multiply. Thus compound interest follows a growth law. Money loaned at compound interest can increase to double the principle amount in a relatively short time at modest rates of interest. It permits the lender to get much more income from the use of his money over extended periods of time at low annual rates of interest.

The Law of Growth

A bacterium may divide every half-hour or so to become two bacteria. Every bacterium in the culture does not divide on the half-hour, but a given percentage of the total divides and doubles at any instant of time with the result that all have divided and doubled in any half-hour period. Their numbers follow a growth law under a favorable environment.

The population of a city may follow a growth law also. For instance, the population of Houston, Texas, has doubled every eighteen years, approximately, since early in the nineteenth century. It may continue to follow the established growth law as long as the environment remains equally favorable. The demand for a petroleum product may be said to increase by a certain percentage annually. If conditions do not change appreciably over the next several years, the time required for the demand to double could be calculated by the growth law. The compound interest or growth law has many applications in oil valuations. The derivation of the basic equation is as follows, where

S = the sum of the principal of a loan or debt plus the periodic increases due to earned and credited interest at the end of anytime t, in years,
r = the interest rate or growth rate per year,
P = the initial principal of the loan or debt,
m = the number of conversion periods, that is, the number of times per year interest charges start on uncollected earned interest.

At the end of the first conversion period:

$$S_{m=1} = P + P\frac{r}{m} = P\left(1 + \frac{r}{m}\right)$$

At the end of the second conversion period

$$S_{m=2} = P\left(1 + \frac{r}{m}\right) + P\left(1 + \frac{r}{m}\right)\frac{r}{m} = P\left(1 + \frac{r}{m}\right)^2$$

By inspection:

$$S_{m=m} = P\left(1 + \frac{r}{m}\right)^m$$

and if there are m conversion periods per year

$$S_t = P\left(1 + \frac{r}{m}\right)^{mt} \tag{6-4}$$

The amount to which one dollar would grow in 10 years at 8 per cent compound interest, converted semiannually, would be, according to Eq. 6-4:

$$\begin{aligned} S_{t=10\text{ yr}} &= \$1.00\left(1 + \frac{0.08}{2}\right)^{2\cdot 10} \\ &= \$1.00(1 + 0.04)^{20} \\ &= \$2.191 \end{aligned}$$

This growth may be compared with that at simple interest whereby the earned interest is not reinvested during the period of the loan.

The demand for gasoline in the United States might be cited as an example of the growth law. If the demand for gasoline increases at a rate of 3.5 per cent annually, the time in years required for the demand to double can be calculated through use of Eq. 6-4. Taking the principal amount P at the present time equal to 1, and the future demand S_t equal to 2, and substituting,

$$2 = 1\left(1 + \frac{0.035}{1}\right)^{1 \cdot t}$$

$$\log 2 = t \cdot \log 1.035$$

$$t = \frac{0.301030}{0.014940} = 20.15 \text{ yr}$$

The number of conversions per year may be any positive number up to and including infinity. Growth becomes continuous with an infinite number of conversions per year and Equation 6-4 takes on a separate form. It states:

$$S_t = P\left(1 + \frac{r}{m}\right)^{mt}$$

Let

$$\frac{r}{m} = \frac{1}{n}, \qquad \text{then } m = nr$$

where

r = rate of increase of P per period t, percentagewise.

$$S_t = P\left(1 + \frac{1}{n}\right)^{nrt} \equiv P\left[\left(1 + \frac{1}{n}\right)^{n}\right]^{rt}$$

The exponent rt always will have a small finite value, so as n becomes infinitely large, that is, as growth becomes continuous, the value of S_t becomes the upper limit of

$$P\left[\left(1 + \frac{1}{n}\right)^{n}\right]^{rt}$$

The expression $(1 + 1/n)^n$ is of the same form as $(x + y)^n$. When n is any positive integer the expression may be expanded through use of the binomial theorem.

$$(x + y)^n \equiv x^n + nx^{n-1}y + \frac{n(n-1)}{1 \cdot 2}x^{n-2}y^2$$

$$+ \frac{n(n-1)(n-2)}{1 \cdot 2 \cdot 3}x^{n-3}y^3 + \frac{n(n-1)(n-2)(n-3)}{1 \cdot 2 \cdot 3 \cdot 4}x^{n-4}y^4$$

$$+ \cdots + nxy^{n-1} + y^n$$

Substituting for x and y

$$\left(1+\frac{1}{n}\right)^n \equiv 1^n + n\frac{1}{n} + \frac{n(n-1)}{1\cdot 2}\left(\frac{1}{n}\right)^2 + \frac{n(n-1)(n-2)}{1\cdot 2\cdot 3}\left(\frac{1}{n}\right)^3$$

$$+ \frac{n(n-1)(n-2)(n-3)}{1\cdot 2\cdot 3\cdot 4}\left(\frac{1}{n}\right)^4 + \cdots + n\left(\frac{1}{n}\right)^{n-1} + \left(\frac{1}{n}\right)^n$$

When $n = \infty$,

$$\left(1+\frac{1}{\infty}\right)^\infty \equiv 1 + 1 + \tfrac{1}{2} + \tfrac{1}{6} + \tfrac{1}{24} + \cdots + 0 + 0$$

The sum of all the terms approaches a finite limit, therefore the series of terms is convergent and the limiting value of $(1 + 1/n)^n$ as n approaches infinity may be carried out to any desired degree of accuracy. Its value to six significant figures is 2.71828. This limiting value is a constant denoted in mathematics by the letter e and is used as the base for the Napierian system of logarithms.

Substituting the limit of $(1 + 1/n)^n$ when $n \rightarrow \infty$,

$$S_t = P\, e^{rt} \tag{6-5}$$

which is the equation expressing the law of continuous growth.

The usual number of conversions per year with respect to savings accounts, loans, and debts is two, that is, semiannual conversions. Sometimes savings and loan institutions may offer more conversion periods per year in an effort to attract more depositors under the impression that much more interest will be earned on their savings accounts. Actually any increase over two conversions per year does not increase the value of S_t appreciably as shown by the following table based on the deposit of one dollar for a period of ten years at 8 per cent compound interest:

Number of conversions per year	Value of S_t at end of 10 years
1	\$2.1589
2	2.1911
4	2.2080
6	2.2138
12	2.2196
360	2.2253
∞	2.2255

Nominal and Effective Interest Rates

When compound interest is converted more than one time per year, the stated annual rate of interest is called the "nominal annual rate." The compounded interest actually earned during the year represents an annual return or "effective annual rate." Where

S = the sum of the principal and the interest.
P = the principal amount.
r = the nominal annual rate of interest.
r_e = the effective annual rate of interest.
m = the number of conversions per year.

According to Eq. 6-4:

$$S_t = P\left(1 + \frac{r}{m}\right)^{mt}$$

For one year:

$$S = P\left(1 + \frac{r}{m}\right)^{m}$$

and

$$S = P(1 + r_e)$$

Equating,

$$P(1 + r_e) = P\left(1 + \frac{r}{m}\right)^{m}$$

and

$$r_e = \left(1 + \frac{r}{m}\right)^{m} - 1 \qquad (6\text{-}6)$$

For example, a nominal annual rate of 8 per cent compounded semiannually is equivalent to an effective annual rate of

$$r_e = \left(1 + \frac{0.08}{2}\right)^{2} - 1 = 0.0816 = 8.16 \text{ per cent annually}$$

Similarly, when m becomes infinity, according to Eq. 6-5,

$$S_t = P\,e^{rt}$$

$$P(1 + r_e) = P\,e^{r}$$

$$r_e = e^{r} - 1 \qquad (6\text{-}7)$$

Values of r_e at various conversion rates per year when the nominal rate r is 8 per cent are given in the following table:

m	r_e
1	0.0800
2	0.0816
4	0.0824
12	0.0830
52	0.0832
360	0.0833
∞	0.0833

Present Worth

Present worth or present value may be defined in a broad sense as the value now of a sum of money which is payable with interest at some future date. Specifically, from an oil property valuation viewpoint, present worth is that amount of money which earning compound interest from the present will total a specified sum at a given future date. The engineering valuation is built around the concept of present worth because income from oil properties may be spread over many years in the future and future income must be discounted to the present.

Present worth can be calculated readily through use of the compound interest laws. According to Eq. 6-4 and redefining P as P_w, the present worth of S_t,

$$S_t = P_w\left(1 + \frac{r}{m}\right)^{mt} \tag{6-8}$$

Transposing,

$$P_w = \frac{S_t}{(1 + r/m)^{mt}} \tag{6-9}$$

If the interest rate is compounded semiannually,

$$P_w = \frac{S_t}{(1 + r/2)^{2t}}$$

if compounded quarterly,

$$P_w = \frac{S_t}{(1 + r/4)^{4t}}$$

and if compounded monthly,

$$P_w = \frac{S_t}{(1 + r/12)^{12t}}$$

The amount S_t usually is taken equal to *one dollar*, which makes for convenience in preparing tables of present worths. Thus equations for the present worths of one dollar compounded semiannually, quarterly, and monthly become, respectively,

$$P_w = \frac{1}{(1 + r/2)^{2t}} \tag{6-10}$$

$$P_w = \frac{1}{(1 + r/4)^{4t}} \tag{6-11}$$

$$P_w = \frac{1}{(1 + r/12)^{12t}} \tag{6-12}$$

Present-worth values of one dollar calculated through use of Eqs. 6-10, 6-11, and 6-12 are given in the Appendix in Tables 2, 3, and 4, respectively.

The present worth of one dollar with infinite conversions per year would become, according to Equation 6-5:

$$P_w = \frac{1}{e^{rt}} \tag{6-13}$$

Present-worth values of one dollar calculated through use of Eq. 6-13 for various rates of interest are given in Table 5 of the Appendix. Banks and other lending institutions sometimes use present worths calculated on either monthly or infinite conversions per year to check security values of oil loans because they given minimum present-worth values.

Present Worth of Oil Property Income. Compound interest converted semiannually usually is used with respect to payments on oil loans and calculations of present worths of production income from oil properties. Equations 6-4 and 6-8 are not considered to be entirely satisfactory for these purposes because of the incremental nature of the production income. For example, an oil loan might be extinguished over a period of years by monthly production payments made directly to the lender by the pipeline purchaser through a special division order signed by the debtor. These production payments may be of either equal or decreasing increments depending on the agreement and the productivity of the wells. Through custom the monthly production payments are accumulated on the books and upon receipt of the twelfth payment simple interest is calculated on the total of the payments for the year at one-half the normal interest rate. Thereafter compound interest converted semiannually is used. The same procedure is used in calculating the present worths of future yearly production income from oil properties for both working and royalty interests. This method is particularly convenient and sufficiently accurate in calculating the present worths of properties for sale or trade. It may also be used to check the security values of loans, where

P = sum of the monthly production payments for one year,
r = the nominal interest rate to be converted semiannually,
S = the sum of production payments plus accumulated interest,
t = years.

$$S_{t=\frac{1}{2}\text{yr}} = \frac{P}{2}$$

$$S_{t=1\text{yr}} = \left(\frac{P}{2} + \frac{P}{2}\right) + \left(\frac{P}{2} + \frac{P}{2}\right)\frac{r}{2} = P\left(1 + \frac{r}{2}\right)$$

$$S_{t=1-\frac{1}{2}\text{yr}} = P\left(1 + \frac{r}{2}\right) + P\left(1 + \frac{r}{2}\right)\frac{r}{2} = P\left(1 + \frac{r}{2}\right)^2$$

$$S_{t=2\text{yr}} = P\left(1 + \frac{r}{2}\right)^2 + P\left(1 + \frac{r}{2}\right)^2\frac{r}{2} = P\left(1 + \frac{r}{2}\right)^3$$

and by inspection,

$$S_{t(\text{yr})} = P\left(1 + \frac{r}{2}\right)^{2t-1} \tag{6-14}$$

By substituting P_w for P, present worth of the sum S at the end of t years, and transposing terms:

$$P_w = \frac{S_t}{(1 + r/2)^{2t-1}} \tag{6-15}$$

If S_t is taken equal to $1,

$$P_w = \frac{1}{(1 + r/2)^{2t-1}} \tag{6-16}$$

Many oil property valuation engineers favor the use of Eq. 6-16 to estimate present worths of all properties where the production income comes in month by month. It is also favored where monthly production payments are to be made to a lending institution. Others believe there is no reason for its preferential use at any time because other errors due to necessary approximations used in the valuation study may more than offset its intended advantages. Perhaps the use of the equation should be thought of as being more realistic with respect to incremental income or payments than with respect to absolute accuracy, keeping in mind that any present-worth calculation is *accurate* only arithmetically.

Present-worth values of one dollar for various rates of interest calculated according to Eq. 6-16 are given in Table 1 of the Appendix.

The Law of Decay

The compound interest law also may be adapted to represent the opposite of a growth curve, namely a *decay curve*. It represents a constant percentage decrease with time in the numerical value of some starting value. For instance, the trade-in value of an automobile might decline at a constant rate of depreciation of 20 per cent per year. Knowing the original cost, the value can be calculated as of the end of any future year. Or the productivity of an oil well might decline at a constant percentage rate per unit of time. The production rate and the remaining recoverable oil at any future date can be calculated. The development of the equation for the decay curve is as follows. Let

P_0 = the initial amount.
$P_1, P_2, \ldots P_n$ = subsequent periodic discounted or decreased value of P_0.
r = constant rate of decline or discount expressed as a percentage per time period t.
m = number of conversions per period t.
n = number of conversions.

$$P_{n=1} = P_0 - P_0\left(\frac{r}{m}\right) = P_0\left(1 - \frac{r}{m}\right)$$

$$P_{n=2} = P_0\left(1 - \frac{r}{m}\right) - P_0\left(1 - \frac{r}{m}\right)\frac{r}{m} = P_0\left(1 - \frac{r}{m}\right)^2$$

by inspection,

$$P_n = P_0\left(1 - \frac{r}{m}\right)^n \tag{6-17}$$

$n = mt$, and

$$P_n = P_0\left(1 - \frac{r}{m}\right)^{mt} \tag{6-18}$$

If m becomes infinity, indicating continuous decline, and using Eq. 6-5 transposed:

$$P = \frac{S_t}{e^{rt}}$$

and letting

P_0 = the initial amount or the equivalent of S_t.
P_t = amount to which P_0 would decline or be discounted in time t.
r = rate of decline or discount, percentagewise, per unit of time t.
t = units of time.

$$P_t = \frac{P_0}{e^{rt}} \tag{6-19}$$

In dealing with the decline in the purchasing power of the dollar P_0 represents \$1 at some initial time, and r is the rate of decline percentagewise per unit of time t. In the case of declining production rates, as from oil wells, P_0 represents an initial rate of production per unit of time t and r is the rate of decline percentagewise per unit of time t.

AMORTIZATION

The basic meaning of *amortization* is the incremental extinction of a debt over a period of time, regardless of the method of payment. The principles of amortization can be applied equally well, however, to accumulate a stated sum of money by making equal deposits over a period of time. Both applications are widely used in the oil industry. A company may repay a banker's loan in equal periodic installments whereby each payment is applied first to accumulated interest charges and the remainder is used to reduce the principal. It is the same process by which so many home mortgages are paid off month by month. The same company could accumulate a planned sum of money at the end of a specified period of time by making equal periodic deposits that accumulate compound interest. Such a plan might be used to provide funds for replacement of certain facilities at the end of their useful life.

The Annuity

Basically an *annuity* is a fund, presumably provided through savings and investment, which is to be paid in installments at stated times to the annuitant, the remaining part of the fund drawing interest and prolonging the payout time. Some annuities are payable in cash upon maturity. They may be purchased over a period of time or by a single cash payment. The simplest kind of annuity is the one payable in cash upon maturity and purchased by one cash payment at some prior time. Insurance companies offer many types. They are related to an amortization in that the basic methods of calculating payments, interest, and maturity values are quite similar, both being based on the compound interest laws.

The purchase of an annuity by a single payment may be figured through use of Eq. 6-8, or if the interest is to be compounded semiannually, through use of Eq. 6-9 which states:

$$P_w = \frac{S_t}{(1 + r/2)^{2t}}$$

where t represents the years until the annuity matures, S_t is the desired amount of the annuity upon maturity, r is the nominal rate of interest paid, and P_w represents the single payment cost.

For example, if one wished to know the amount of money that must be deposited today at 8 per cent compound interest, converted semiannually, to equal \$10,000 payable at the end of ten years, then according to Eq. 6-9:

$$P_w = \frac{10{,}000}{(1 + 0.08/2)^{20}}$$

$$= \$4563.87$$

The more frequent method of purchasing an annuity is to make equal periodic payments, each of which accumulates compound interest up to the time of maturity of the annuity. Irrespective of the method of collecting the annuity, the method of calculation is essentially the same as for an amortization. Usually a specific number of equal periodic payments are to be made whose amount plus accumulated compound interest equals a desired total.

The first payment or deposit on the purchase of an annuity by equal periodic installments may be made at either the beginning or the end of an interest conversion period. The first of the equal payments to be made on a debt, an account, or an amortization, and the amount of one of the equal periodic production payments on certain kinds of oil loans usually becomes due and payable at the end of the first interest conversion period. The last payment falls due on the date of maturity. The amount of each equal periodic payment can be calculated through use of the laws of compound interest. Where:

R = one of the equal periodic payments due, to be paid and credited at the end of each conversion period.
S = sum of all equal periodic payments R, plus all earned interest.
n = number of equal payments R to be made.
m = number of conversion period per year.
r = nominal interest rate.

The last payment R_n in this kind of transaction will be equal to the first payment R_1. The last payment R_n will earn no interest because it will be made upon the date of maturity. The next to the last payment R_{n-1} will earn interest over only one conversion period.

$$S_{R_n} = R$$

$$S_{R_{n-1}} = R\left(1 + \frac{r}{m}\right)$$

The next preceding payment R_{n-2} will earn interest over two conversion periods, etc. The first payment R_1 will earn interest over $n - 1$ conversion periods.

$$S_{R_1} = R\left(1 + \frac{r}{m}\right)^{n-1}$$

$$S = S_{R_1} + S_{R_2} + \cdots + S_{R_n}$$

Let b equal the constant $(1 + r/m)$

$$S = Rb^{n-1} + \cdots + Rb^3 + Rb^2 + Rb + R$$

Reversing the order of the terms

$$S = R + Rb + Rb^2 + Rb^3 + \cdots + Rb^{n-1}$$

Multiply by b

$$Sb = Rb + Rb^2 + Rb^3 + \cdots + Rb^n$$

Subtracting

$$Sb - S = Rb^n - R$$

$$S(b - 1) = R(b^n - 1)$$

$$R = \frac{S(b - 1)}{b^n - 1)}$$

Substituting the value of b

$$R = \frac{S[(1 + r/m) - 1]}{(1 + r/m)^n - 1} \tag{6-20}$$

or

$$R = \frac{S(r/m)}{(1 + r/m)^n - 1} \tag{6-21}$$

When payments are made twice each year and interest is converted semiannually

$$R = \frac{S(r/2)}{(1 + r/2)^n - 1} \tag{6-22}$$

Equation 6-22 states that R is one of the n equal payments earning compound interest converted semiannually which would provide an annuity of the amount of S dollars.

For example, to provide an annuity of \$10,000 at the end of ten years at 8 per cent interest would require twenty semiannual payments, each equal to

$$R = \frac{10{,}000 \cdot 0.08/2}{(1 + 0.08/2)^{20} - 1}$$
$$= \$335.82$$

Amortization of a Debt by Equal Periodic Payments

Payments to a building and loan company in buying a home are based on the principle of amortizing a debt. Loans on oil properties are usually of this type.

Equation 6-4 stated:

$$S_t = P\left(1 + \frac{r}{m}\right)^{mt}$$

which, letting i equal r/m, is equivalent to

$$S = P(1 + i)^n \tag{6-23}$$

And Equation 6-20 is equivalent to

$$S = \frac{R[1 - (1 + i)^n]}{1 - (1 + i)}$$

and

$$\frac{R[1 - (1 + i)^n]}{1 - (1 + i)} = P(1 + i)^n$$

Transposing,

$$R = \frac{[P(1 + i)^n][1 - (1 + i)]}{1 - (1 + i)^n}$$

$$R = \frac{[P(1 + i)^n](-i)}{1 - (1 + i)^n} = \frac{Pi(1 + i)^n}{(1 + i)^n - 1}$$

Multiplying by

$$\frac{1}{(1+i)^n}$$

$$R = \frac{Pi(1+i)^n/(1+i)^n}{(1+i)^n/(1+i)^n - 1/(1+i)^n}$$

Simplifying and substituting the equivalent of i,

$$R = \frac{P(r/m)}{1 - 1/(1+r/m)^n} \tag{6-24}$$

and if compound interest converted semiannually is applicable,

$$R = \frac{P(r/2)}{1 - 1/(1+r/2)^n} \tag{6-25}$$

Equation 6-25 states that a debt or a loan P drawing compound interest at a rate of r converted semiannually will be extinguished by n equal semiannual payments, each of an amount R.

For example, a loan or a debt of $10,000 drawing compound interest converted semiannually at a rate of 8 per cent would be extinguished in ten years by 20 equal payments of

$$R = \frac{(10{,}000)(0.08/2)}{1 - 1/(1+0.08/2)^{20}}$$

$$= \$735.82$$

The amortization of a bank loan on a producing oil property may be handled as an amortization whereby equal monthly payments are to be made to the bank through a division order to the pipeline purchasing company. For example, an oil loan of $500,000 is to be extinguished by 36 equal monthly payments in oil worth $9 per bbl. Interest is to be charged at 8 per cent compounded monthly (an effective rate of 8.2999 per cent). To find the number of barrels of oil for which the bank must be paid each month, according to Eq. 6-24,

$$R = \frac{P(r/m)}{1 - 1/(1+r/m)^n}$$

Substituting,

$$R = \frac{500{,}000 \cdot 0.08/12}{1 - 1/(1+0.08/12)^{36}}$$

$$= \$15{,}668.18$$

$$= 1740.91 \text{ bbl per month}$$

REFERENCES

Raymond W. Dull, *Mathematics for Engineers*, 2nd ed., McGraw-Hill, New York, 1941.

Max Kurtz, *Engineering Economics for Professional Engineers' Examinations*, McGraw-Hill, New York, 1959.

Walter R. VanVoorhis and Chester W. Topp, *Fundamentals of Business Mathematics*, 2nd ed., Prentice-Hall, Englewood Cliffs, N. J., 1955.

7 INCOME, EXPENSES, AND PROFITS

The capital needs of the industry, as in all businesses, must originate through savings—self-generated savings and those of others. Self-generated savings are that portion of net income retained by a company after paying dividends. Net income is often synonymously referred to as "profits" by the news media. It has a derrogatory connotation when used in reporting annual increased net incomes by the petroleum industry. The industry's survival as a private industry depends upon ever larger "profits," i.e., net incomes.

INCOME

From a business viewpoint the word *income*, somewhat like the word *value*, has many shades of meanings. It too needs modifiers to provide correct connotations. The more important kinds of income from the valuation viewpoint are the following:

1. Gross income.
2. Net income.
3. Net income before taxes.
4. Net income after taxes.
5. Discounted net income.
6. Real income.

From a valuation viewpoint *gross income* may be limited to those amounts received in the immediate vicinity of the well or the property from the sale of production less royalty and other payments to participants in the production. The payments usually are made directly to them by the purchaser of the production. The term *net income* from a property might be thought of as gross income minus all applicable expenses, in which case it is sometimes referred to as *net income after taxes* for clarity. All taxes–local, state, and federal–are meant and included in expenses. Lending institutions are interested in that income primarily. Evaluation tabulations also need to show yearly taxable income or *net income before taxes*; taxes in both cases refer to federal income taxes because all other taxes on production or a property usually are quite small in comparison. *Discounted net income* is synonymous with the present worth of future net income after taxes, although present-worth tabulations in valution reports are usually the sum of the discounted net incomes and applicable depreciation charges. Depreciation charges were expensed previously in arriving at net taxable income.

The term *real income* is sometimes used within the industry and by economists in referring to the purchasing power of the net income after taxes at the end of the accounting period in comparison with that of some previous period. Real income is particularly meaningful during periods of inflation.

DIRECT COSTS AND EXPENSES

Most of the multitudinous kinds of costs and expenses chargeable against the business income of an integrated oil company for income tax and profit-computing purposes are not of particular interest in oil property valuation work. The more important larger items of interest include the following:

1. Leasehold costs.
2. Development costs.
3. Production costs.
4. General and administrative costs.
5. Interest, if actually paid out.
6. Taxes.

Depreciation on producing oil properties is chargeable against the gross income from the property for tax computation purposes. Depletion also may be chargeable under certain conditions depending largely upon yearly volumes of production. Neither is a direct or out-of-pocket cost; both represent book-keeping charges and are treated separately.

Leasehold Costs

Leasehold expenses vary widely depending largely upon location, the lessor, and the long periods some leases must be carried before the properties provide any income. They may be broken down into the following:

1. Options and bonuses.
2. Purchase price (if a fee property).
3. Leasing expense.
4. Surveys.
5. Title and other legal fees.
6. Rental (first year's and delay).
7. Exploration costs (geological, geophysical, and exploration drilling).

These costs occur before any production and income are established on a property. Delay rentals may be considered as an expense. No development or equipment costs should be included in leasehold costs, even though production is not yet being sold from the property. Such a policy permits easy transfer from undeveloped to developed status on the books. In a purchased producing property the leasehold cost is the cost minus an inventoried estimate of the value of all depreciable equipment and materials.

Some development wells may actually be semiexploratory wells in that they may be located on drilling units outside the definitely proven limits of the field.Generally one or more unproductive wells are drilled into every reservoir. Unless such wells can be made productive in higher or lower reservoirs they must be paid for entirely out of the income from other productive wells on the same or other properties. The question naturally arises as to whether such expenditures for development-well failures should be considered as capital costs or current expense to the property or to the company as a whole for income tax purchases. Also, unless pipelines or other carrier facilities are readily available, enough wells must be drilled to prove up sufficient productive capacity to warrant the necessary investment in a pipeline outlet. Until the pipeline is provided there may be no income from the property irrespective of mounting costs.

The competitive needs within the industry for added domestic productive capacity and previous rejections of many realistic bids gradually led to some unbelievably high bonus offers for well located off-and onshore tracts. A bonus of $28,233 per acre was paid for a block of approximately 2,700 acres in the Prudhoe Bay area of Alaska in 1969. As of mid-1976, the highest bonus paid was $39,108 per acre for a 3,000-acre tract off the coast of Louisiana in the Fall of 1974. Such high bonuses for prospective but untested acreages could be offered only by the largest companies, and even they for some time had been forced to resort to joint bidding.

Total leasehold costs usually amount to only a small percentage of a company's total development and operating costs for onshore properties. Normally the exploratory wells are the major expense, especially if drilled to 20,000 ft or more, and no production is found. The opposite is usually true for leased offshore properties because the very high bonus costs may be many times the total of all other leasehold costs. The geophysical, leasing, rental, and subsequent exploratory drilling costs are very small in comparison with what might be the bonus bid. For example, the Florida, Alabama, Mississippi offshore lease sale in 1973 drew total bonus offers of $1.4 billion for 89 tracts. Two of the major companies and one larger independent company jointly bid $632.4 million for six 5740-acre tracts, one of which in the Fort Walton Beach area was bid in at $36,803 per acre. Seven dry test holes on the seven tracts at a total cost of approximately $15 million were drilled on them leaving very little if any chance of finding commercial production through added drilling.

A new policy was tried by the federal government at the Louisiana Fall sale of 1974 in order to permit smaller companies to participate in the costly offshore operations. Joint bidding by the major companies was restricted to those more distant and deeper offshore tracts which presented extremely high risks and very costly exploration, development, and production operations. Flat bonuses of $25 per acre were set on ten tracts with bidding limited

Table 7-1

Lease Acquisitions from U.S. Federal Government

Year	Millions of Dollars: Public and Acquired Lands[1]	Offshore[2]
1965 and prior	31.77	1,219
1966	.57	209
1967	.76	510
1968	.97	1,346
1969	.76	111
1970	.48	945
1971	1.16	96
1972	1.12	2,251
1973	2.20	3,082
1974	2.30	5.023
1975	8.20	1,088
1976	4.93*	2,242
Totals	55.22	18,122

*Bureau of Land Management, U.S. Department of Interior, Personal Communication.

Table 7-2
Oil Shale Lease Acquisitions

Lessor	Acres	Bonus	
		Total	Per Acre
Gulf Oil Corp. and Standard Oil of Indiana	5,089	$210,305,600	$41,326
Atlantic Richfield Co., Oil Shale Corp., Ashland Oil Co., and Shell Oil Co.	5,094	$117,000,000	$22,968
Phillips Petroleum Co. and Sun Oil Co.	5,120	$ 75,600,000	$14,766
Totals and Average Cost per Acre	15,303	$402,905,000	$26,328

to the royalties offered. Even the major companies approved of the method for future use as being far less costly initially as well as later if a tract should prove to be unproductive.

A total of 13.74 million acres of the Outer Continental Shelf had been leased from the Federal Government as of December 31, 1976. Total bonuses amounted to $18.122 billion as shown in Table 7-1. The Federal Government had also leased 1.23 million acres of Public and Acquired Lands to December 31, 1976, for a total of $55.22 million. Bonuses per acre for the Public and Acquired Lands averaged $45 per acre. In marked contrast, bonuses on the Outer Continental Shelf tracts averaged $1319 per acre.

In addition to the bonuses paid for oil and gas leases, the industry acquired oil shale leases, also from the Federal Government, in northwestern Colorado and adjoining Utah totalling $403 million as shown in Table 7-2. The lowest bonus paid for the shale leases was $14,766 per care. In as much as no commercial method of extracting the oil is anticipated prior to 1980, such high bonuses emphasize the industry's dire needs for future domestic production.

Development Costs

In general, development costs include all expenditures made on the property after production has been established. These costs include the drilling of development wells, completing them for production purchases, laying of gathering lines, installation of separators, tanks, and all other lease equipment. Actually development costs can occur any time during the productive life of the property, including period of secondary and tertiary recovery operations. Basically not all costs on a producing property are maintenance and operations. Development costs as well as some repairs may be divided into *tangible* and *intangible* costs. The cost of any item or piece of equipment that may depreciate with time or use and have a salvage value after being installed and/or used is a tangible cost. Cost of things that can have no salvage value are intangible costs. Both costs are applicable for income tax purposes on a property once commercial production has been proven.

Surface installations and equipment are tangible cost items. Wellheads, pumps, separators, tanks, and pipelines are examples. They have salvage value and may be depreciated with time and use. Wages, fuel, hauling, supplies, repairs, road building, drilling and tank-site preparation, and clean-out operations are intangible costs and may be treated as current expenses and deducted in the year they are rendered or they may be capitalized at the option of the operator and recovered through depreciation and/or depletion. The election must be made the first year the expenditures are made and it is not subject to change later.

Generally it will be found convenient and preferable to capitalize all the intangible costs for financial reasons, in which case they can be amortized over the life of the property.

Production Costs

Production costs or current expenses, exclusive of development costs, may be broken down into several categories. For valuation purposes operating costs may be conveniently limited to intangible costs such as wages and salaries, fuel and power, minor repairs and maintenance, cleanouts, extraction costs, and treating costs, all items that are not subject to depreciation. Extraction costs would be applicable in some gas-cycling operations and when liquids are extracted from marketed gas by other than wellhead separators. Treating costs are those related primarily to removal of water production.

General and Administrative Costs

General and administrative costs are usually small yet important on a per unit of production basis. Losses that might be occasioned through fires, pipeline breaks, and damage suits are usually omitted because of uncertainities of occurrence and amounts. Insurance costs may be carried if anticipated, but they too would be quite small on a unit of production basis.

Interest and Taxes

Interest cannot be charged as a cost on any part of the investment in the property unless there is an actual loan from an outside source on which interest is paid. Taxes are discussed separately later in the chapter.

DEPRECIATION

Depreciation and depletion often are used synonymously by those not fully acquainted with taxation of mineral production. Both are deductible charges or costs applicable in the calculation of profits before income taxes. Both are intended to permit the return of certain forms of "used-up" capital tax free. Neither entails a transfer of actual cash; only accounting procedures are involved. Both are related in many ways, yet their meanings are different and distinct with respect to the extractive mineral industries.

Depreciation is the older term. It is better understood and more widely applicable, yet was once considered to be of lesser importance with respect to profitable producing properties. More and more attention is now being given to it in valuation work to make up in part for losses in depletion allowances.

Depreciation charges are usually made annually against all tangible assets which have more than one year of useful life and, if used to produce income, decrease in value through obsolescence, deterioration, wear, tear, and usage. The intended purpose of depreciation charges is to provide for an accumulation of income tax-free funds plus salvage to replace a production facility at the end of its estimated useful life. Charges begin when the facility is placed in service and may end long before the facility is retired from service. Land may be productive, but it cannot be depreciated for income tax purposes.

Depreciation from a business viewpoint is based on four fundamental concepts:

1. The decline in value concept.
2. The appraisal value concept.
3. The impaired utility concept.
4. The amortization concept.

The *decline in value concept* merely recognizes the fact that the usefulness or market value of a production facility decreases with time. Neither the exact reasons nor the rate of decline, whether constant or variable, is pertinent to the concept.

The *appraisal value concept* is based on an actual appraisal of an existing facility in comparison with the most economical new facility for performing the same service. The concept may be applied at times to determine whether certain production facilities that are perhaps obsolete or costly to operate could be replaced by facilities more efficient in the same service. The difference between the cost of the replacement and the appraised value of the old would be a measure of the depreciation.

The *impaired utility concept* refers to facilities that no longer serve their intended purposes efficiently or economically. For example, pumping units may be handling the volumes of oil required, but maintenance and repair charges are perhaps excessive. Or pumping units may be operating satisfactorily from a maintenance viewpoint but may be of insufficient volumetric capacity to handle both the oil and water production.

The *amortization of cost concept* is actually the accounting concept whereby the cost of a facility less salvage value is apportioned over the useful life of the facility. The system of apportionment used is only an accounting procedure and has no intended relationship with the market value of the facility during its useful life. A number of methods of computing depreciation under the amortization concept are acceptable for computations of income tax liability. Under periods of a "stable dollar" this method of providing for depreciation enables a company to accumulate sufficient income-tax-free capital to replace a facility having served its estimated useful life.

Depreciation should be taken into account in every production operation in arriving at production expense. It should be accounted for in valuation computations of oil properties for the following reasons:

1. Equipment investments per unit of production are steadily increasing because of inflation and ever increasing well depths.
2. Depreciation is an important cost dollarwise.
3. The profitability of an oil production operation depends in part on the allowable depreciation charges.
4. Income tax obligations are based on profits.
5. The value of a property is determined largely by anticipated profits.

Depreciation accounts for each property usually are separately posted in a depreciation ledger. Two basic methods for computing the periodic charges are permitted. One is a function of time; the other a function of usage. A lastingly wise choice of method is not always possible. In general the choice depends considerably on the cost, newness, and expected life of the production facilities, the productive life of the property, and the desired rate of profitability of the property. Excessively high annual depreciation charges reduce earnings and income tax obligations, yet they provide added cash for expansion or other purposes. Very low annual depreciation charges increase earnings and income tax obligations, but decrease available cash for expansion and other purposes.

Computation Methods

A company may keep two depreciation accounts on its books—one for internal financial purposes and one for income tax purposes. Actual company costs and profits are based on the income tax procedures irrespective of the many ways in which an accounting department may wish to keep the charges. That is, some items may be appreciating in asset value due to inflation or scarcity while being depreciated for income tax purposes. The oil property valuation engineer is concerned only with those methods acceptable to the Internal Revenue Service. The three methods most often used are the following:

1. The straight line method.
2. The declining balance method.
3. The sum of the years-digits method.

All are based on the concept of depreciation with time as contrasted with actual use. The selection of one of these methods, or any other reasonable method, depends on the newness of the production facility and the allowable or desired rate of amortization. Production facilities cannot be depreciated to less than reasonable salvage values. The salvage value of a peice of production equipment or of a production facility is the amount that the taxpayer estimates at the time the asset is acquired, or put into use, will be realized on sale or other disposition of the asset after it has fulfilled its estimated useful life.

One may elect to deduct an extra 20 per cent of the cost of certain tangible facilities in addition to the regularly calculated first-year depreciation. The extra deduction is limited to facilities having a useful life of six or more years and a cost limit of $10,000.

Straight Line Method. The straight line method is the oldest, simplest, and most generally used method of computing depreciation. Intangible assets can be depreciated only by this method. It is permissible for either new or used production facilities. It permits periodic deductions for deprecication equal to the cost or other basis of value, less salvage value, divided by the number of equal periods in its estimated remaining useful life. The period is usually taken as one year, where

d_{slm} = the annual allowable depreciation charge by the straight line method,
n = estimated useful life of the facility in years,
v_s = salvage value,
v_c = original cost or other basis.

$$d_{slm} = \frac{v_c - v_s}{n} \tag{7-1}$$

Declining Balance Method. The declining balance method provides for annual depreciation charges up to twice that allowed by the straight line method. Salvage value is not deducted from the cost or other basis prior to making the calculation.

The declining balance method is more realistic with respect to oil production facilities than the straight line method in that it provides for the maximum depreciation during early useful life with gradually smaller charges each succeeding year. Annual depreciation charges, each of which is a fixed percentage of the remaining value, are successively subtracted from the original cost or other value.

The maximum rate of depreciation that may be used under this method must not exceed twice the straight line rate. Twice the straight-line rate may be used for all tangible property having a useful life of three years or more providing it (1) was acquired new by the taxpayer after December 31, 1953; or (2) was contructed, reconstructed, or erected by the taxpayer after December 31, 1953. Accelerated depreciation rates on facilities meeting these requirements can be changed to the straight-line method at any time upon option of the taxpayer unless prohibited by a written agreement as to useful life and depreciation rates.

A maximum rate one and one half times the straight-line rate can be used under this method on tangibles not meeting the double-rate requirements, if acquired (1) before January, 1, 1954, as new or used property; or (2) after December 31, 1953, as used property. These depreciation rates cannot be changed from the the declining balance method without consent of the Internal Revenue Service. Depreciation rates once changed from the declining balance method to the straight line method can never be changed back without consent of the Internal Revenue Service, where

v_c = original cost or other basis of value,
v_s = salvage value,
t = years in use,
v_t = depreciated value at end of year t,
d_{dbm_t} = annual depreciation for year t,
r = constant annual rate of depreciation expressed as a fraction of v_t (must not exceed one and one-half to two times d_{slm}).

$$v_t = v_c(1 - r)^t \tag{7-2}$$

and

$$d_{dbm_t} = v_{t-1}(r) = v_{t-1} - v_t \tag{7-3}$$

Annual depreciation d_{dbmt} must be discontinued when v_t equals v_s.

Sum of the Years-Digit Method. The sum of the years-digit method also permits accelerated depreciation rates during the early life of tangible property. The method may be used only on assets meeting the requirements for twice the straight-line rate under the declining balance method, as already stated. Under the sum of the years-digits method, annual depreciation is a decreasing fraction of the original cost or other basis of value reduced by estimated salvage value. In general it may be applied only to single asset accounts.

The numerator of the fraction changes each year to a number representing the years of future life of the asset as of the beginning of the year for which the computation is made. The denominator of the fraction remains constant. It is the sum of the digits representing the years of future life of the facility, or its equivalent, calculated by squaring the life of the facility in years, adding the life, and dividing by two, where

d_{sdm} = annual allowable depreciation,
v_c = original cost or other basis,
v_s = salvage value,
n = estimated useful life of the facility in years,
n_r = remaining useful life in years at beginning of taxable year.

By the first method

$$d_{sdm_{n_r}} = \frac{n_r(v_c - v_s)}{[n + (n-1) + (n-2) \cdots (n-n)]} \qquad (7\text{-}4)$$

By the second method

$$d_{sdm} = \frac{n_r(v_c - v_s)}{(n^2 + n)/2} \qquad (7\text{-}5)$$

The annual allowable depreciation by both methods is the same. For example, for the third year

$$d_{sdm_3} = \frac{8(v_c - v_s)}{55}$$

for a facility having a useful life of ten years.

Other methods of computing depreciation are acceptable to the Internal Revenue Service providing the deduction under any such method does not result in accumulative allowances at the end of any tax year, during the first two-thirds of the useful life of the facility, greater than the total which could have been deducted under the declining balance method. The *remaining life method* meets this requirement. It is applicable to group, composite, or classified accounts not acceptable under the sum of the years-digit method.

The allowable depreciation is a changing fraction of the unrecovered cost or other basis less salvage value. The numerator of the fraction is the years of remaining useful life of the facility at the beginning of the year for which the computation is made. The denominator is equal to the sum of the digits representing the estimated remaining useful life of the facility. The unrecovered cost is cost or other basis of value minus estimated salvage value, and minus allowable depreciation by this method for the previous year. Annual depreciation charges computed according to this plan will be equal to those computed under the sum of the years-digits plan.

Another alternative method is known as the *unit of production method* which is particularly adaptable to any production facilities serving a single field and whose useful life is a function of the productive life of the operation. Reasonably accurate estimates of the recoverable reserves are a prerequisite. The cost of the facility less salvage value divided by the recoverable reserves gives a "unit of depreciation." Depreciation for the year is the product of the units of production and the unit of depreciation. The method provides a means of tying depreciation to both use and income.

Comparative Write-offs for Depreciation. Assume that a new asset with an estimated useful life of ten years (n), cost \$11,000 (v_c), at the beginning of the first year. Salvage value (v_s) is estimated to be \$1,000.

Table 7-3
Straight Line Method

Year	Cost less salvage $(v_c - v_s)$	Useful life (n)	Depreciation (d_{slm})	Reserve (end of year)
1	\$10,000	10	\$1,000	\$ 1,000
2	10,000	10	1,000	2,000
3	10,000	10	1,000	3,000
4	10,000	10	1,000	4,000
5	10,000	10	1,000	5,000
6	10,000	10	1,000	6,000
7	10,000	10	1,000	7,000
8	10,000	10	1,000	8,000
9	10,000	10	1,000	9,000
10	10,000	10	1,000	10,000

Table 7-4
Declining Balance Method

Year	Unrecovered cost (v_t)* (end of year)	Rate (r)	Depreciation (d_{dbm})	Reserve (end of year)
1	$8,800	0.20	$2,200	$2,200
2	7,040	0.20	1,760	3,960
3	5,632	0.20	1,408	5,368
4	4,506	0.20	1,126	6,494
5	3,604	0.20	901	7,396
6	2,884	0.20	720	8,116
7	2,307	0.20	577	8,693
8	1,846	0.20	461	9,155
9 †	1,476	0.20	369	9,524
			(423)	(9,578)
10	1,182	0.20	295	9,818
	(1,000)		(422)	(10,000)

* Original cost v_C ($11,000).

† Change to straight line method in ninth year would completely depreciate the asset by end of tenth year, as shown by figures in parentheses.

Table 7-5
Sum of the Years-Digit Method (First Method)

Year	Cost less salvage ($v_c - v_s$)	Fraction $n_r/[n + (n-1) + \cdots (n-n)]$	Depreciation (d_{sdm})	Reserve (end of year)
1	$10,000	$\frac{10}{55}$	$1,818	$ 1,818
2	10,000	$\frac{9}{55}$	1,636	3,454
3	10,000	$\frac{8}{55}$	1,455	4,909
4	10,000	$\frac{7}{55}$	1,273	6,182
5	10,000	$\frac{6}{55}$	1,091	7,273
6	10,000	$\frac{5}{55}$	909	8,182
7	10,000	$\frac{4}{55}$	727	8,909
8	10,000	$\frac{3}{55}$	545	9,454
9	10,000	$\frac{2}{55}$	364	9,818
10	10,000	$\frac{1}{55}$	182	10,000

Table 7-6
Sum of the Years-Digit Method (Remaining Life Method)

Year	Unrecovered cost less salvage (beginning of year)	Fraction	Depreciation	Reserve (end of year)
1	$10,000	$\frac{10}{55}$	$1,818	$ 1,818
2	8,182	$\frac{9}{45}$	1,636	3,454
3	6,546	$\frac{8}{36}$	1,455	4,909
4	5,091	$\frac{7}{28}$	1,273	6,182
5	3,818	$\frac{6}{21}$	1,091	7,273
6	2,727	$\frac{5}{15}$	909	8,182
7	1,818	$\frac{4}{10}$	727	8,909
8	1,091	$\frac{3}{6}$	545	9,454
9	546	$\frac{2}{3}$	364	9,818
10	182	$\frac{1}{1}$	182	10,000

Both of the sums of the years-digit method and the remaining life method give the same annual depreciation and reserves. There is merely a choice in the method of calculation. The annual rates of depreciation and the depreciation reserves by the different methods of calculation are shown graphically in Figure 7-1. The declining balance method provides the highest annual rates during the early life of the facility and the highest depreciation reserve during the first half of the life.

Depreciation Practices in Other Nations. Allowable costs for depreciation of oil property facilities are neither well understood nor adequately provided for in some of the nations with post-World War II petroleum industries. This may be especially true and a cause of misunderstandings, when profit sharing is a part of the petroleum contract. Each concessionaire must work out procedures with the national government.

Depreciation allowances on oil properties in Canada are treated in nearly the same manner as in the United States. They are called *capital cost allowances*, and, as in the United States, may vary considerably from those carried on the books for purely company accounting purposes. In general the method of calculation[3]

> . . . roughly coincides with the group theory of depreciation applied on the declining balance method. The allowance is based on accumulated costs less capital cost allowance claimed to date and proceeds of disposals. The groups and maximum rates allowed are defined by regulation; however, less than the maximum allowance may be claimed if desired. Any allowance claimed

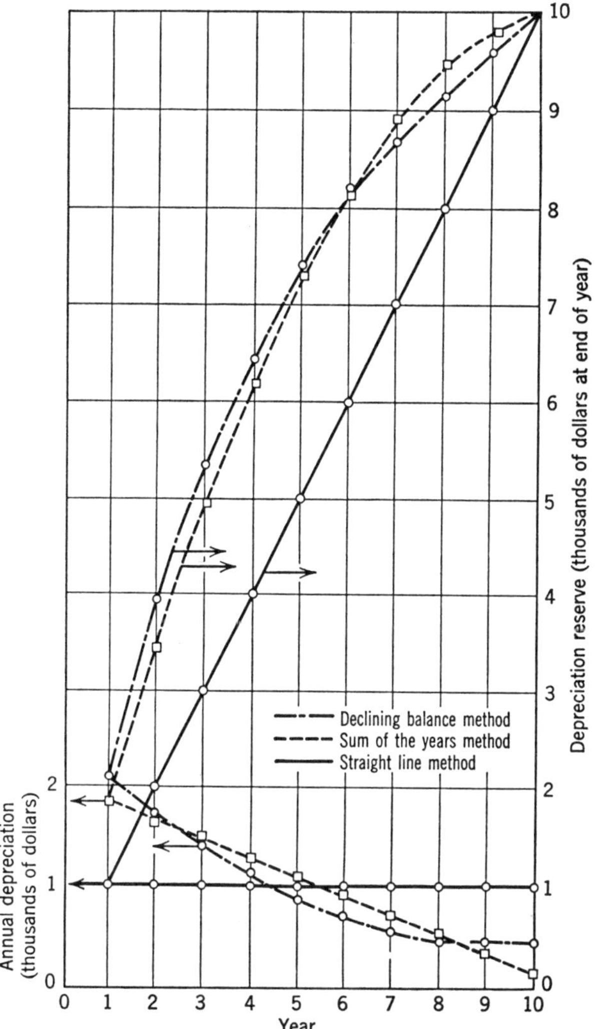

Figure 7-1. Graphical comparison of rates of annual depreciation and depreciation reserves as calculated by the three common methods.

in excess of income earned during the year (before deduction of exploration and development expenses therefrom) results in a business loss, and as such may be carried back one year and forward five years to apply against the profits of those years.

Casing is considered to be an exploration and development expense and as such may be deducted to the extent of 100% in the year acquired. Such costs should not be classified with equipment for capital cost allowance purposes.

DEPLETION

The term depletion is applicable only to those natural resources that become progressively exhausted in the process of production for utilization by man. Metallic mineral deposits, coal beds, oil shales, accumulations of oil and natural gas, and stands of virgin timber are examples. Zimmermann[4] called such exhaustible natural resources from a conservation viewpoint nonrenewable, fund, or stock resources. In contrast, such natural resources as surface waters, ground waters, soil, and soil fertility are renewable or flow resources.

Nonrenewable or fund resources, exclusive of virgin timber resources, can be created only by geological processes over geological time. Virgin timber resources such as those of redwoods and douglas fir in the Pacific northwestern states and northern California have required from 200 to 800 years for their present-day growth. Renewable or flow resources may be sustained and/or periodically renewed within Man's lifespan through his efforts, rainfall, and solar energy.

There is a question as to whether fund resources should be thought of as exhaustible. Man will never exhaust the total stock of oil and gas resources. There will always be those accumulations believed to exist yet uneconomic to explore for. And some of every discovered accumulation will never be recovered for economic reasons. Thus a fund resource only becomes exhausted from an economic viewpoint. Improvements in extraction methods or higher prices may bring about renewed operations for a time, yet there will be no increase in the original stock in place. There will only be an increase in the percentage of the original stock recovered.

Depletion is often discussed from three viewpoints, namely: physical depletion, economic depletion, and depletion for tax purposes. Physical depletion refers to the exhaustion of a mine or an oil or gas reservoir by actual extraction of the mineral, the oil, or the gas. Economic depletion refers to the reduction in the value of any wasting asset through production operations. Depletion for tax purposes refers to the removal and sale of the production of a wasting asset. The tax concept differs in that both removal and sale or involved. Depletion in the oil industry is seldom discussed from other than the income tax viewpoint. It still has to do with the exhaustion through production of a fund resource.

Depletion allowances were intended to provide in part, if not wholly, income-tax-free replacement in place of extracted and marketed mineral production. Oil and natural gas are clearly fund resources. So are deposits of sand, gravel, sulfur, salt, coal, uranium, metallic ores, etc. Considerable disagreement has always existed within the legal profession, and even more so among economists and the legislative branches of our government, with respect to the justification of the depletion allowance, and especially, if justified, to that permitted the petroleum industry. Misinterpretations, disagreements, and accusations started of course upon inclusion of a depletion allowance for income tax purposes under U. S. tariff and revenue acts, 1913-1926.

Taxation of income in the modern sense began with adoption of the 16th Amendment and passage of the Tariff Act of 1913. Taxation of income was tried for a time during the Civil War period and a 1 per cent tax upon annual incomes of corporations above $5000 was authorized under the Tariff Act of 1909. The 1909 law was vague and omissive in defining what was meant by taxable income and was particularly so with respect to the mineral extractive industries. The phrase "depreciation of production facilities" was used and interpreted very broadly. A treasury decision of 1910 permitted estimated depreciation in oil and gas wells, buildings, and machinery. The law stated that details were to be provided by the taxpayer if depreciation exceeded 5 per cent of the value as previously inventoried. The income tax law under the Tariff Act of 1913 allowed domestic corporations:[5]

> . . . in the case of mines a reasonable allowance for depletion of ores and all other natural deposits, not to exceed 5 percentum of the gross value of the output for the year for which the computation is made. . . .

The Senate Finance Committee had struck out a provision of the House providing for depletion on the basis of original cost in cash, or its equivalent, and substituted the 5 per cent limitation. The Act also imposed a 1 to 6 per cent surtax in addition to the normal tax of 1 per cent in the 1909 law. Such a progressive tax prompted serious questioning as to the equity of the fixed 5 per cent depletion allowance. It was considered to be an arbitrary and fictitious rule and too low and inequitable according to spokesmen for the Senate Finance Committee in 1916.

The Revenue Act of 1916 reverted back in part to the version originally passed by the House in 1913 which provided for a cost depletion basis. The 1916 law did away with the inequitable 5 per cent depletion allowance and in its place provided a "reasonable depletion allowance" not to exceed (1) the original cost, or (2) the value as of March 1, 1913. The terms "in the ground market value" and "market cost" came into general use. In-ground market value had several interpretations. A seller would have one viewpoint; a buyer and government, having different aims, might have other viewpoints. Market cost, on the other hand, was a fixed figure corresponding to the amount of money paid in cash in an outright purchase of a producing property. Both industry and government were still faced with ambiguous and inequitable expressions of value for tax purposes. Matters were complicated more by the inability of either industry or government to estimate accurately underground amounts of oil and gas. Both were working diligently and cooperating on the problem particularly with respect to what had become known as "production decline curve methods" to estimate original and future recoverable oil and gas.

J. O. Lewis and Carl H. Beal, both outstanding governmental authorities at the time on the behavior of oil and gas wells, played a very important part in making the laws workable. They spent several years trying to develop methods, and especially production decline curve methods, for estimating future production from wells. Beal, as Federal Oil and Gas Inspector for the "Five Civilized Tribes" in Oklahoma, started in 1916 to collect production statistics on oil wells. He studied the production data of some 20,000 wells in the United States during the next two years. His resulting publication included a comprehensive discussion with respect to depletion. He stressed the differences between depreciation and depletion by emphasizing that depreciation is a loss through exhaustion and wear, whereas depletion covers the decrease in the amount of natural deposits through production, and as long as there is production there will be depletion. He stated with respect to methods of computing depletion for purposes of taxation of producing oil and gas properties:[6]

. . . Individuals and corporations owning oil and gas properties are authorized by this act to deduct from gross income "a reasonable allowance . . . for the actual reduction in flow and production, . . . provided that when the allowance authorized . . . shall equal the capital originally invested, or in case of purchase made prior to March 1, 1913, the fair market value as of that date, no further allowance shall be made."

The reason for this provision is that capital, being returned out of profits, shall not be subject to tax. Return of capital comes out of gross profit. Hence a deduction must be made from the gross profit in order to find the true profit (taxable income) made over and above all costs. It is intended as a relief of tax upon such part of the gross profits as represents a return of the capital invested. . . .

The method first set forth [T. D. 2447, Office of Commissioner of Internal Revenue, February 8, 1917] provided that the annual deduction authorized by the provision quoted above must be reasonable and not in excess of such a percentage of the cost or value, as the case may be, and as herein defined, of the oil and gas producing properties as indicated by the reduction in the original flow or "settled" production of one year as compared with that of the proceeding year.

If the decline in the flow and production during the year of, say, 10 wells, costing $100,000, has been 5 percent, as compared with the production and flow as indicated by a test made at the beginning of the period, then 5 percent of $100,000, or $5,000, will, for the year for which the compensation is made, constitute an allowable depletion deduction in favor of the individual or corporation owning and operating the property.

Beal emphasized certain fallacies in that method of determining depletion by stating the following:

> There are two inherent fallacies in this method of determining depletion. The first is that "settled" production is a comparative term only and not susceptible to a specific definition; the second is that if the aggregate flow of all the wells in a district for which depletion is to be determined is greater than the aggregate flow of the wells in the same district a year before (this often happens on account of new drilling), the book value will show an *appreciation* instead of a *depreciation*. Obviously, this is absurd, for production means depletion, and wherever there is production the operator should have the privilege of deducting from his gross income a certain portion of his invested capital. Thus, at a time when depletion is greatest no deductions from gross income are allowed.

Beal also called attention to other fallacies in depletion methods when using such concepts as book value, unit cost methods, etc., and recommended the use of decline curves for calculating ultimate recoveries from oil wells, or the "appraisal curve" in those cases where one had lots of wells in a field or a district. Unfortunately, at the time few within the industry understood decline curve methods of estimating future recoveries.

The problems and inequities of the 1916 law and amendments to it in 1917 with respect to the mineral extractive industries, and particularly with respect to the oil and gas industry, were studied thoroughly by House and Senate committees in 1918. Their studies centered around nine unique problems and conditions, as follows:[7]

1. The exhaustion of raw materials.
2. The conversion of capital values.
3. The difficulties of measuring inground values.
4. Abnormally high risks.
5. Extreme variations in productive flow.
6. Declining production with rising costs.
7. High initial costs.
8. Abnormal time interval between initial investment and return of capital.
9. A market value which has no stable or predictable relationship to direct finding costs.

The Revenue Act of 1918, as passed in February 1919, was thought of as an entirely new law. The "original cost" criterion was retained and broadened to include certain costs of development. Now, with respect to properties acquired prior to March 1, 1913, the principle of fair market value might be used as a basis for depletion instead of original cost. Also, discovery value could be used instead of original cost, providing: (1) the discovery occurred after March 1, 1913; (2) the property had not been acquired by purchase or lease after having been proven; (3) the fair market value had been determined on the date of discovery, or within 30 days thereafter; and (4) the fair market value was found to be disproportionate to the direct cost. These limitations also proved to be unsound. Under them depletion allowances were tied into market values which in turn were subject to considerable fluctuation.

Congress is said to have intended through inclusion of the "discovery value" criterion:[8]

> First and foremost . . . to provide a tax-free return of capital. Such a tax-free return of capital was not always assured to a high-risk enterprise when depletion allowances are limited to original cost.
>
> Secondly, the Congress deliberately aimed to provide *something more* than mere cost depletion. It was the deliberate intent of Congress to provide a depletion allowance liberal enough to cover indirect costs, such as dry holes in oil and gas exploration.
>
> Thirdly, the Congress deliberately aimed to provide a depletion allowance that would be a positive stimulus to further exploration and discovery.

The only developments of importance between 1919 and 1926 with respect to the depletion allowance were increasing unhappiness on the parts of producers and both proponents and critics with the Revenue Act of 1918, and, in the Revenue Act of 1924, a limitation to the allowance. The latter stated in part:[9]

> . . . in the case of mines, oil and gas wells, discovered by the taxpayer after February 28, 1913, and not acquired as the result of purchase of a proven tract or lease, where the fair market value of the property is materially disproportionate to the cost, the basis for depletion shall be the fair market value of the property at the date of discovery of within thirty days thereafter; but such depletion allowance based on discovery value shall not exceed 50 per centum of the net income (computed without allowance for depletion) from the property upon which the discovery was made, except that in no case shall the depletion allowance be less than it would be if computed without reference to discovery value.

A Select Senate Committee, sometimes called the Couzens Committee, was appointed to 1924 to study the overall problem of depletion allowances to recommend changes particularly with respect to the costly and inequitable concept of "discovery value." Its scathing criticism as well as studies carried out by the Senate Finance Committee, whose chief spokesman was Senator Reed of New York State, were largely responsible for a complete change in acceptable methods of computing depletion allowances.

The Revenue Act of 1926 provided for a uniform statutory depletion allowance of 27½ per cent annually for all petroleum and natural gas production. The rate was a compromise between the House and the Senate who had proposed rates of 25 and 30 per cent, respectively. The new rate of 27½ per cent was intended to approximate depletion rates claimed by industry after 1918 under the "discovery value" concept.

Revenue Acts of 1932 and 1954

The Revenue Act of 1926 was amended in 1932 to permit every operating owner and every owner of an "economic interest" in the production from oil and gas wells to claim a depletion deduction in his share of the proceeds from the marketed production from the property. It also provided for revisions of previous reserves estimates and the basing of future allowances upon such revised estimates. No other important depletion provisions were passed until 1954 when "discovery value" as a basis for depletion allowances was eliminated from the law.

Revenue Acts of 1969 and 1975

The 27½ per cent rate remained in effect for 43 years. It was arbitrarily reduced in 1969 to 22 per cent under the overall "Tax Reform Act." For the industry as a whole the reduction amounted to an equal decrease in needed net income that could be offset only by increased sales, higher prices, or a combination of both. The effects of the reduced allowance were barely overcome when the major companies were confronted with increased demands for majority control of many of their foreign operations by host nations. The "Arab oil embargo" in 1973 quickly followed the meeting of these demands. In meeting these Middle East price demands, domestic reserves of crude oil and products in storage and in transportation facilities were suddenly of much higher value in terms of world markets. Compensatory increases in prices of products resulted in increased net incomes, particularly for the major companies in 1973 and 1974 in comparison with those of 1972. Congress retaliated against the industry's so-called "obscene profits" by repealing the percentage depletion allowance of 22 per cent for all the larger producing companies as of January 1, 1975.

The original justification of a depreciation allowance on anything subject to exhaustion when used to produce business income was incorporated in the early concepts of a "cost depletion" allowance with respect to the production of minerals including oil and natural gas. A cost depletion allowance was never eliminated for oil and natural gas operations. It still can be used subject to certain limitations. Every owner of an economic interest in the production from a property must calculate his depletion allowance for income tax purposes separatelyfor each property by both methods and, if an allowance is permissible, deduct *the larger of the two amounts*.

An individual has an economic interest in a property in which he owns or has acquired by investment any interest in the oil and gas in place and secures, by any form of legal relationship, income solely from its extraction and sale, and to which he must look for a return of his capital. It may be an ownership in fee, an oil landowner's royalty, overriding royalty, working interest, net profits interest, or an oil payment. Generally, a "property" is each separate interest one owns in each separate tract or parcel of land. Allowances must be computed separately by each partner in a partnership, each allocating his proportionate share of production and costs. Intangible drilling and development costs may be deducted as expenses, or capitalized providing one holds a working or an operating interest in the development and operation of the property. Such intangible costs include wages, fuels, supplies, repairs, transportation, etc.; i.e., items not having a salvage value. Intangible costs also include any contract drilling and development work, exclusive of depreciable items. The election to capitalize or to expense such costs must be made in the income tax return for the first year in which the expenditures were made.

Cost Depletion. Cost depletion is computed by dividing the total number of recoverable units (barrels of oil, condensate, other hydrocarbon liquids, and cubic feet of gas, determined in accordance with prevailing industry methods) ascribed to the property into the adjusted basis of the property, and multiplying the resulting rate per unit by (1) the number of units for which payment is received during the tax year if the cash receipts and disbursements method is used, or (2) the number of units sold if an accrual method of accounting is used.

The procedure for calculating the allowable cost depletion for any tax year may be expressed as

$$d = b\left(\frac{p}{r + p}\right) \tag{7-6}$$

where

d = annual cost depletion allowance.
b = the "adjusted basis" of the property.

p = units of production sold or for which payment was received during the tax year.

r = "recoverable units" of production remaining at the end of tax year.

The adjusted basis of the property (b) is the original cost of the property less all the depletion (cost and percentage) allowed on the property since its acquisition by the taxpayer. Obviously the adjusted basis can never be less than zero. Capital investments in a property after acquisition, such as tangible expense on added wells or secondary recovery installations, may increase the adjusted basis of the property during the tax year such investments are made.

Recoverable units (r) used in figuring cost depletion are the remaining units of recoverable production as of the end of the tax year. These figures are always estimates and subject to upward or downward adjustments when warranted, the burden of proof falling on the taxpayer. Percentage depletion still may be allowed after the taxpayer has recovered his costs under cost depletion methods.

Cost depletion is generally most applicable and more favorable for the taxpayer when he has purchased a producing property for cash or its equivalent based on a given amount per unit of estimated reserves at the time of the acquisition. This form of acquisition is frequently most favorable to both seller and purchaser from an income tax viewpoint.

Percentage Depletion. Percentage depletion, unlike depreciation and cost depletion, needs no starting basis, even if all costs have been recovered. It is simply a percentage deduction based upon gross income from the sale of production from a property, such income being considered as a measure of use and exhaustion. It is not necessarily allowed on the proceeds from an entire tract or property, but only on the gross income received from the sale of the oil, gas, and/or other hydrocarbons produced from the immediate vicinity of the well or wells. If any production is not sold on the property, but is manufactured or converted into a refined product before sale, the market or field price of the production at the wellhead as of the date of the sale, before conversion and transportation, is to be used in calculating the gross income.

Only small independent oil producers can still claim a percentage depletion allowance. Under the revision of 1975 the allowance periodically declines to 15 per cent in 1984 and remains at that rate thereafter. Dollarwise, percentage depletion under the revision is subject to a limitation of 65 per cent of ones taxable income from all sources computed without regard to the depletion allowance and certain carrybacks.

Percentage depletion for gas wells was given separate attention. Excluded from the act was all natural domestic gas sold under a fixed contract in effect

February 1, 1975, and at all times thereafter before the sale. The price must not have been adjusted to reflect the increase in the liability of the seller for tax because of the change in the law regarding percentage depletion for gas. Natural gas sales regulated by the Federal Power Commission were to be excluded until July 1, 1976. After that time regulated prices were supposed to include adjustments reflecting liability for increased taxes because of the changes in the law.

Percentage depletion is allowed only to relatively small independent producers and royalty owners as indicated in Table 7-7. Their exemption under primary recovery methods allows for the 22 per cent depletion rate through 1980 with 2 per cent annual decreases until 1984. A rate of 15 per cent is allowed in 1984 and years thereafter. Limitations in production rates started at 2,000 bbl/day of oil or 12 mm cf/day of natural gas and declined 200 bbl/day of oil anf 1.2 mm cf/day of gas until 1981. Production allowables remain at 1,000 bbl/day of oil or 6 mm cf/day of gas after 1980.

Where the taxpayer has production of both oil and natural gas, the allowable production must be allotted between the two types of production. One barrel of crude oil is the equivalent of 6,000 cf of gas. The taxpayer's average daily production rates are determined by dividing his totals for the taxable year by the number of days therein. Taxpayers whose secondary or tertiary production does not exceed 2,000 bbl/day or its equivalent in natural gas are allowed the 22 per cent depletion allowance until January 1, 1984. No taxpayer's percentage depletion allowance may exceed 65 per cent of his taxable income computed without regard to the depletion allowance and any carrybacks for operating and capital losses. Small producer exemptions do not apply to the transferee in the case of a transfer of any interest in a property.

Application of the Depletion Allowance. Assume an operator owns three drilled-up producing oil properties. There is no market for the small volume of produced gas which is used in the production operations. Properties A and B are primarily recovery operations subject to 1/6 royalty. Property C is a secondary recovery operation purchased outright at $3 per bbl estimated recoverable oil. Operating costs for the properties are $3.00, $6,00, and $5.00 per bbl, respectively. All the produced oil is sold at $8.00 per bbl from properties A and B, and at $10.00 per bbl from property C. Table 7-8 summarizes the operating data and shows the taxable income for each. A comparison of the depletion allowances for properties A and B shows that the 65 per cent depletion limitation may induce premature abandonment of the more costly production operations.

Originally the depletion allowance was intended to permit the return of invested capital in mineral production operation income-tax free. In the case

Table 7-7
Allowable Percentage Depletion Schedule

Year	Average Daily Production of Oil	or	Average Daily Production of Gas	Rate (Percent)
1975	2,000 bbl/day		12,000,000 cf/day	22
1976	1,800 bbl/day		10,800,000 cf/day	22
1977	1,600 bbl/day		9,600,000 cf/day	22
1978	1,400 bbl/day		8,400,000 cf/day	22
1979	1,200 bbl/day		7,200,000 cf/day	22
1980	1,000 bbl/day		6,000,000 cf/day	22
1981	1,000 bbl/day		6,000,000 cf/day	20
1982	1,000 bbl/day		6,000,000 cf/day	18
1983	1,000 bbl/day		6,000,000 cf/day	16
1984 and years thereafter	1,000 bbl/day		6,000,000 cf/day	15

Table 7-8
Examples of Depletion Allowances

	Property		
	A	B	C
Gross production–bbl	146,000	146,000	146,000
Net production to operator–bbl	121,667	121,667	146,000
Gross income	$973,336	$973,336	$1,460,000
Production costs	$438,000	$876,000	$730,000
Income before taxes	$535,336	$97,336	$730,000
Depletion allowances			
Cost depletion	–	–	$438,000
22 percent of gross income	$214,134	$214,134	$321,200
65 percent of net income	$347,968	$63,268	$474,500
Allowable depletion	$214,134	$63,268	$438,000
Taxable income	$321,202	$34,068	$292,000

of oil and natural gas the owner of an economic interest or an operating owner might recoup some of his capital investment tax free to drill for or to obtain more producing properties in order to stay in the production business. Those who search for oil and natural gas, but neither find nor produce any, never benefit through it. Neither do those whose production expenses exceed 66 per cent of their gross income from primary production operations through 1980 and from secondary recovery operations through 1983. The depletion allowance with respect to oil and natural gas has been called a "tax gimmick," a "tax loop-hole," and an "oil production subsidy." It is no more any of these than is depreciation where applicable. The depletion allowance promotes efficiency and successful accomplishment, both of which in turn promote national self-sufficiency. The original aim of the allowance is well illustrated by the percentage allowances shown in Table 7-9 for other commonly produced minerals as of 1976.

Table 7-9
Depletion Allowance in Mining Industries

Minerals	Percentage Depletion Allowance
Bauxite, manganese, molybdenum, nickel, platinum, sulphur, tantalum, tin, titanium, tungsten, uranium	22
Copper, gold, iron, oil shale, silver	15
Magnesium	14
Coal, sodium chloride	10
Clay and shale used for bricks, etc.	7½
Sand, stone, gravel, clays used for drain pipe, roofing tiles, etc.	5

Criticisms of the Depletion Rationale

The minerals depletion law of 1926 with respect to oil and natural gas production has been studied, questioned, and attacked by economists and our legislators repeatedly since its passage. Although the entire nation and every state that produces any kind of minerals such as gold, silver, copper, coal, iron ore, sand, and clay benefit through it, only the petroleum industry has been denounced for having that privilege. Some have never seen any justification for it at all with respect to the petroleum industry; others have merely thought it too liberal.

Lentz concluded in 1961 after months of studying the oil depletion rationale, that:[10]

1. Percentage depletion allowances are not mere loopholes of tax avoidance.
2. The concepts and principles underlying percentage depletion are sufficiently sound to justify their retention as a basis of future Federal tax policy.
3. The failure to provide legal recognition for the unique problems and circumstances of extractive production would have a punitive and confiscatory effect upon capital and enterprise.

Nevertheless, Lentz believed that three obvious weaknesses and vulnerable points of concern existed:[11]

1. The rate of depletion for royalty owners.
2. The empirical data and method by which Congress established percent age depletion in 1926.
3. The option to expense intangible drilling and development costs.

He thought that the first two weaknesses were clearly minor, but that the third was quite serious, not necessarily as an error on the part of Congress, but largely due to a change in modern political thinking and our government's continual need for additional revenue.

The oil depletion law of 1926 had withstood the repeated attacks of Senators,[12] of economists, and of the courts. Responsible daily press and financial publications periodically attacked it, directly and indirectly, and the industry for profiting by it.[13-15] The same was true to some extent in Canada.[16] Naturally these critical attitudes, and particularly those originating in the United States, helped to influence unfavorable thinking in all those nations in which the companies were operating. Obviously the industry always strongly supported the original concept of the depletion allowance, but not very convincingly. The same was true with respect to the national need to maintain original rates.[17-19]

Industry had expected some changes in the depletion allowance during the administration of President Kennedy. The Revenue Act of 1964 did not change the amount of the percentage depletion allowance, but it did include one related minor change in the laws of 1954.

Beginning in 1954 operators were allowed to combine properties for income tax purposes, even though scattered, as long as they were a part of a "single operating unit." Operators supposedly could combine high income properties and wells with low income properties and wells to increase the overall allowable depletion. The 1964 law forbid such aggregations of wells and properties. The taxpayer now had to compute and report each lease or property separately unless it was a part of an actual "unitization agreement."

The future market price of all our hydrocarbon products has always depended in part upon any adverse action that Congress might take with respect to the depletion allowance. Up until late 1975 the consensus within the industry was that the depletion allowance as such would never be eliminated; only less favorable methods of calculation from an industry viewpoint might be in the offing. As criticism of the high allowance for the petroleum industry increased within Congress and the news media some high officials within the major companies were said not to be against its elimination. If so, they undoubtedly did so in order to help eliminate criticism knowing that its loss would have to be eventually offset by increased retail prices for petroleum products. Otherwise a healthy industry could not be maintained.

Many estimates indicated net overall increases in the industry's net income taxes of 4 to 5 percent annually of gross revenue from production operations. Other estimates quoted amounts of perhaps $2 billion per year, which represented equal future shortages of capital so badly needed to maintain future domestic adequacy. The net incomes within the industry quickly showed marked declines which indicated that the former "obscene profits" were almost entirely limited to increased profits on crude and products in storage or in transit. These reductions in needed self-generated capital reduced producing property values with respect to obtaining needed loans as well as providing less cash for debt service. The real beneficiaries at the time were the Government and the American consumer. Its real abuse has been limited to a great extent to those small companies organized apparently for the sole purpose of bilking those persons in high income tax brackets. The new regulations do very little to remedy that practice. The depletion allowance was aptly compared in the late sixties to the "seed corn" that the farmer sets aside each Fall in order to grow his next crop. In similar light oil percentage depletion is a way of letting the legitimate oil producer retain, free of income tax as with the farmer, funds that represent his "seed corn."[20]

Depletion Practices in Other Nations. It was always difficult for many of the overseas oil producing nations to understand the justification for a depletion allowance to foreign operators. In general allowances were provided for production from properties and concessions held and operated by American companies. Following cognizance of attitudes in the United States and the fact that most foreign nations owned all the oil and gas in place, many took over partial to complete ownership and operation of their national properties, irrespective of concession or other unexpired contracts.

Canada had always been quite generous in its depletion allowance irrespective of operating ownership. Early provisions gave all owners of working oil and gas interests an allowance equal to one-third of the aggregate of net profits reasonably attributable to the production of domestic and foreign oil and gas. A royalty owner was allowed 25 per cent of his gross royalty as a depletion allowance prior to 1977. Thereafter royalty income was classified as production income eligible for earned depletion allowance.

Beginning with the year 1977, upon expiration of the former depletion allowance,:[21]

. . . all taxpayers will be required to earn depletion at the rate of $1.00 of depletion allowance for very $3.00 of eligible expenditure. For purposes of computing earned depletion in 1977 and subsequent years, taxpayers may accumulate all eligible expenditures incurred after November 7, 1969. Eligible expenditures will include exploration and development expenses incurred in Canada, but will exclude resource property acquisition costs, foreign exploration and development expenses, and any interest capitalized in the exploration and development account.

Earned depletion allowance may be deducted in 1977 and subsequent taxation years at a maximum annual rate of 33 1/3% of net Canadian production profits (including royalty income); after recovery of all Canadian exploration and development expenses and deduction of certain oother prescribed items.

A doubled depletion allowance and additional tax write-offs were granted for wells drilled for frontier oil and gas during the period March 31, 1977 and April 1, 1980, and meeting certain cost requirements. The provision of a 20 per cent depletion allowance on stockholders dividends was permanently eliminated at the close of 1971.

TAXES AND INCOME

Taxes can be one of the most important costs determining the income and profit picture of an oil property valuation. It is income after taxes that determines the value of the property, that provides funds for reinvestment, expansion, and growth, and that attracts outside capital and investment. Our state and federal governments are plagued by problems of raising additional new revenue each year and the oil industry sometimes appears to be a natural tax target. Federal income taxes, of all the many direct, indirect, and hidden taxes, are the most important to the indsutry. The federal government may take nearly half of what is left from income after deducting all other taxes and applicable costs. The value of a property is often quoted "before taxes" and "after taxes," meaning federal income taxes in both cases, because the value of a property to either the seller or to the prospective buyer may depend in part on which income tax bracket he is in. Thus, next to field operating costs, taxes are probably the most important items in all valuations and they must be allowed for in every valuation. Today's rates usually are used. Yet there is a danger that they may be higher in the future and thus reduce the present-worth value of a property unless one can count on compensatory increases in market prices for lease production. Statements such as that of Representative Curtis following a 1955 congressional review of income taxation are cause for concern. He stated in part:[22]

> . . . I am impressed with the fact that "taxation for revenue only" seems to have pretty well disappeared from modern-day thinking, whether, it be of the conservatives or the liberals. Using the taxing powers to produce specific economic results and to operate as the enforcing power behind government regulations seems implanted in modern political thinking.

Changes in the depletion allowances in 1975 for oil and natural gas producers well illustrated that trend. The public was led to believe that the industry long deserved such punishment. Government and indirectly the public were to benefit through increased tax revenues. The fact that prices of all petroleum products would have to be increased in time accordingly in order to maintain a healthy industry was overlooked entirely.

Our larger oil companies seldom approve of oil property investments until their tax experts have studied the proposals from every angle to determine tax benefits or detriments. This is especially true with respect to the rates at which depreciation may be written off, the time it will take to recover that part of the investment, and whether or not the expenditure should be capitalized entirely or partially expensed. In some cases it appears preferable to amortize certain intangible costs over the productive life of the property, especially if one is convinced that the depletion allowance may be reduced before the property reaches the abandonment stage. The tendency seems to be gaining favor within the industry whereby a company prefers to rent, lease, or otherwise contract for use of equipment and services, particularly with respect to the drilling of wells and to a lesser extent with respect to all servicing operations in production.

The oil property valuation engineer will find it very difficult to keep up with all aspects of taxation having to do with producing oil properties. Many rely on "tax experts" or simply lump all taxes exclusive of federal income taxes as some percentage of gross income. A total of 10 to 15 per cent is often used.

Taxes are generally classified into *direct* and *indirect* taxes. Basically a direct tax is supposed to be actually borne by the party that pays, whereas an indirect tax is one that the taxpayer can supposedly pass on to the consumer and others in his business dealings. Obviously, there can be no clean-cut distinction in modern times between the two. Yet those taxes on interests in oil property valuations are considered on the whole as direct taxes. There are many kinds, the more important of which fall on the following:

1. The corporate structure.
2. Its physical property.
3. Its production operations.
4. Its earnings as well as those of its employees.

Taxes on Corporate Structure

Taxes on the corporate structure are quite small with respect to operations and income in the modern established corporation and normally are not considered in valuations of separate oil properties. Some may be thought of as fees if paid only once and as a tax if paid periodically. Examples would be fees paid for the privilege of a franchise to do business and to sell capital stock in a company. The total rarely amounts to more than a few dollars per $1,000 of capitalization and is seldom considered in valuation studies.

Taxes on Physical Property

Taxes on physical property are familiarly known as *ad valorem* or *property taxes*. They are primarily applicable to tangible things used in production operations such as physical equipment and depreciable improvements. Ad valorem taxes usually are intended to be proportional to the assessed value of the tangible property, that is, exclusive of oil and gas in place. It is difficult to tax oil and gas in place on an ad valorem basis because the tax would have to be based on (1) an accurate estimate of the recoverable reserves in place and (2) an estimate of their value. Naturally, where oil and gas in place are taxed, there would be a hesitation to prove-up a property completely unless production was needed. Basically the purchase of a producing property at a unit value per recoverable unit of production establishes a base for an ad valorem tax. Even so, experience in production operations shows that estimates for purposes of sales or purchases can be very much in error.

The assessed valuation of physical property is normally some fraction of its "fair market value." The needs for revenue usually determine rates as well as assessments which vary from state to state and from county to county within a state. Rates up to what may be the legal limit depend upon assessed value. If rates cannot be legally increased, assessed valuations may be increased.

Oftentimes legal limits on both must be increased. Oil and gas in place usually are not subject to ad valorem taxes. There are too many conflicts, not only as to who should pay the tax, but also questions concerning the volume of reserves, recovery factors, and unit values.

Rates are usually somewhat less than one per cent of assessed valuation and are expressed in mills per $100 valuation. Oil companies may pay more than one-half of all ad valorem taxes in sparsely populated oil-producing counties. At one time the Texas State Tax Commission estimated that 37 per cent of all ad valorem taxes paid in the state were collected from the oil and gas industry.

Taxes on Production

All oil producing states, and most counties and many municipalities within them, have one or more separate taxes on all production from oil and gas wells. They are sometimes in place of, and often in addition to ad valorem taxes. Production taxes are usually on a unit of production basis, and sometimes on both. Some production taxes are levied for specific purposes; others are simply a source of revenue. They go by various names such as gross production taxes, conservation taxes, severence taxes, excise taxes, school taxes, gathering taxes, pipe-line taxes, etc. Examples of production taxes levied by various states are as follows:[23]

Alabama

Oil – 2 per cent conservation tax and a 4 per cent production tax on the gross value at the point of production.

Gas– Same as for oil.

Oil and gas production, producing leases, rights in producing properties, and unrecovered oil and gas under producing properties are exempt from ad valorem taxes. Additional assessments can not be added to the surface value of such lands.

Alaska

Oil – 1/8 cent per bbl conservation tax on all oil removed or sold and credited to the operator's account.

Production taxes of 5, 6, and 8 per cent of the value of the first 300 bbl, next 700 bbl, and all over 1,000 bbl, respectively, daily average production per well per month of 27° API gravity crude. The percentages increase or decrease 2 per cent for each variance of 1° API gravity with no variance above 40° API gravity. The value taxed fluctuates with changes in the U. S. Department of Labor Wholesale Price Index for domestic crude petroleum.

A related tax of 2 per cent on real and tangible personal properties used or committed for production purposes.

Municipalities may levy property taxes up to certain limitations.

Gas– 4 per cent of the value at the well head on gas and related liquid products.

Oil and Gas–A reserve tax not to exceed 20 mills for the two years ending December 21, 1977, subject to certain limitations.

Exempt from local taxation are severance taxes, oil and gas reserves, oil and gas production, leases, etc.

Arkansas

Oil – Severance tax of 4 per cent of production from wells producing 10 bbl per day or less, 5 per cent on wells producing more than 10 bbl per day. Lease use is excluded.

An assessment tax of not more than 10 mills per bbl.

The severance tax does not exclude nor eliminate general property, county, municipal, district, or special taxes.

Gas– Severance tax of 3 mills per 1,000 scf.

An assessment tax of not more than 1 mill per 1,000 scf.

As with oil the severance tax does not exclude other taxes on gas production.

California

The state sets and collects in addition to any and all other charges, taxes, assessments, or licenses, an annual levy on oil and gas production in an amount sufficient to support the office of the oil and gas supervisor and to provide for a reserve fund at the end of each fiscal year of $100,000, plus an annual charge for each operating well in the subsidence area of the Wilmington field.

Oil – $0.008353 per bbl conservation tax (1977 rate).

Gas– $0.008353 per 10 mcf conservation tax (1977 rate).

Individual localities in oil producing areas also impose fees upon production, licenses, drilling permits, etc.

Colorado

Oil – $0.001 per dollar of market value at the well conservation tax and a production (or privilege) tax as follows on annual gross income:

Up to but not over $25,000, 2 per cent.

Over $25,000 but not over $100,000, $500 plus 3 per cent of excess over $25,000.

Over $100,000 but not over $300,000, $2750 plus 4 per cent of excess over $100,000.

All over $300,000, $10,750 plus 5 per cent of excess over $300,000.

Under $25,000, 2 per cent.

$25,000 and under $100,000, 3 per cent.

$100,000 and under $300,000, 4 per cent.
$300,000 and over, 5 per cent.
Credit for ad valorem taxes to be deducted from the above taxes. (Producers or first purchasers must withhold 3 per cent from royalty oil.)

Gas– $0.001 conservation tax on market value at well and a production (privilege) tax at same rates as for oil.

Kansas

Oil – 2½ mills conservation tax per bbl produced.
1 mill per bbl to be paid to State Department of Health and Environment.

Gas– $0.0008 conservation tax per 1,000 scf.
$0.00005 per 1,000 scf to be paid to State Department of Health and Environment.

Kentucky

Oil – 0.5 per cent of market value of all production.
1 per cent of market value per bbl serverance tax imposed by 60 counties and ½ per cent imposed by one county.
County fiscal courts cannot levy excise taxes for revenue.

Louisiana

Oil – 12½ per cent of value severance tax, to be suspended to the extent it exceeds 3⅛ per cent of the value of the production from wells unable to produce an average of 10 bbl per day, and 6¼ per cent of the value of the production from wells unable to produce an average of 25 bbl per day if also producing an average of 50 per cent of salt water per day.
12½ per cent of gross value severance tax on distillates, condensates, and similar liquids.
10 cents per bbl severance tax on natural gasoline, casinghead gasoline, ethane, methane, and other natural gas liquids recovered through processing separator gas.
5 cents per bbl on butane and propane similarly recovered.

Gas– 7 cents per 1,000 cf measured at 15.025 psia and 60° F.
3 cents per 1,000 cf if produced from an oil well with 50 psi well head pressure or less.
1.3 cents per 1,000 cf if produced from a gas well unable to produce an average of 250,000 cf/day/month.
Adjustments are made under certain Federal Power Commission's allowable pricing rules, to municipal electric generating plants.

Oklahoma

Oil – 7 per cent of gross value, the first $150 gross value being exempt from 2 per cent of the tax if the well produces an average of less than 3 bbl per day per calendar month.

0.085 of 1 per cent of gross value per barrel excise tax.

Gas– 7 per cent of gross value of natural and/or casinghead gas, the first $150 gross value being exempt from 2 per cent of the tax if the well produces less than 1½ mmcf per month.

0.085 of 1 per cent excise tax per 1,000 cf of natural and/or casinghead gas which are subject to the gross production tax.

Texas

Oil – 4.6 per cent of market value occupation tax, also applicable to recovered liquid hydrocarbons from separator.

Gas– 7½ per cent of market value providing such tax shall never be less than 121/1500 of 1 cent per 1,000 cf.

Gas and oil used for production purposes is exempt.

Taxes on Earnings and Income

Taxes on earnings and income can be major items of expense in production operations. Actually the more efficient the production operation the greater the tax burden. Some operators appear to avoid paying income taxes by incurring large annual tangible and intangible expense items or other "write-offs" which result in small net taxable incomes for the year. There may be an appreciable loss in allowable depletion under such practices. It is spending of that so-called "tax money," especially in exploration, that may turn out profitable not only to the individual but to the nation with respect to needed petroleum reserves. Unless federal income tax rates are lowered considerably in the future, however, future income taxes still can catch up quickly once exploratory or other large deductible expenditures are discontinued. In other words one cannot purposefully operate any producing property forever at a loss for income tax purposes.

Federal Income Taxes. Federal income tax rates for corporations in 1976 were 20 per cent of the first $25,000 of taxable income, 22 per cent of the next $25,000, and 48 per cent of all in excess of $50,000. Rates were considerably higher in 1963 at 30 per cent of the first $25,000 and 52 per cent of all in excess of $25,000. The rates were reduced in 1964 and more so in 1965 to rates only slightly higher than for 1976. Comparisons are given in Table 7-10.

The reductions in the lower income tax rates were sizable for the corporations with very small incomes. Individuals owning royalty interests are taxed at regular rates, which after taking advantage of the depletion allowance, are still higher than for the corporate operator. Royalty incomes of $25,000, $50,000, and $100,000 after depletion allowances, would be taxed $6,020,

$17,060, and $45,180, respectively, on a joint income basis. These higher rates are not out of line because the tax savings through the depletion allowance are rarely used by the royalty owner to find and to develop new reserves.

Table 7-10
Comparison of 1965 and 1976
Corporate Federal Income Tax Rates

Taxable Income	1965	1976
$ 10,000	$ 2,200	$ 2,000
25,000	5,500	5,000
50,000	17,000	10,500
100,000	41,500	34,500
500,000	233,500	226,500
1,000,000	473,500	466,500
10,000,000	4,793,500	4,786,500
100,000,000	47,993,500	47,986,500

State Corporation Income Taxes. State corporation taxes where collected on oil and gas operations are small in comparison with federal income taxes. They are usually within the 5 to 6 per cent range and often included in an overall tax item, rather than separately.

Capital Gains Taxes. Both the federal and the state governments tax short-term capital gains as full income. Long-term capital gains are given preferential treatment. State laws vary with respect to federal laws as to what constitutes long and short terms. The federal tax rules specify that the income tax on capital gains shall be at regular income tax rates on 50 per cent of the capital gain, but in no case shall it exceed 25 per cent of the capital gain. The federal government taxes capital gains at regular income tax rates unless the asset has been owned for a long term period. Prior to 1977 any period of ownership of six months or more was called long term. It was changed to nine months for the 1977 tax year and to twelve months after 1977. Many states follow the rules set up by the federal government, exclusive of those for common stocks and securities. As much as a two-year holding period may be required for them.

Corporations are permitted an aternative method of computing federal income tax on the capital gains portion of taxable income. It permits worthwhile savings if the corporation's taxable income exceeds $50,000. The capital gains portion of the total taxable income is taxed at only 30 per cent.

PROFITS

Any man on the street can define business profit. It would be said to be the difference between sales and costs, or the difference between the total income and the total expense. Either of these definitions is fundamentally correct yet interpretations and applications vary. An economist goes to great length to explain the many ramifications of the many meanings. Marshall[24] described profits as follows:

> When a man is engaged in business, his profits for the year are the excess of his receipts from his business during the year over his outlay for his business. The difference between the value of his stock of plant, material, etc. at the end and at the beginning of the year is taken as part of his receipts or as part of his outlay, according as there has been an increase or decrease of value. What remains of his profits after deducting interest on his capital at the current rate . . . is generally called his *earnings of undertaking or management.* The ratio in which his profits for the year stand to his capital is spoken of as his *rate of profits.*

Marshall's "earnings of undertaking or management" is called *true profit* by Collery[25] who refers to the book difference between sales and costs as the *accounting profit*. He stated:

> To obtain true profit from an accounting profit, you must subtract something for the wages an owner could have earned if he had worked for someone else (economists call this implicit wages). You must also subtract something for the interest an owner could have earned if he had lent his capital to some other firm (economists call this implicit interest).
>
> . . . Therefore, much of the profit reported by accountants is in reality interest and does not represent a true profit.

Drucker probably would agree that much of reported profits are in reality interest and criticize the term "true profit."[26] He believes that the very businessmen who complain about the economic illiteracy throughout our society seem to know nothing about profit and profitability. In fact, they may be among the worst offenders of the the "free enterprise system." He stated:

> . . . what they say to each other as well as to the public inhibits both business action and public understanding.
>
> . . . businessmen owe it to themselves and owe it to society to hammer home that there is no such thing as "profit." There are only "costs": costs of doing business and costs of staying in business; costs of labor and raw materials,

and costs of capital; costs of todays jobs and costs of tomorrow's job and tomorrow's pensions.

There is no conflict between "profit" and "social responsibility." To earn enough to cover the genuine costs which only the so-called "profit" can cover, is economic and social responsibility—indeed it is the specific social and economic responsibility of business. It is not the business that earns a profit adequate to its genuine costs of capital, to the risks of tomorrow and to the needs of tomorrow's worker and pensioner, that "rips off" society. It is the business that fails to do so.

The general public, the press, and even most shareholders would find it difficult to understand the correctness of Drucker's thesis, especially if used in an annual report. The use of the term "profits" with its many connotations, right, wrong, or prejudiced, is too firmly established in the public mind.

Misunderstandings on the part of the general public with respect to profits of oil companies often arise on account of failure to realize the following:

1. The industry as a whole has suffered a profit squeeze since the mid-1950's, that is, the ratio of net income to gross income has declined considerably.

2. Only a fraction of the net income shown on a balance sheet is passed on to the stockholders or other owners of the company at the end of the year.

3. Net income for a single year, particularly if high, is far less meaningful than the average over a period of years.

4. Quantitatively, neither the net income nor the payments made to the stockholders and other owners of the company are true measures of the return on the investment because dividends and income, like wages, must be measured with the same inflated dollar.

The net income for the year shown in the annual reports of our oil companies corresponds to the "accounting profit" or simply the net profit for the year. The term "net profits" is not often used in the reports because the term has come to connote something unjust or unfair to the consuming public. These amounts, dollarwise, may seem large and even excessive especially to critics of the industry, economists, and those who must frequent the filling stations. They are sometimes portrayed as dollars literally stolen, at least in part, from an almost helpless consuming public. Yet a little serious thought would show that these profits are the backbone of the free enterprise system of which this nation is so proud.[27]

Profits (after taxes) provide a firm with funds for reinvestment and help attract new capital.

Reinvested funds and new capital are the essential ingredients for expansion.

Expansion creates new facilities and tools, more employment.
More employment means greater purchasing power and a dynamic economy.

These essentials for a dynamic economy, especially with respect to the petroleum industry, are a national necessity. For at least the past two decades progress has been at the expense of the shareholder. Shareholders' earnings are included in Drucker's costs of doing business and must be paid out of accounting profits. A mid-1970's study[28] showed that more than 14 million persons were shareholders of the six largest U. S. oil companies, either directly through purchases or indirectly in financial institutions such as mutual funds. Other owners included 91 colleges and universities, nearly 200 mutual insurance companies, and approximately 1000 charitable institutions and educational foundations. As of 1975, dividends paid to the oil companies' shareholders usually amounted to only one or two cents out of every dollar of revenue. Governmental taxes, exclusive of income taxes, amounted as a rule to several times dividend payments for domestic operators, and from two to five times dividend payments for international operators. Net incomes per dollar of sales and other operating income and cash dividends paid as a percentage of net earned income for two of the representative international oil companies are given in Table 7-11 for the years 1967-1976. The industry's averages over the period amounted to only $0.08 and 55.65 per cent, respectively. The industry's rate of return, i.e., the ratio of net income to net worth for 1976 was exceeded by six of the other leading industries.

Corporate surveys have shown that the vast majority of the public during the mid-seventies believed corporate profits for the oil industry ranged from 25 to 60 per cent per dollar of sales or revenue. The real figure had rarely exceeded five per cent for several years. As a result, profits or net incomes during the mid-seventies were only half as large a proportion of wages and salaries as they were twenty-five years earlier. Congressional attitude toward the industry, and particularly toward the largest companies, has been thought to be a major factor. It ignores the overall economics of the industry with respect to its needs for larger and larger annual "profits" that it might accumulate sufficient capital to provide for future energy requirements under high inflationary conditions The industry was literally stunned several times during 1975 upon learning that a surprisingly large number of Congressmen were pledged to break up the major companies, in part by horizontal divestiture, and particularly through vertical divestiture.

The retained part of the accounting profit may result in still larger accounting profits in the future, which when true may in turn result in increased dividends and/or increased value of ownership shares. Investors in oil company shares are attracted primarily by possibilities of appreciation in share values through company growth, not the small dividend payment alone.

Table 7-11
Ratios–Net Incomes, Sales, Dividends[29,30]

	Net Income per dollar of sales and other operating income*		Cash dividends paid as a percentage of net earned income	
Year	Exxon	Gulf Oil Corp.	Exxon	Gulf Oil Corp.
1967	$0.079	$0.138	64.4%	44.8%
1968	0.082	0.137	61.5	46.5
1969	0.074	0.123	65.3	51.1
1970	0.071	0.102	63.4	56.7
1971	0.074	0.094	55.4	55.6
1972	0.069	0.032	54.6	158.0**
1973	0.086	0.095	38.9	37.0
1974	0.066	0.065	36.9	28.8
1975	0.051	0.049	44.7	47.3
1976	0.050	0.050	46.2	41.2
Ave.	0.070	0.089	53.1	56.7

*Consumer excise taxes excluded.
**Dividends exceeded net income.

The discount factor used by valuation engineers to obtain the present worth of an oil property corresponds in part to what may be called the desired accounting profit expressed percentagewise with respect to the future net income from the property. The percentage is difficult to decide upon with any degree of confidence despite its being a measure of the value of a property. And the value of property, being a measure of estimated future income, results only through operations, and if sold, through capital gains. Future profits have become quite difficult to estimate due primarily to continued inflation and its rate, as well as growing retaliatory and confiscatory attitudes by all governments toward the petroleum industry. Reduction and/or loss of the depletion allowance, unrealistic price controls for natural gas and crude production from older fields are prime examples. Threats of divestiture, horizontal and vertical, unrealistic environmental controls, sharing of lower cost oils, and national offshore operations add to the confusion. Thus, a basic discount factor certainly should amount to some 8 to 9 per cent compounded semiannually because that much may be earned with safety in many forms of saving accounts. In addition, some percentage discount factor may be included as a safety measure against miscalculations especially with respect to possibilities of increased labor costs and operating expenses, unanticipated well repairs, increased taxes, and lower prices for produced oil.

The concensus is that lower prices could be only short-lived and that ever higher prices will compensate, in part if not wholly, for these uncertainties. Nevertheless, some discount factor must be included to compensate for the need to take some risk. A 10 to 12 per cent discount factor for these contingencies is not unreasonable. Thus many evaluations are figured with a 20 and even 25 per cent discount factor. Such discount factors bring the present worth of a property down to the approximate going exchange value, either in the case of ordinary sales and trades or with respect to a forced sale value in case of a foreclosure by a lending institution.

Others prefer to use a going interest rate for the discount factor to calculate the present worth of a property and then add another overall discount to the present worth so obtained. Such a procedure provides a dollar value for a property of perhaps 60 to 80 per cent of the present worth when calculated at current bank loan rates, a figure for which one might hope to buy, sell, or trade the property, voluntarily or under a foreclosure.

The profit on a producing oil property is never known until the property has been abandoned.

REFERENCES

1. *Federal and Indian Lands Oil and Gas Production, Royalty Income, and Related Statistics*, U. S. Department of Interior, Conservation Division, Washington, D. C., June 1976, p. 17.
2. *Outer Continental Shelf Statistics*, U. S. Department of Interior, Geological Survey, Conservation Division, Washington, D. C., June 1976, pp. 74-84 (compiled from).
3. *Oil and Gas–A Guide for Oil and Gas Operators in Canada*, Bank of Montreal, Oil and Gas Department, Calgary, Alberta, Canada, Revised ed., 1974, p. 41.
4. Erich W. Zimmerman, "Conservation in the Production of Petroleum," *Petroleum Monograph Series No. 2*, Yale University Press, New Haven, Conn., 1957, p. 6.
5. Oscar H. Lentz, "Mineral Economics and the Problem of Equitable Taxation," *Quart. Colorado School of Mines*, Vol. 55, No. 2, April 1960, p. 12.
6. Carl H. Beal, "The Decline and Ultimate Production of Oil Wells with Notes on the Valuation of Oil Properties," *USBM Bull. 177*, 1919, pp. 97-98.
7. Oscar H. Lentz, "The Depletion Rationale and Recent Political Pressures of Erosion," *J. Petr. Tech.*, June 1961, p. 524.
8. *Ibid.*
9. Lentz, "Mineral Economics . . . ," *op. cit.*, p. 20.
10. Lentz, "The Depletion Rationale . . . , *op. cit.*, p. 524.
11. *Ibid.*
12. "Depletion Will Be Cut," *Oil and Gas J.*, Feb. 6, 1961, pp. 78-79.

13. "The Tax Beater," *Forbes*, Sept. 1, 1960, p. 20.
14. "Should Oil 27½% Be Cut?," *Newsweek*, March 4, 1963, pp. 66-68.
15. "The Coming Battle over Depletion," *Forbes*, April 15, 1959, pp. 17-21.
16. "Depletion Allowance Claimed 'Give Away' for Oil Industry," *Oilweek*, March 2, 1964, pp. 13-14.
17. "The Case for 27½% Depletion," *Oil and Gas J.*, Feb. 23, 1959, pp. 99-114.
18. Samuel T. Williamson, "Why Oil Depletion Allowance Is Necessary," *The Orange Disc*, Gulf Oil Corp., May-June 1959, pp. 2-7.
19. "Oil and Gas Conservation and Depletion, A Committee Study," pamphlet, *Interstate Oil Compact Commission*, March 13, 1959.
20. "Percentage Depletion–Seed Corn for Minerals," *The Lamp*, February 1968, pp. 6-9, (Condensation of a presentation before the House by Congressman, A. Sydney Herlong, Jr.)
21. *Oil and Gas* . . . op. cit., p. 40.
22. Lentz, "Mineral Economics . . . ," *op. cit.*, p. 75.
23. *State and Local Oil and Gas Severance and Production Taxes, January 1, 1977* (condensation of data furnished through courtesy of the American Petroleum Institute, New York).
24. Alfred Marshall, *Principles of Economics*, 8th ed., Macmillan, London, 1922, p. 74.
25. Arnold Collery, "Primer on Profits," *Petroleum Today*, Summer 1962, American Petroleum Institute, New York.
26. Peter F. Drucker, "The Delusion of Profits," *Wall Street Journal*, February 5, 1975, Editorial Page.
27. Don Lambert, "What You Should Know about Profits," *World Oil*, October 1963, pp. 11-25.
28. "Who Owns the large Oil Companies?" *The Lamp*, Fall 1975, Exxon Corporation. (Original source–American Petroleum Institute).
29. *Financial and Statistical Supplement to the Annual Report, 1976*, Exxon Corporation, New York, N. Y.
30. *Annual Reports*, Gulf Oil Corporation, 1970, 1973, and 1976, (calculated from Five-Year Annual Financial and Statistical Summaries), Pittsburgh, Pa.

Additional References

The Future of American Oil–The Experts Testify, American Petroleum Institute, Washington, D. C., 1976. (Testimony by 16 educators, lawyers, and bankers before Congress dealing with the effects of vertical and horizontal divestiture on the petroleum industry).

"Who Needs Multinational Oil Companies," *The Lamp*, Exxon Corporation, Summer 1976.

"The Multinationals and the Oil Crisis–Petroleum Industry Profits," *Orange Disc*, Gulf Oil Corporation, March-April 1975.

"The Profit Situation, A Special Petroleum Report," Energy Economics Division, The Chase Manhattan Bank, New York, N. Y., April 1974.

Clark W. Breeding and John R. Herzfeld, "Effect of Taxation on Valuation and Production Engineering," *J. Petr. Tech.*, September 1958, pp. 21-25.

J. Waddy Bullion, "Tax Considerations in Various Oil Transactions," *J. Petr. Tech.*, August 1956, pp. 12-18.

E. L. Dougherty and J. Lohrenz, "Statistical Analyses of Bids for Federal Offshore Leases," *Jour. Petr. Tech.*, November 1976, pp. 1377-1390.

Lawrence T. Harbeck, "Some Plain Truth about Profit." *Wall Street Journal*, editorial page, January 20, 1972.

Jim West and Steven Kelley, "U.S. Oils' Profits Skid Sharply from Stellar '74" *The Oil and Gas Journal*, May 5, 1975, pp. 143-148.

8 RESERVOIRS AND RESERVOIR ROCKS

The oil property valuation engineer will find himself handicapped without some knowledge of what has developed over the past thirty to forty years into the "science" of reservoir engineering. Pirson[1] has called reservoir engineering an "art"—

. . . the art of forecasting the future performance of a geologic oil and/or gas reservoir from which production is obtained according to probable and pre-assumed conditions.

Oil property valuations are based on the recoverable contents of oil and gas reservoirs. The term *reservoir* is relatively new in the production branch of the industry. It may be defined as any body of underground rock with a continuously connected system of void spaces which are filled with hydrocarbon fluids to such extent that they move toward wells drilled into the rock for recovery purposes under the influence of either natural or artificial forces.

Every geological formation irrespective of age or composition must have two physical characteristics in order to be a reservoir rock, namely, porosity and permeability. Although each is important in itself, neither is of value without the other. It must not only have a complex of void spaces to provide for storage of reservoir fluids, but these voids must be interconnected in order to serve as passageways through which the reservoir fluids may flow. Flow capacity or permeability depends on porosity to some extent, but porosity does not depend on permeability. That is, reservoir rocks of high porosity are not necessarily of high permeability and those of low porosity are not necessarily of low permeability. The porosity of reservoir rocks was studied long before their permeability, hence, "old timers" in the industry may still use the term porosity to describe both properties.

THE OIL FIELD

During the very early days of the industry not more than one stratigraphic horizon was found to be productive in a given area at any one time. The discovered productive accumulation of oil became known as a *pool* largely due to a misconception as to how it occurred underground. Several wells tapping the same pool outlined an *oil field*. Over the years the two expressions became somewhat synonymous and are often so used today. Strictly speaking, pool is not an all-inclusive term; pool and reservoir are more nearly synonymous terms today.

It is better to think of the term field as being the all-inclusive term. A field would consist of all pools or reservoirs underlying a continuous geographic area with no large enclosed areas being considered as unproductive. All fields have been given names since the early days of the industry. Field names may become confusing in time, especially if a later discovery is given the same name. A new field may be assigned to an area surrounding a new producing well. As development drilling progresses the newly found reservoir may join or partly underlie or overlie another productive reservoir in an adjoining named field. The production from a multicompletion well or dual wells producing from two or more reservoirs might be designated as coming from two or more overlapping fields. The Thompson fields, Fort Bend County, Texas, are examples. The number of reservoirs in a single field are virtually unlimited. Most salt dome fields and some of the highly faulted fields of California are examples. The Timbalier Bay field, located over a salt dome off the coast of Louisiana, is said to have 434 separate reservoirs in 81 different productive sand horizons at depths of 2000 to 17,500 ft.[2] The Ventura field, California, is another example of a highly faulted multireservoir field.

THE RESERVOIR

If the hydrocarbon fluids move through the reservoir toward recovery wells in volumes sufficient to permit economic recovery, the accumulation is a commercial reservoir and usually referred to as a *proven reservoir*. The void spaces of proven reservoirs normally contain some *interstitial water*, commonly called *connate water*, in addition to the hydrocarbons. All or most of the interstitial water is held in an immobile state by capillary forces. Thus proven reservoir rocks are saturated with *reservoir fluids*, namely, liquid hydrocarbons, gas, and interstitial water. The amount of each reservoir fluid in the void spaces is expressed percentagewise.

Reservoirs, pools, and fields are found or discovered only by drilling to sufficient depths to verify someone's scientific recommendation, or occasionally someone's hunches. Holes drilled for such probing purposes are called *wildcat wells, exploratory wells,* or *test wells*. A successful wildcat well is called a *discovery well*, that is, it proves one or more prospective formations to be commercially productive reservoirs. Subsequent wells drilled into proven reservoirs for production purposes are called *development wells*. An unsuccessful wildcat or an unproductive development well is known as a *dry hole* even if it would have produced water. Dry holes do not necessarily rule out future discoveries in an area. Prolific fields have been found in areas and at depths previously condemned by drillers of dry holes. The great East Texas oil field is an outstanding example.

Development wells for either oil or gas used to be located haphazardly and close together on a property. Since about the late 1930's, they have been located according to definite spacing patterns, uniform distances apart, and at uniform positions within land-subdivision units. Oil wells drilled in the United States today usually are widely spaced and located at the centers of 40-acre tracts, at like ends of 80-acre tracts, at alternate ends, at centers of quarter sections, etc. The customary spacing for gas wells is now on a more or less uniform pattern allotting from 160 to 640 acres per well. Under certain favorable conditions 1280-acre spacing for gas wells is not considered excessive. Alberta Province, Canada, will permit spacing of one gas well per 1440 acres under ideal reservoir conditions.[3] Well spacing in Canada since World War II has been greater on the average than in the United States. That acreage assigned to each development well is known as a *drilling unit* prior to completion of the well and as a *production unit* upon successful completion. The sizes and shapes of drilling units and the position of development wells on them are usually established and enforced by state regulatory bodies shortly after completion of the discovery well. A tendency since World War II to go to ever larger drilling units in order to reduce field development expense came about largely as an outgrowth of ever increasing required drilling depths and higher well completion costs combined in some states with low allowable production rates. However, closely spaced wells are not considered as necessary as once thought to be for efficient and maximum recovery.

Classification of Reservoirs

Wilhelm[4] has classified oil and gas reservoirs from a geological viewpoint by combining their structural, depositional, and stratigraphic characteristics into five major groups which will be found satisfactory for valuation purposes.

Group A. *Convex trap reservoirs* which are completely surrounded by edgewater as the porosity extends in all directions beyond the reservoir areas. The reservoir peripheries are therefore defined by uninterrupted edgewater limits. The trap is due to convexity alone.

Group B. *Permeability trap reservoirs* with the periphery partly defined by edgewater and partly by the barrier resulting from loss of permeability in the reservoir layer. In the extreme case a reservoir may be entirely surrounded by such a barrier.

Group C. *Pinchout trap reservoirs* with the periphery partly defined by edgewater and partly by the margin due to the pinchout of the reservoir bed.

Group F. *Fault trap reservoirs* with the periphery partly defined by edgewater and partly by a fault boundary.

Group G. *Piercement trap reservoirs* with the periphery partly defined by edgewater and partly by a piercement contact.

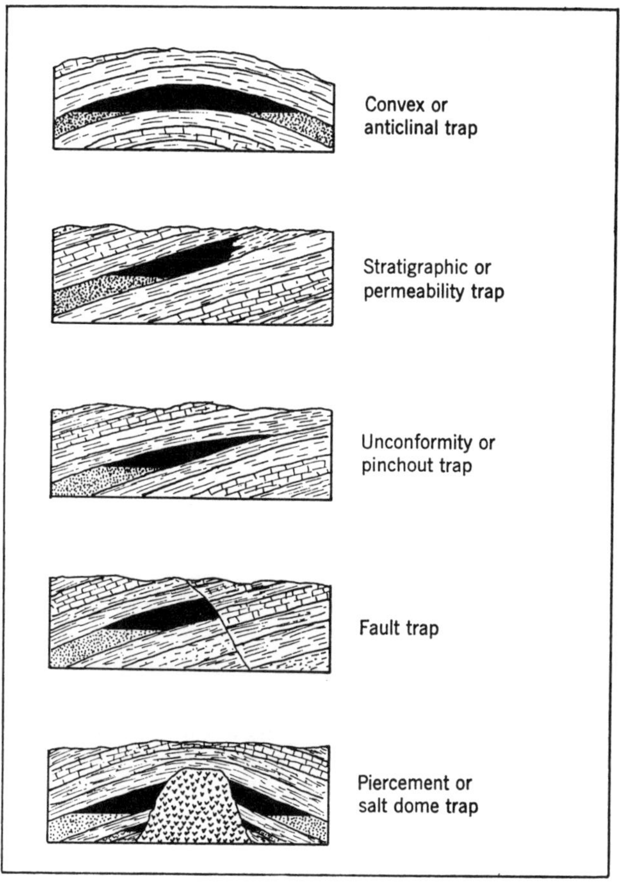

Figure 8-1. Structural classification of oil and gas reservoirs.

These types are commonly called anticlinal, stratigraphic, unconformity, fault, and salt dome traps, respectively, and are illustrated in Figure 8-1. Levorsen[5] has described many kinds of reservoirs in detail, yet he includes them all under three general classes, namely: structural traps, stratigraphic traps, and combination traps. Heald[6] used only two overall classes of reservoirs: those closed by local deformation of strata and those closed because of varying permeability of the rock.

A reservoir may be bounded by connected pore spaces filled with water, a physical change in the body of the rock, particularly with respect to a lack of connected pore spaces, a change in hydrocarbon saturation, or a sealing fault.

A reservoir may prove to be noncommercial upon additional drilling and testing; or it may become so through exhaustion as a result of production operations. Some exhausted reservoirs may become commercial again upon an increase in the market prices for oil and gas, the application of modern well stimulation techniques, adoption of cheaper and more efficient production practices, and other advances in oil recovery technology.

RESERVOIR ROCKS

All geological formations making up reservoirs are often called *reservoir sands*. The preferable and all-inclusive term is *reservoir rocks*. Both terms are used rather loosely to include all consolidated and unconsolidated sandstones, limestones, dolomites, and shales, as well as those productive igneous and metamorphic rocks making up some unique reservoirs. Only a few commercially productive shale reservoirs have been found and still fewer igneous and metamorphic reservoirs.

Wilhelm[7] summarized the many different kinds of reservoir rocks with their relative frequency of occurrence—common, frequent, infrequent, rare—as follows:

1. *Sand, conglomeritic sand and gravel*, in varying state of consolidation, porosity due to fragmental texture; common.

(a) Clean sand, etc.; pore space between sand grains uncontaminated.

(b) Argillaceous sands, etc.; pore space partially filled with argillaceous matter.

(c) Silty sands, etc.; pore space partly filled with silt.

(d) Lignitic sands, etc.; pore space partly filled with lignitic matter.

(e) Bentonitic sands; pore space partly filled with volcanic ash.

2. *Porous calcareous sandstone and siliceous sandstone*, porosity due to incomplete cementation; frequent.

3. *Fractured sandstone and fractured conglomerate*, porosity due to fracturing in tight sandstones or hard conglomerates caused by faulting or sharp folding; infrequent.

4. *Arkosic feldspathic sand, arkose, arkosic conglomerate (granite wash)*, porosity due to fragmental texture; infrequent.

5. *Detrital limestone* (calcitic and dolomitic), porosity due to fragmental texture and frequently increased by solution; common.

6. *Porous crystalline limestone* (calcitic and dolomitic), porosity due mainly to solution; common.

7. *Cavernous crystalline limestone* (calcitic and dolomitic), porosity due to strong solution effects; common. Note: 5, 6, and 7 are not sharply separable.

8. *Fractured limestone* (calcitic, dolomitic, and siliceous), porosity due to open fissures along fracture pattern; frequent.

9. *Sugary dolomite*, "saccharoidal" porosity possibly due to volume shrinkage in the process of formation of dolomite from calcite sediment; common.

10. *Oölitic limestone*, porosity due to oölitic texture with uncemented or partially cemented interstices; frequent.

11. *Coquina and shell breccia*, porosity due to fragmental texture; infrequent.

12. *Crinoidal limestone*, a variety of coquina, porosity to fragmental texture; infrequent.

13. *Porous cap rock* on shallow salt plugs, porosity due to solution; infrequent.

14. *Honey-combed anhydrite*, porosity due to leaching; rare.

15. *Fractured shale*, porosity due to fracturing of brittle siliceous shale under sharp folding; rare.

16. *Fractured chert*, porosity due to fracturing under sharp folding, rare.

17. *Porous tectonic breccia*, formed along fault and thrust zones, porosity mainly due to incomplete cementation or subsequent solution; rare.

18. *Contact metamorphic shales*, porosity due to volume shrinkage after "baking"; rare.

19. *Porous igneous rock*, porosity primarily as in tuffs, or due to fracturing as in basalt, or due to decomposition; rare.

Krynine[8] has shown that less than 20 different mineral species form 99 per cent of the detrital sediments and that many times no more than three to six minerals make up the bulk of a sediment. Some 160 different sediment-forming minerals from reservoir rocks have been identified. Krynine also pointed out that sediments differ from igneous rocks in that they are always mixtures of two series of end members. The first series makes up the detrital fraction and the second series forms the soluble chemical fraction. The first series usually consists of quartz, feldspar, detrital chert, micas, and clay minerals. The second or chemical series is usually in the form of carbonates, sulphates, authigenic chert, and secondary quartz.

The Detrital Rocks

Krynine[9] divided the detrital rocks into the following three major groups:

1. The quartzite group.
2. The graywacke group.
3. The arkosic group.

The quartzite group is represented by "clean sands." The graywacke group may be called the "dirty sands," that is, sands containing considerable percentages of clays and micas. The arkosic group may be defined as ashy, light gray (sometimes red), dirty sands with much feldspathic material. Shales and siltstones are really the very fine-grained members of these three groups.

The graywacke group is the widest spread and most productive group on a worldwide basis. The arkosic group is second, particularly in the United States. Detrital limestones and dolomites occur and should be good reservoir rocks. They are difficult to identify except from core material with the result that recognized production from them is relatively small.

The Chemical Rocks

Limestones and to a lesser extent dolomites are the most widespread of the productive chemical reservoir rocks. Due to difficulties in identifying dolomites, except in cored wells, the all-inclusive term of limestone reservoirs predominates. Chemical reservoir rocks also may be composed wholly or in part of silica. Such domestic reservoirs are uncommon except in West Texas and parts of California. "Limestone" reservoirs are important producers in the Permian Basin of West Texas, adjoining New Mexico, Illinois, Kansas, Michigan, Indiana, locally in the Rocky Mountain region, and especially in the Golden Lane area of Mexico and the Persian Gulf area of the Middle East.

POROSITY

Reservoir rocks may have two values of porosity, namely, *total* or *absolute porosity* and *effective porosity*. Total porosity includes all voids whether sealed off or interconnected; effective porosity is a measure of only the interconnected voids. Industry used to rely on laboratory determinations of effective porosities of cored material from representative wells. Within recent years there has been a tendency to rely almost entirely on porosity interpretations of electric logs which theoretically provide an average reliable effective porosity value.[10, 11] Values so obtained should be checked carefully against all other available information before using them in reservoir volume calculations.

Effective porosity is generally less, but never greater than, absolute porosity. Obviously, effective porosity values should always be used. There is some doubt as to the representativeness of values obtained from electric logs, but the savings made by eliminating coring and laboratory determinations has influenced thinking with respect to the need for exact values. Porosity of a reservoir rock may be estimated much more closely than permeability and percentage saturations.

Effective porosity may be of either *primary* or *secondary* origin depending on whether it developed before or after lithification of the reservoir rock. Secondary porosity is usually negligible to absent in sand reservoirs and occurs most frequently in limestone and dolomitic reservoirs as a result of solution by circulating groundwaters, fracturing, jointing, and recrystallization. Such pores and openings in limestones and some dolomitic limestones may vary from microscopic size to large spongelike voids. Occasionally limestone reservoirs may be cavernous which nullifies the usual concept of porosity, but as a general rule limestone porosities are quite low. Large pores, large openings, vugs, and caverns are quite indicative of one or more periods of solution after deposition and consolidation. Hohlt[12] has classified primary porosity in limestones as follows:

1. Voids between the individual constituent particles of detrital carbonate rocks, i.e., conglomerates, sands, conquinas, and oölites.
2. Voids between individual crystals, in crystalline limestones, and along individual cleavage planes of crystals.
3. Voids along bedding planes due to differences of material deposited and differences of crystal size and arrangement.
4. Voids within the skeletal structure of invertebrates and within the tissue of calcareous algae.

Piper[13] stated that secondary porosity in limestones is due to

1. Fractures, including
   (*a*) joints caused by contraction of the sediment during consolidation,
   (*b*) joints and faults resulting from crustal movement,
   (*c*) joints due to mineralogic changes.
2. Solution openings related to present or former erosional surfaces.
3. Intercrystalline voids produced by mineralogic change.

Fracture porosity is limited neither to limestones nor to crystalline rocks. It frequently occurs in shales. The high-grade oil production from the fractured shallow Mancos shale in the Rangely field, Colorado, is a well-known example. Productive fractured shales are usually relatively shallow.

There is often some question as to the percentage of porosity in limestone and dolomite reservoirs and experience is required to arrive at an acceptable value. On the other hand porosities of sandstone reservoir rocks are easily measured quite accurately. One study of a California sand reservoir showed that porosity values are likely to be rather narrowly distributed about a mean value.[14] For example, 240 equally spaced porosity samples from 200 ft of cores showed a maximum deviation of only 0.97 per cent porosity from the mean value.

Crawford[15] evaluated porosities of well-known reservoir rocks in the Rocky Mountain area as follows:

Negligible porosity	0–5 per cent
Poor porosity	5–10 per cent
Fair porosity	10–15 per cent
Good porosity	15–20 per cent
Excellent porosity	20–25 per cent
Exceptional porosity	25–35 per cent

The ranges and averages of porosities of reservoir rocks from other producing areas are given in Table 8-1.

Representativeness of Porosity Measurements

Present-day laboratory measurements of effective porosities of reservoir rocks are sufficiently accurate if carefully made. Yet the use of such data has been questioned with respect to their applicability under actual reservoir conditions. The question was called to industry's attention by Hall[16] in 1953 after he measured sizable decreases in porosities of reservoir rocks when subjected to internal pressures up to 1500 psig with simultaneous external pressures up to 3000 psig.

Later Fatt[17] found that there was no overall correlation between compressibility and porosity. He agreed with Geertsma,[18] another investigator of the problem, that pore compressibilities measured in the laboratory under simulated reservoir conditions may be as much as twice those occurring under actual reservoir conditions. Wyble[19] independently obtained decreases in porosities with respect to pressures comparable with those reported by Fatt. All such reductions are in part a function of the differences between external rock pressures and internal pore pressures. These reductions would be most pronounced in poorly consolidated sands and low porosity reservoir rocks. Reductions probably do not exceed 5–10 per cent in most reservoirs, irrespective of depth, but considerably more research is needed to provide

Table 8-1

Range of Porosities and Permeabilities in U.S. Oil Fields*

Area	Formations represented	Porosity (per cent)			Permeability (md)		
		Range	Range of averages	Mean of averages	Range	Range of averages	Mean of averages
Arkansas	14	1.1–42	14.2–39.4	25.7	0.1–13,700	61 –2,100	871
East Texas	18	1.5–40.9	11.7–27.9	19.1	0.1–13,840	1.2–1,185	200
North Louisiana	17	3.4–41	12.9–31.4	20.9	0.1– 7,350	0.7– 760	256
California	10	14 –41	19.5–35.6	27.7	4 – 9,400	107 –1,410	777
Texas Gulf Coast							
Corpus Christi	8	14 –38	19 –30	26	1 – 9,000	50 –7,000	364
Houston	13	14.5–38.5	19.6–35.9	29.4	3 –13,100	86 –2,970	990
Louisiana Gulf Coast	4	15.7–39.0	27.3–30	28.5	18 – 9,470	920 –1,630	1,230
Oklahoma-Kansas	105	0.6–39.3	4.0–22.3	13.5	0.1– 2,690	0.7– 372	66.5
West Texas-S.W. New Mexico	56	1.3–27.7	3.3–21.0	11.1	0.1– 9,410	0.1– 609	652

* Summarized from data in *Petroleum Production Handbook* by Frick (1962), with permission of McGraw-Hill Book Co., Inc.

acceptable corrections usable in volumetric calculations. Porosities calculated from electric well logs supposedly eliminate the compressibility factor.

PERMEABILITY

The unit used to express the permeability of fluid conductivity of reservoir rocks is called the *darcy*. It was named in 1933 after a Frenchman, Henri Darcy, who in 1856 studied and reported on rates of flow of water through sand filter beds. A reservoir rock has a permeability of one darcy when[20]

> . . . a single-phase fluid of one centipoise viscosity that completely fills the voids of the medium will flow through it under "conditions of viscous flow" at a rate of one cubic centimeter per second per square centimeter of cross-sectional area under a pressure or equivalent hydraulic gradient of one atmosphere (76.0 cm of Hg) per centimeter.

Most reservoir rocks have permeabilities of much less than one darcy and are reported in *millidarcys* (thousandths of a darcy). Permeabilities of low darcy values are usually quoted in millidarcys also, rather than decimal fractions. In general oil reservoir rocks of low permabilities show the most rapid declines in production rates and as a consequence accumulative productions are equally low.

The first actual measurements of the fluid-flow capacity or permeability of an actual oil sand were made by Newell in 1885. Some fifty years later Fancher, Lewis, and Barnes[21] laid the foundations of our present concepts of this important property of reservoir rocks. They devised laboratory methods for its measurement which were consistent with known theories of fluid flow through porous media. At the time the permeability of a given reservoir rock was believed to be a constant and that any gas or liquid could be used for the determination as long as the proper correction was made for its viscosity and, in the case of a gas, for its flow volume at mean pressure and standard temperature.

Since that time many research workers have questioned both the theory and the laboratory methods with the result that now five different measures of permeability are recognized. Each has its particular value and usefulness in reservoir calculations. They are usually spoken of as the following:

1. Specific permeability.
2. Klinkenberg permeability.
3. Effective permeability
4. Relative permeability.
5. Directional permeability.

Specific Permeability

Specific permeability is the measure of the fluid conductivity of a porous medium when no other fluid than the one used in the measurement occupies pore space. It is said to be analogous to the electrical conductivities of metals. Air is used most often in making the measurement because of its convenience. Other gases, water, alcohol, oil, etc., could be used, but permeabilities to different gases and liquids do not always agree with each other. Such discrepancies were once explained by probable lack of thorough wetting of pore surfaces and by chemical or other reactions. Laboratories were advised to report the fluid used in making the permeability measurements and such values became known as specific permeabilities.

Klinkenberg Permeability

Discrepancies in permeability measurements when using different gases, or when using the same gas under different mean pressures, were unexplainable for several years. Klinkenberg[22] solved the problem by demonstrating that permeability to a gas varies inversely with the mean pressure used. From a theoretical viewpoint those molecules of a flowing liquid in direct contact with the pore surfaces are stationary. In the case of a gas flowing at a mean low pressure, however, the molecules in contact with the pore surfaces are not stationary. Their mean free paths vary inversely with the mean pressure and under conditions of flow they will have a net forward movement. Hence at low pressures larger volumes of gas than anticipated flow through the pore openings. Klingenberg showed that a theoretical permeability corresponding to that of any nonreactive liquid could be obtained when using a gas by finding the specific permeabilities to the gas at two or more mean pressures. The values so found may be plotted on Cartesian coordinate paper against the reciprocals of the mean pressures used and extrapolated to zero—the reciprocal of infinite mean pressure. The extrapolated value theoretically corresponds to the permeability of the medium to any nonreactive liquid.

Such a measurement is reported as a "Klinkenberg permeability" and is preferable to any single gas permeability measurement, especially if made at a low mean pressure. Klinkenberg permeabilities may be as much as 50 per cent less than the usual single air permeability measurement in the case of low permeability sands. Table 8-2 illustrates some typical differences. Any permeability measurement made with a nonreactive liquid should correspond to the Klinkenberg permeability. Either measurement is often reported as the *absolute permeability* of the reservoir rock.

Table 8-2
Illustrative Values of Regular Permeability Measurements Using Low Pressure Air and Corresponding Klinkenberg Values

Reservoir rock	Air permeability (md)	Klinkenberg permeability (md)
Bradford sand, Pa.	3.1	2.1
	8.0	6.2
	19.0	15.2
	46.0	38.6
	130	110
Wilcox sand, Texas	237	204
	513	474
	1980	1860
Louisiana reservoirs	26.8	19.5
	97	80
	202	176
	561	521
	1036	965

Effective Permeability

A reservoir rock saturated completely with two or more reservoir fluids will have an *effective permeability* to each fluid at any given combination of fluid saturations. Hence effective permeability is a measure of the conductivity of a reservoir rock to one fluid in the presence of another fluid at a particular state of saturation.

The sum of the percentages of effective pore space occupied by each fluid must equal 100 per cent. Effective permeability of a reservoir rock will always be less than either the Klinkenberg permeability or the specific permeability to that particular fluid. For example, a reservoir rock might have a Klinkenberg permeability of 320 md. When saturated 60 per cent with oil; 30 per cent with interstitial water, and 10 per cent with free gas, the effective permeabilities to oil; water, and gas might be of the order of 200 md, 0 md, and 40 md, respectively. The effective permeabilities of the reservoir rock to the reservoir fluids at specific times determine the producing characteristics of the reservoir to a considerable degree.

The effective permeability of the reservoir rock about the well bore to a produced fluid can be determined through application of the proper "radial-flow equation" to production and pressure data taken at the well. A value so determined represents an overall average for the productive section of the reservoir rock and will provide more reliable data for certain purposes than laboratory measurements using cores.

Relative Permeability

The relative permeability of a reservoir rock to a fluid is the ratio of its effective permeability at a given condition of saturation to its Klinkenberg or absolute permeability. Thus the relative permeability to oil, water, and gas in the above example of effective permeability would be 0.625, 0, and 0,125, respectively. All values of relative permeabilities lie between zero and 1.00. A typical relative permeability graph is shown in Figure 8-2.

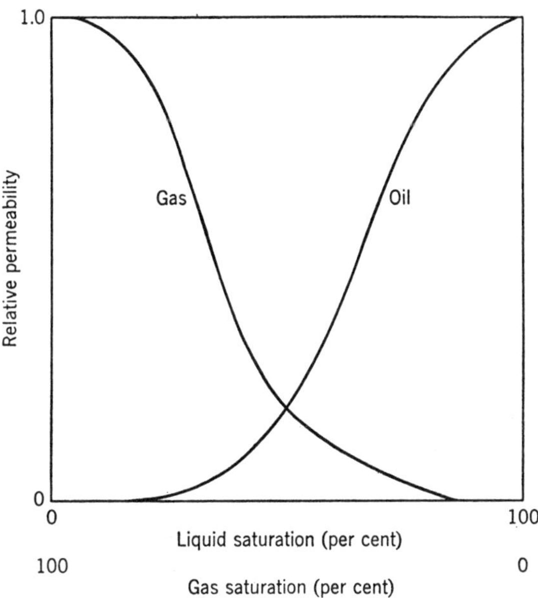

Figure 8-2. Relative permeabilities of a reservoir rock.

Directional Permeability

Permeability measurements of reservoir rocks may not be isotropic, that is, the same in all directions. Almost all reservoir rocks of sedimentary origin will show higher permeabilities parallel to the bedding planes than across them.

Permeabilities measured parallel to the bedding may vary depending on the general direction of fluid flow. Such directional irregularities are most common in clastic sediments. The Bradford field, Pennsylvania, has its maximum permeability essentially parallel to the bedding planes in a direction of N35° W.[23, 24] Permeabilities at right angles to the maximums and also essentially parallel to the bedding planes may be as much as 40 per cent less.

Other examples of directional permeability may occur in formations whose porosity results partially, if not wholly, from a system of fractures. The Spraberry field, Texas, is an outstanding example.[25, 26] Fluid injection tests in connection with intended widespread secondary recovery operations proved that the trend of maximum permeability is in a southwest-northeast direction, somewhat askew to the drilling pattern for primary recovery. The ratio of permeability along the fracture trend to that perpendicular to it ranges from approximately 6 to 144, or higher, over an area of approximately 15 by 17 miles.

Representativeness of Permeability Measurements

The representativeness of either individual or average permeability values of cores of reservoir rocks, due to the usual wide range of values and the probable unreported method of measurement, is much more questionable than for porosities. Crawford's[27] evaluation of permeabilities of reservoir rocks in the Rocky Mountain area is as follows:

Negligible and unproducible	below 0.1 md
Poor but may produce	0.1–1 md
Poor but will produce	1.0–10 md
Fair and will produce	10–50 md
Good and producible	50–200 md
Excellent and producible	200–700 md
Exceptional	above 700 md

Ranges and averages of permeabilities of reservoir rocks from other producing areas in the United States are given in Table 8-1.

Permeability values measured by the same general procedure may be averaged by arithmetic, exponential, logarithmic, and harmonic methods. Law[28] studied core data mostly from the Dominquez field, California, and concluded that permeability assemblages usually give satisfactory logarithmic and arithmetic normal frequency distributions. Cardwell and Parsons[29] believed that qualitative reasoning indicated equivalent permeability lies nearer the upper, arithmetic limit than to the lower, harmonic limit.

Fatt and Davis[30] measured reductions up to 50 per cent in permeability of cores when subjected to simulated overburden pressures up to 10,000 psi. Fatt[31] recommended that the effective permeability to be used in reservoir calculations be obtained by multiplying the relative permeability by the permeability at the calculated overburden pressure of the reservoir. McLatchie, Hemstock, and Young[32] reported with respect to selected core material from Western Canada when subjected to overburden loads up to 8000 psi that reduction of permeability in clean sands was relatively small and the results agreed with published data. Very large reductions in permeability with increasing overburden pressure was found in sandstones containing large amounts of clay.

Effective permeabilities to the reservoir fluids being produced are of primary interest yet no method of laboratory measurement or averages will give representative data at all times. Interpretations of pressure buildup tests of producing wells provide data from which effective permeability values can be calculated for the producing section as a whole. Operating values so obtained during the early productive life of a field should prove most valuable despite the methods of well completion.

REFERENCES

1. Sylvain Pirson, *Oil Reservoir Engineering*, 2nd ed., McGraw-Hill, New York, 1958, p. 1.
2. C. A. Lipari, "An Engineering Challenge–Development of South Louisiana's Giant Timbalier Bay Field," *J. Petr. Tech.*, February 1963, pp. 127-132.
3. "Prorationing: What Effects?" *Oilweek*, Sept. 7, 1964, p. 25.
4. O. Wilhelm, "Classification of Petroleum Reservoirs," *Bull, AAPG*, Vol. 29, No. 11, November 1945, pp. 1537-1579.
5. A. I. Levorsen, *Geology of Petroleum*, Freeman, San Francisco, 1956, p. 142.
6. K. C. Heald, "Essentials for Oil Pools," *Elements of the Petroleum Industry*, AIME, New York, 1940, pp. 47-55.
7. Wilhelm, *op. cit.*, p. 1541.
8. Paul D. Krynine, "Sediments and the Search for Oil," *Producers Monthly*, January 1945, pp. 12-22.
9. *Ibid.*
10. Sylvain J. Pirson, *Handbook of Well Log Analysis*, Prentice-Hall, Englewood Cliffs, N. J., 1963.
11. J. M. Edwards and A. L. Simpson, "A Method for Neutron Derived Porosity Determination of Thin Beds," *Trans. AIME*, Vol. 204, 1955, pp. 132-136.
12. Richard B. Hohlt, "The Nature and Origin of Limestone Porosity," *Quart. Colorado School of Mines*, Vol. 43, No. 4, October 1948, p. 9.
13. A. M. Piper, "Ground Water in North-Central Tennessee," USGS *Water Supply Paper 640*, 1932, pp. 69-70.
14. "Effect of Core Sampling Interval upon Mean Porosity and Permeability," *API, Proc. 26th Annual Meeting*, Sec. IV, 1946, pp. 128-135.

15. James G. Crawford, "Core and Water Analyses Speeded up in Rocky Mountains and Western Canada by New Flying Service," *Oil and Gas J.*, June 23, 1949, pp. 178-186.
16. H. N. Hall, "Compressibility of Reservoir Rocks," *Trans. AIME*, Vol. 198, 1953, pp. 309-311.
17. I. Fatt, "Pore Volume Compressibility of Sandstone Reservoir Rocks," *J. Petr. Tech.*, March 1958, pp. 64-66.
18. J. Geertsma, "The Effect of Fluid Pressure Decline on Volumetric Changes of Porous Rocks," *Trans. AIME*, Vol. 210, 1957, pp. 331-340.
19. D. O. Wyble, "Effect of Applied Pressure on the Conductivity, Porosity, and Permeability of Sandstones," *Trans. AIME*, Vol. 213, 1958, pp. 430-435.
20. *Recommended Practice for Determining Permeability of Porous Media*, 3rd ed., RP 27, American Petroleum Institute, New York, 1956.
21. G. H. Fancher, J. A. Lewis, and K. B. Barnes, "Some Physical Characteristics of Oil Sands," Pennsylvania State College, *Mineral Ind. Exp. Sta. Bull 12*, 1933, pp. 65-171.
22. L. J. Klinkenberg, "The Permeability of Porous Media to Liquids and Gases," *API Drill. and Prod. Pract.*, 1941, pp. 200-213.
23. W. E. Johnson and R. V. Hughes, "Directional Permeability Measurements and Their Significance," *Producers Monthly*, November 1948, pp. 17-25.
24. J. C. Griffiths, "Directional Permeability and Dimensional Orientation in Bradford Sand," *Producers Monthly*, June 1950, pp. 26-32.
25. "Flood Pattern Is Key to Spraberry Oil," *Petr. Week*, Sept, 2, 1960, pp. 18-20.
26. Lincoln F. Elkins and Arlie M. Skov, "Determination of Fracture Orientation from Pressure Interference," *Trans. AIME*, Vol. 219, 1960, pp. 301-304.
27. Crawford, *op. cit.*, p. 183.
28. Jan Law, "A Statistical Approach to the Interstitial Heterogeneity of Sand Reservoirs," *Trans. AIME*, Vol. 155, 1944, p. 219.
29. W. T. Cardwell, Jr., and R. L. Parsons, "Average Permeabilities of Heterogeneous Oil Sands," *Trans. AIME*, Vol. 160, 1945, p. 38.
30. I. Fatt and D. H. Davis, "Reduction in Permeability with Overburden Pressure," *Trans. AIME*, Vol. 195, 1952, p. 329.
31. I. Fatt, "The Effect of Overburden Pressure on Relative Permeability," *J. Petr. Tech.*, October 1953, Sec. 1, p. 15.
32. A. S. McLatchie, R. A. Hemstock, and J. W. Young, "The Effective Compressibility of Reservoir Rock and Its Effect on Permeability," *J. Petr. Tech.*, June 1958, pp. 44-51.

Additional References

H. W. Hardy, "Continuous Velocity Logger Yields Excellent Results," *World Oil*, April 1959, pp. 111-117, 137.

C. C. Miller, A. B. Dyes, and C. A. Hutchinson, "The Estimation of Permeability and Reservoir Pressure from Bottom-Hole Pressure Build-up Characteristics, *Trans. AIME*, Vol. 189, 1950, pp. 91-104.

R. L. Perrine, "Analysis of Pressure Build-up Curves," *API Drill. and Prod. Pract.*, 1956, p. 482.

H. E. Pickhard and L. M. Holley, "Sonic Log Proves Valuable Porosity Tool," *World Oil*, September 1959, pp. 65-68, 120.

H. J. Ramey, Jr., "Interference Analysis for Anisotropic Formations–A Case History," *J. Petro. Tech.*, October 1975, pp. 1290-1298.

9 RESERVOIR FLUIDS

Reservoir fluids occupying the pores of reservoir rocks are commonly referred to as oil, gas, and interstitial water. All reservoir rocks contain interstitial water. It is usually a minor constituent and seldom exceeds 40 per cent of the pore space at the time of discovery of the reservoir. The oil and gas may occur in separate reservoirs, in contact with each other in a single reservoir manifested as a free gas cap overlying the oil phase, as gas dissolved in the oil phase, and occasionally as oil dissolved in a gas phase.

The terms oil and gas are used rather loosely in referring to the hydrocarbon fluids in specific reservoirs. Their appearances and properties in the reservoirs may be very different from those observed under atmospheric conditions. The mode of occurrence in the reservoir depends on composition, reservoir temperature, and pressure. The liquid phase may vary from a very viscous, heavy nonvolatile black oil to fluids resembling gasoline in appearance, volatility, and odor. These highly volatile fluids are referred ro rather broadly as condensates which may be difficult to recover from reservoirs except under carefully controlled production methods. Produced gas may vary from a dry gas coming from a 100 per cent gas reservoir to a wet gas carrying condensable hydrocarbon liquids in widely varying proportions.

Researchers have made rapid progress since 1930 in learning the physical and chemical properties of reservoir fluids under various conditions of temperature and pressure and their modes of occurrence. Despite today's wealth of research and literature on the subject, the origin of oil and gas is still quite speculative.

ORIGIN AND ACCUMULATION OF RESERVOIR FLUIDS

A knowledge of the many theories dealing with the origin and migration of oil, and more particularly with its accumulation, will prove to be helpful in evaluating undrilled structures and very helpful in evaluating structures not completely drilled. The oil property valuation engineer is often faced with one or the other of these problems which are usually referred to the geologists.

The interstitial water of reservoir rocks is thought of as ancient sea water which filled all voids in the sediments at the time of their deposition. The interstitial water found in reservoir rocks today is thought to still be sea water, sometimes concentrated and other times diluted, but not necessarily that which was trapped in the porous rock at the time of deposition. Constant compaction of sediments and structural changes in the formations may have caused movement and gradual replacement of the interstitial water originally trapped, at least up to the time a porous rock became a reservoir rock. Interstitial water is held in place largely through capillary forces. Thus its abundance in reservoir rocks is controlled by its composition, the sizes and shapes of the pore spaces, temperature, and the surface properties of the water and the pore walls.

Oil and gas are not believed to be original "pore fillings" as in the case of interstitial water. Oil and gas probably collected in the pores long after the reservoir rocks had undergone compression, consolidation, and structural deformation. Both oil and gas must have had a related origin as to source and methods of accumulation.

The consensus since 1913 has favored an organic origin of oil. Yet earlier theories of an inorganic origin persist in addition to some modern versions including one or cosmic origin. Hoyle[1] is one of the most recent proponents of an inorganic or cosmic origin. He suggested that the so-called oceans on Venus may be oceans of oil and the clouds may well be hydrocarbon gases. If so, oil could have been an original or early component of the earth also. The origin of oil is much more speculative than how it accumulated into the traps or reservoirs where found today.

Two of the more recent theories of accumulation are of particular interest. One proposed by Hubbert,[2] called the "Hydrodynamic Theory," requires a porous and permeable rock saturated with a moving water phase in which very small globules of oil and minute bubbles of gas are carried along until trapped in the pores of certain geological structures where the forces due to buoyancy exceed the dynamic vector forces of the moving water. Hubbert's theory offers a logical explanation for tilted interfaces between accumulated oil and surrounding edgewater such as are prevalent in the Rocky Mountain region.

Another recent theory of particular interest was proposed by Gussow.[3] He stated that the vector forces of buoyancy alone can account for the lateral and vertical movements of very small droplets of oil and bubbles of gas. Oil and gas are said to spill up dip from one structural trap to another depending on regional dips, pore capacity of successful traps, and the abundance of the migrating hydrocarbons. Gussow's theory needs only a continous porous, permeable, and water-saturated rock with regional dip under a hydrostatic condition, but it does not rule our the added benefits of a hydrodynamic condition. Both Hubbert's and Gussow's theories allow for migrations of tens of miles or more from the places of origin to the places of commercial accumulation.

CLASSIFICATION OF CRUDE OILS

The value of a crude oil to a purchaser is largely a function of its composition as measured by the amounts and values of the products refined from it. The specific gravity and viscosity of a crude oil is related to its composition. Thus the "gravity" of a crude oil can be used as an approximation in determining the purchase price of crude oils.

Gravity of a Crude Oil

The American Petroleum Institute adopted the Baumé Tagliabue gravity relationship for crude oils which became known as the API gravity. A high API gravity oil would have a low specific gravity and vice versa. An oil of a 1.0 specific gravity at 60°F has a "gravity" of 10°API. A few fields produce oils of less than 10°API gravity. Such oils will not float on water. The gravity relationship is:

$$^{\circ}\text{API gravity} = \frac{141.5}{\text{specific gravity}} - 131.5$$

The specific gravities of both the oil and water are taken at 60°F.

Base of a Crude Oil

Our early-day refiners found that crude oils from Pennsylvania, West Virginia, New York, and southeastern Ohio precipitated a waxy to vaseline-like material upon slow chilling. These oils which were of low specific gravity and opalescent green colors became known as *paraffin-base* oils. Oils discovered later in other regions and particularly in California and along the Gulf Coast were usually darker to black in color and left asphalt-like residues upon distillation. Waxy-like material might or might not be a constituent. Also, many fields produced crudes that were not exactly one or the other. These crudes were called asphaltic-paraffinic or mixed-base crudes.

The U. S. Bureau of Mines had analyzed crude oils from all the United States by 1928 and from fields throughout the world by 1935. It had recommended use of "naphthene base" instead of asphalt base and "intermediate base" instead of mixed base, and had proposed a seven-base classification in 1935.[4] Two more bases were added in 1937 with the explanation that most of the crude oils of the world would fall within the previous classification.

The Bureau's classification is based on analyses of 800 crudes from all over the world by its modified Hempel distillation method.[5] Two distillations are made: one at atmospheric pressure and the other at an absolute pressure of 40 mm of mercury. The method is as follows:

1. Note the API gravity of "key fraction no. 1," the fraction distilling between 250° and 275°C (482° and 527°F) at atmospheric pressure.
2. Note the API gravity of "key fraction no. 2," the fraction distilling between 275° and 300°C (527° and 572°F) at 40 mm absolute pressure.

If the gravity of key fraction no. 1 (the lighter or lower boiling fractions of the oil) is 40.0° API or lighter the lower boiling fractions are paraffinic; if it is 33.0°API or heavier they are naphthenic; and if it is between 33.0° and 40.0° (33.1° to 39.9°API, inclusive) they are intermediate.

Similarly, if the gravity of the key fraction no. 2 is 30.0°API or lighter the higher boiling fractions of the oil are paraffinic. They are naphthenic if the key fraction no. 2 is 20.0°API or heavier, and intermediate if the key fraction is between 20.0° and 30.0°API (20.0° to 29.9°API, inclusive). Thus, there are nine possible classes of crude oil.

1. Paraffin base: Distillates paraffinic throughout.
2. Paraffin-intermediate base: Light fractions paraffinic; heavy fractions intermediate.
3. Intermediate-paraffin base: Light fractions intermediate; heavy fractions paraffinic.
4. Intermediate base: Distillates intermediate throughout.
5. Intermediate-naphthene base: Light fractions intermediate; heavy fractions naphthenic.
6. Naphthene-intermediate base: Light fractions naphthenic; heavy fractions intermediate.
7. Naphthene base: Distillates naphthenic throughout.
8. Paraffin-naphthene base: Light fractions paraffinic; heavy fractions naphthenic.
9. Naphthene-paraffin base: Light fractions naphthenic; heavy fractions paraffinic.

Average analyses of 315 worldwide samples are given in Table 9-1. Sulfur is a worldwide component.

Table 9-1
Average Analyses of 315 Worldwide Crude Oils*

Class	Sulfur, %	Gravity, key fraction no. 1, API	Gravity, key fraction no. 2, API	Gasoline, light ends, %	Total gasoline and naphtha, %	Total gasoline and naphtha, °API gravity
Paraffin, wax-bearing	0.22	43.0	32.7	6.1	22.6	60.1
Paraffin-intermediate, wax-bearing	0.50	40.6	28.0	5.8	28.9	58.4
Intermediate-paraffin, wax-bearing	0.19	37.8	31.0	5.6	21.9	52.0
Intermediate, wax-bearing	1.00	37.0	25.2	5.4	25.0	55.3
Intermediate, wax-free	0.36	34.9	23.4	4.5	25.1	52.5
Intermediate-naphthene, wax-bearing	...	37.1	19.0	9.3	38.1	56.0
Intermediate-naphthene, wax-free	...	34.3	17.3	...	11.2	50.9
Naphthene-intermediate, wax-bearing	0.35	31.5	23.4	3.3	24.7	49.4
Naphthene-intermediate, wax-free	0.80	31.1	22.6	...	6.8	48.6
Naphthene, wax-bearing	1.50	29.4	16.7	4.7	21.5	51.0
Naphthene, wax-free	1.07	28.8	16.0	2.9	15.1	50.3

* From *USBM Bull.* 401, p. 164.

FLUID SATURATIONS

Fluid saturations of reservoir rocks are expressed as percentages of effective porosity. Initially, saturations may vary over wide limits and of course always change during production operations. For example, *initial saturations* of a reservoir rock, that is, at the time of discovery, may have been 70 per cent oil and 30 per cent interstitial water. Upon completion of a recovery process or upon abandonment, *residual saturations* might be approximately 50 per cent oil, 20 per cent free gas, and 30 per cent interstitial water. Or, in other cases, residual saturations might be in the range of 40 per cent oil, 10 per cent free gas, and 50 per cent water (30 per cent interstitial water and 20 per cent encroached water). Oil saturation changes in a reservoir are functions of the initial saturation, initial pressure, initial composition of the oil, production methods, and sources of reservoir energy. Residual saturation at the end of a primary phase of oil recovery may be thought of as the initial saturation at the beginning of a secondary recovery phase. And a new residual oil saturation value would ensue at the end of the secondary recovery operation. The same terminology would apply in a tertiary recovery operation. The term residual water saturation also is used to designate the percentage of water held in a reservoir rock by capillary forces at any time during the productive life of the reservoir.

Reservoir natural gas may be classified as follows:

1. Associated gas (*a*) Dissolved gas–gas originally in solution in the oil phase of an oil reservoir or in solution in the hydrocarbon-liquids phase of condensate reservoirs. (*b*) Gas-cap gas–free gas in a free gas cap overlying the oil in an oil reservoir.
2. Nonassociated gas–free gas neither associated with nor dissolved in liquid hydrocarbons in a reservoir.

The distribution of reservoir fluids is determined largely by the type of fluid combined with the wetting properties of each fluid, the surface properties of the pores, and the pore geometry. Most students of the problem of oil accumulation believe that the pores of present-day reservoir rocks were filled with water from the time of deposition until displaced in part by accumulating oil and/or gas. If the accumulating oil appears to have adhered preferentially to the pore surfaces, thus expelling most of the pore water, the reservoir rock is said to be preferentially *oil wet*. Water having a higher surface tension than oil is difficult to expel, especially from the sharp-angled interstices at the contacts of individual grains. Consequently, oil-wet reservoir rocks are believed to be rare. The Wilcox sand of the Oklahoma City field, Oklahoma, is thought to be preferentially oil wet. Reservoir rocks with approximately 10 per cent or less of interstitial water, exclusive of natural gas reservoirs, are probably oil wet, at least in part. Any reservoir rocks with 20 to 25 per cent or more of initial interstitial water saturation is probably preferentially *water wet*. Nearly all reservoir rocks are believed to be of this type.

Generally, sands of low permeability have high water saturations initially and high residual oil and residual water saturations upon abandonment, irrespective of their preferential wetting properties.

PHASE BEHAVIOR OF RESERVOIR HYDROCARBONS

The way in which the three states or phases of matter (gaseous, liquid, and solid) exist at any particular time depends on the termperature and confining pressure, either of which may change while the other remains constant; or both may change simultaneously. These changes in matter with pressure and temperature are called *phase behavior*. The three phases of water, namely, ice (solid), water (liquid), and steam (gas), are well known. Pure individual hydrocarbons and to a certain extent hydrocarbon mixtures change their states similarly. Solids are seldom formed or found under reservoir or production conditions. Primary interest from the oil property valuation viewpoint is in the liquid-gas relationships.

The phase behavior of hydrocarbons like all matter is explained by the behavior characteristics of individual molecules. Four factors contribute to their behavior, namely: (1) pressure, (2) molecular attraction, (3) kinetic energy, and (4) molecular repulsion. Molecular attraction, which varies with the mass of the molecules and pressure tends to increase density. Kinetic energy of all molecules increases with temperature. It may become sufficiently high as to throw molecules in the liquid state into the gaseous state. The kinetic energy of molecules in the gaseous state decreases with a lowering of temperature; then they tend to attract each other more and to coalesce into the liquid state again. And if the temperature is sufficiently low the liquid may freeze into the solid state.

All molecules are said to be in a state of equilibrium when the attracting forces and the repelling forces under a given condition of pressure and temperature just neutralize each other. Small molecules such as those of methane and ethane have less molecular attraction for each other than do molecules of propane and hexane which are larger and denser. Thus the molecules of propane and hexane tend to be pulled into the liquid state at pressures and temperatures which, through their higher kinetic energies, tend to keep methane and ethane in a gaseous state.

Oil property valuation engineers are concerned more with the physical properties of reservoir fluids than with the chemical properties. Both properties under reservoir conditions may be quite different from those observed at the surface under conditions of lesser pressures and temperatures. A reservoir rock may be saturated completely with oil and interstitial water under reservoir conditions, yet produced fluids at the surface may consist of only oil and gas. The oil phase under reservoir conditions may be composed of several hundreds of hydrocarbon compounds varying from methane, the simplest, to the complex such as waxes, tars, and asphalts, Under surface conditions the same is true for the overall produced mixture except that the simpler and less dense components probably have evolved to give a separate gas phase. The gas phase, either as a free gas in the reservoir or as an evolved gas at the surface, is always a simpler mixture than the oil phase. Normally under reservoir conditions, each component hydrocarbon is miscible with all other present in the reservoir.

PVT Studies

Phase behavior studies of hydrocarbons are called *PVT studies*, that is, studies of hydrocarbons under changing conditions of pressure and temperature with respect to corresponding changes in physical composition and respective volumes. PVT relationships of single component systems such as ethane are quite simple as shown in Figure 9-1. This single-component

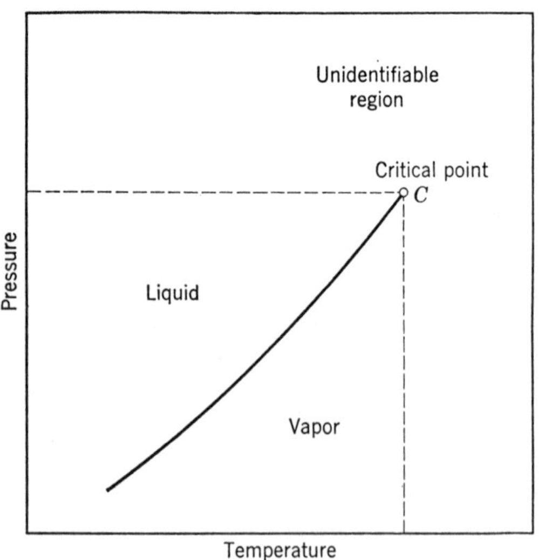

Figure 9-1. Phase behavior curve for a single-phase hydrocarbon.

hydrocarbon may exist either as a vapor or a liquid depending on the combination of pressure and temperature. The locus of points at which it may change from one to the other with small changes in temperature or pressure is called the *vapor-pressure curve*. At some high temperature and corresponding high pressure it is impossible to distinguish whether the ethane is in liquid or vapor form. This pressure and temperature defines its *critical point, C* in the figure. Every component hydrocarbon has a vapor-pressure curve and critical point which may be determined in the laboratory.

PVT studies of mixtures of two or more simple hydrocarbons such as methane and propane may be made. There is still a critical point *C* for all such mixtures, but a sharp boundary between a 100 per cent liquid and a 100 per cent vapor phase has changed to an envelope of continuously varying percentages of the component hydrocarbons as shown in Figure 9-2. Although such studies are of interest and value in reservoir engineering, the valuation engineer is interested primarily in PVT studies of actual *bottom-hole samples*, i.e., samples of reservoir fluids taken at the depth of the productive reservoir sand from closed-in wells and kept at reservoir pressure. At times recombined samples of reservoir fluids as produced at the surface are equally satisfactory.

One or more bottom-hole samples should be taken from the first well in every reservoir shortly after completion in order to provide PVT data needed in making subsequent reservoir calculations including estimates of reserves.

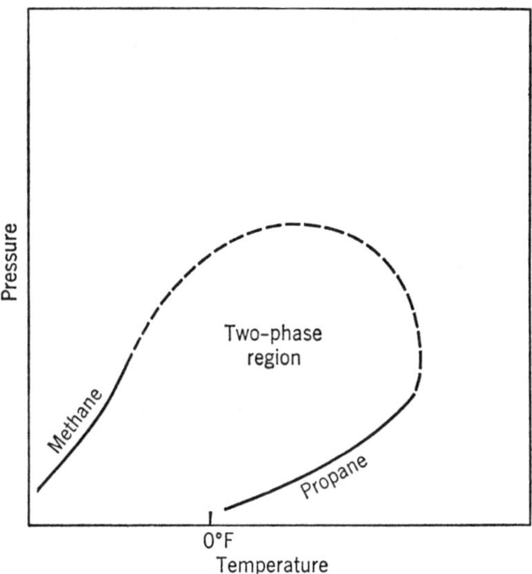

Figure 9-2. Phase behavior diagram of a simple two-component hydrocarbon fluid.

This is true with respect to deep or high-pressure reservoirs and especially true for condensate reservoirs. Periodic sampling and PVT studies of the same and perhaps additional wells are advisable if the reservoir is large.

A virgin oil reservoir normally has gas in solution in the oil phase. The oil under reservoir conditions may be either saturated or undersaturated, depending on the compositions of the oil and the gas, pressures, and temperatures. If the reservoir oil is saturated, some gas in addition to that in solution may be present under original reservoir conditions in the form of a free gas cap. Usually the higher the reservoir pressure the higher the solution gas-oil ratio, depending in part on the compositions of the gas and the oil and the reservoir temperature. PVT studies usually are made at constant temperature, that of the reservoir which remains essentially the same throughout its productive life.

Two kinds of PVT studies are prevalent, namely, *flash vaporization* and *differential vaporization*. A constant volume high-pressure cell partially filled with the bottom-hole sample under a pressure equal to or greater than the reservoir pressure is used for both studies. The cell is kept at constant temperature and is continuously agitated between observations. Each study consists of letting gas vaporize from solution as pressure is lowered. Means are provided for measuring the volume of the evolved gas phase as well as that of the oil phase after each change in pressure. The gas may be removed from contact with the oil phase in the cell as evolved or it may be left in contact over large pressure drops.

Differential vaporization procedures attempt to keep the composition of the system continuously changed by removing the gas phase as liberated under continuous pressure. drop. Actually the procedure must be conducted under small stepwise decreases in pressure, and the volume of the liberated gas after each pressure drop is removed as a unit.

Flash vaporization procedures keep the liquid and vapor phases in contact over large decreases in pressure and until equilibrium is established at the lower pressure. All the vapor is removed from the cell at this time and the pressure may be reduced to some lower value for the next stage of flash vaporization. Often the two phases are kept in contact over pressure drops equaling that of the reservoir down to atmospheric pressure or to representative back pressures at points in the well-flow system.

An attempt is made under both procedures to simulate what takes place in the reservoir and in the well with its surface hookup under operating conditions. Probably neither flash nor differential vaporization approximating a laboratory test occurs as a separate process in the production operation until the surface gas-oil separator is reached. Some combination of the two processes is most likely at all other times. Both the *intensive properties* and the *extensive properties* of such multicomponent fluids are measured. The intensive properties are those that are independent of the amount of material under consideration, and the extensive properties are those that are directly related to the amount of material under consideration.

Series of pressure-volume tests on a given sample can be run at various constant temperatures ranging from temperatures above the reservoir down to atmospheric. The relative volumes of measured liquid and gas at various combinations of pressure and temperature provide data for plotting phase diagrams to be used in making reservoir calculations. Viscosities of the oil and gas phases at various stages of the PVT study may be made simultaneously in auxiliary equipment.

Phase Diagrams of Multicomponent Hydrocarbons

The oil property valuation engineer is much less interested in the exact chemical composition of the reservoir hydrocarbons than in their phase

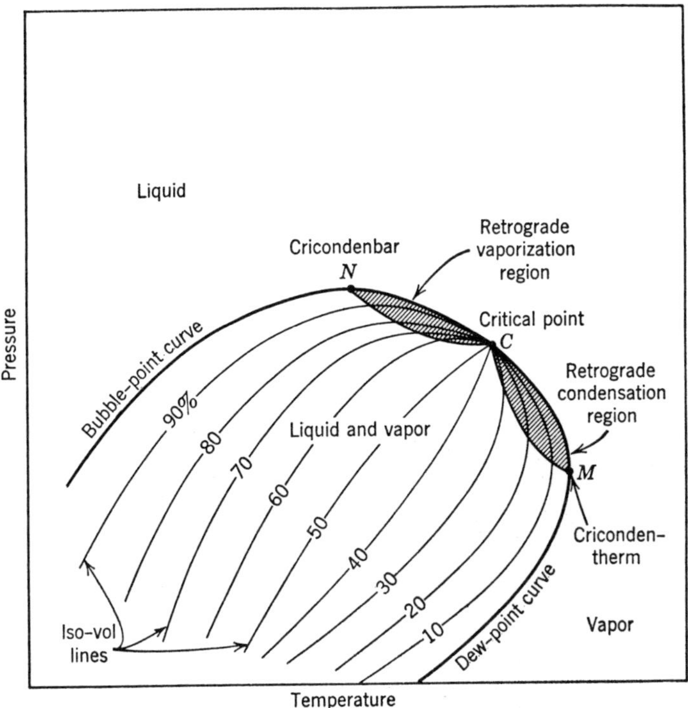

Figure 9-3. Phase diagram of a multicomponent reservoir fluid illustrating retrograde vaporization and condensation.

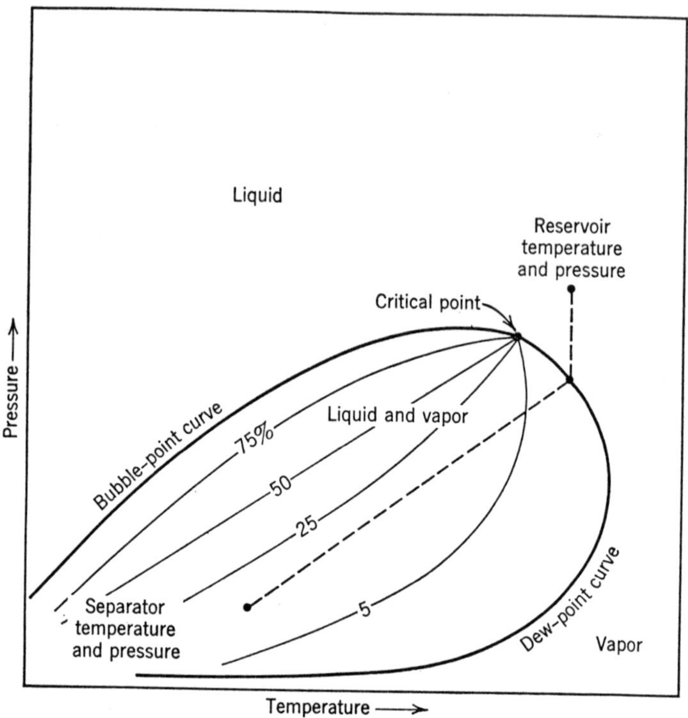

Figure 9-4. Phase behavior diagram of a condensate-type reservoir fluid.

behavior under decreasing reservoir pressure. Phase behavior data will provide some knowledge as to the state of the hydrocarbon contents of the reservoir at various pressures and aid in the selection of best production procedures. Condensate and wet-gas reservoirs cannot be produced to recover the liquid phase efficiently without such knowledge. They are more of academic interest with respect to very low gravity oils.

A phase diagram of a simple multicomponent hydrocarbon fluid is shown in Figure 9-3. The following points, boundaries, and regions are of interest:

Critical point (C), also shown on Figure 9-1 for a single component fluid, that pressure and temperature at which the intensive properties of the gas and liquid phases are indistinguishable.

Cricondentherm (M), the highest temperature at which a hydrocarbon liquid and its vapor can coexist in equilibrium.

Cricondenbar (N), the highest pressure at which a hydrocarbon liquid and its vapor can coexist in equilibrium.

Bubble-point curve, the locus of points of pressure and temperature at which the first bubble of solution gas evolves as the pressure on the hydrocarbon liquid is reduced.

Dew-point curve, the locus of points of pressure and temperature at which the first droplet of liquid is condensed from a hydrocarbon gas upon increasing or decreasing the pressure.

Critical temperature, the temperature at the critical point.

Critical pressure, the pressure at the critical point.

Two-phase region, that part of the phase diagram enclosed by the bubble-point and dew-drop curves wherein gas and liquid coexist in equilibrium.

Iso-vol lines, the loci of points of equal liquid volume per cent within the two-phase region.

Retrograde region, those parts of the two-phase region where condensation occurs upon lowering the pressure at constant temperature or increasing the

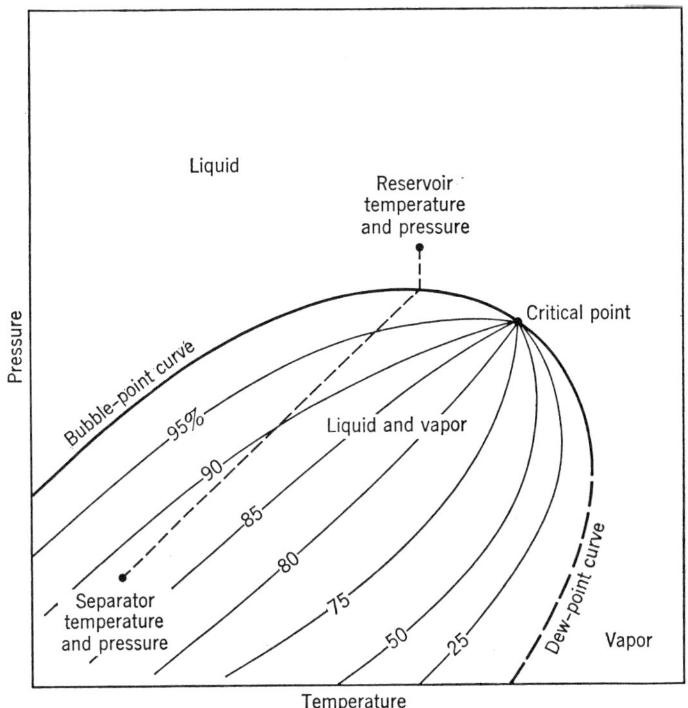

Figure 9-5. Phase diagram of a low-shrinkage reservoir fluid.

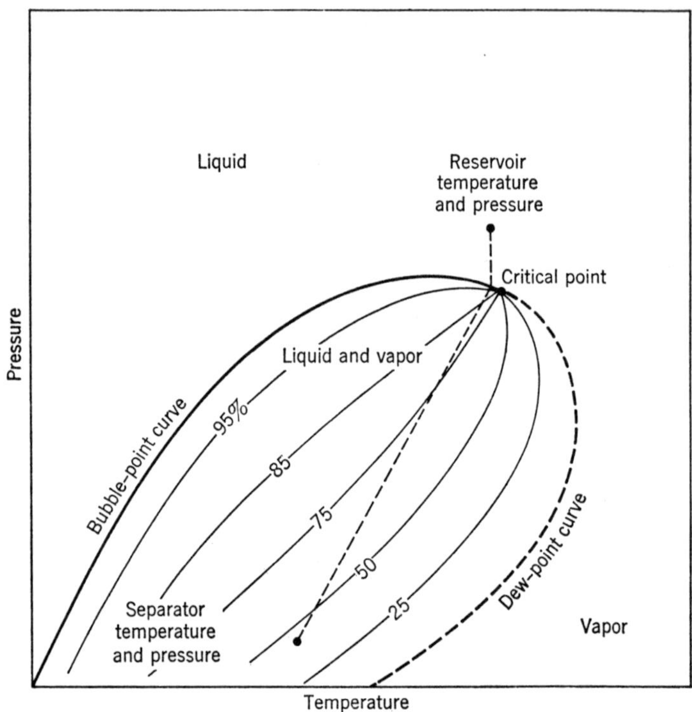

Figure 9-6. Phase behavior diagram of a high-shrinkage reservoir fluid.

temperature at constant pressure, and where vaporization occurs upon lowering the temperature at constant pressure or increasing the pressure at constant temperature. These conditions are called *retrograde condensation* and *retrograde vaporization*, respectively.

Figure 9-4 illustrates a phase diagram of a condensate-type reservoir fluid. Wet-gas reservoirs are sometimes difficult to distinguish from condensate reservoirs without a phase diagram of the reservoir fluid. Ordinarily the reservoir temperature which remains constant during production operations falls outside the dew-point curve on a wet-gas diagram. A small percentage of the production from a wet-gas reservoir may be hydrocarbon liquids if the separator pressure and temperature fall within the dew-point line. Figures 9-5 and 9-6 illustrate phase diagrams for low- shrinkage and high-shrinkage oils, respectively.

OTHER PVT CURVES OF INTEREST

Other curves needed in reservoir calculations and based upon PVT data would include:

1. Shrinkage (B) factor curve for the oil phase.
2. Gas-oil ratio (GOR) curve.
3. Oil viscosity curve.
4. Deviation (Z) factor curves for the natural gas phase.

B-Factor Curves

A unit volume of liquid hydrocarbon in the reservoir will shrink considerably by the time it reaches the stock tank because of the loss of dissolved gas and very light hydrocarbon liquids. The amount of shrinkage depends on whether either or both flash and differential vaporization occur between some production point in the reservoir and the stock tank. There are several ways of expressing these changes in liquid volumes. The most common

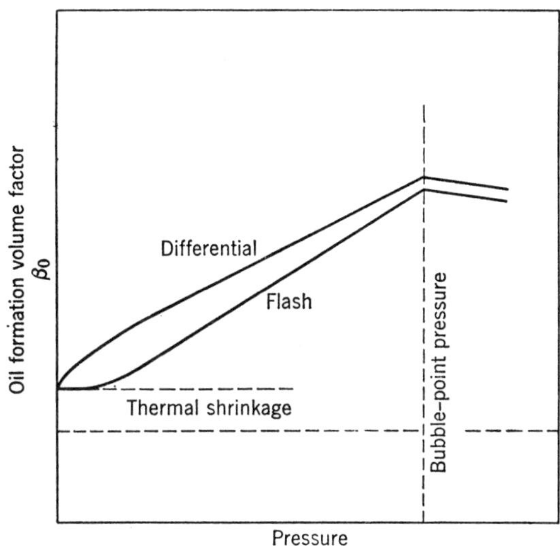

Figure 9-7. Idealized comparison of B-factors for reservoir oils.

method is to calculate and plot the so-called B_o-factor, the reciprocal of its shrinkage. It may be expressed as the ratio of a volume of oil plus its dissolved gas at reservoir temperature and pressure to its volume under stock-tank conditions. The B_o-factor is a maximum at the saturation pressure of the reservoir as shown in Figure 9-7 for both flash and differential vaporization. A differential liberation curve is most often used.

Pure water and brines will dissolve natural gas. The amount in reservoir brines depends on the pressure, amount of dissolved solids, composition of the gas, and temperature. Consequently, in addition to the compressibility B_w of water itself, an added factor due to dissolved gas must be considered. Dodson and Standing [6] made measurements of the factor for mixtures of natural gas and water as given in Table 9-2.

Table 9-2
Formation Volumes of Mixtures of Natural Gas and Water*

Saturation pressure (psia)	Formation volumes—barrel per barrel			
	100°F	150°F	200°F	250°F
5000	0.9989	1.0126	1.0301	1.0522
4000	1.0003	1.0140	1.0316	1.0537
3000	1.0017	1.0154	1.0330	1.0552
2000	1.0031	1.0168	1.0345	1.0568
1000	1.0045	1.0183	1.0361	1.0584

* Dodson and Standing.[6]

Theoretically the total B_{sw}-factor for a reservoir brine at depths of 15,000 to 20,000 ft might approach a value of 1.2, which is a large correction when applied to oil recoveries. Even in shallow reservoirs such as the East Texas reservoir it may be sizable due to very large volume of surrounding edgewater.

Evaluation engineers seldom take the B_{sw}-factor into account. When necessary the reservoir engineers will have already done so for them.

GOR Curves

The GOR curve is a plot of the calculated standard cubic feet (scf) of gas dissolved in one barrel of stock tank oil (STB) at reservoir temperature and varying pressures up to that of the original reservoir pressure. Such a curve is shown in Figure 9-8.

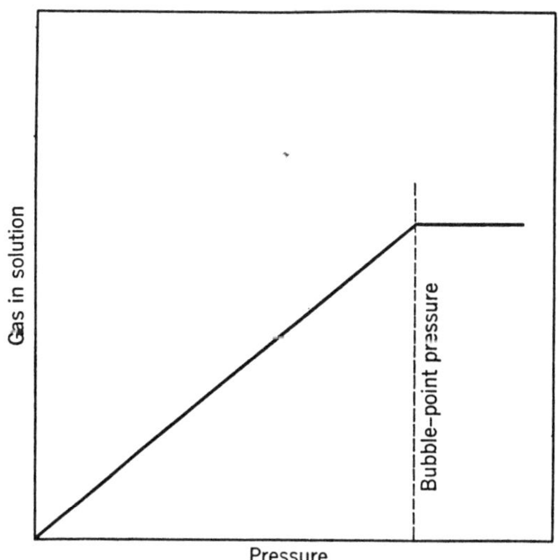

Figure 9-8. Idealized gas in solution pressure curve for a reservoir oil.

Oil Viscosity Curves

The viscosity of the liquid phase is always lower in the reservoir than in the stock tank. The viscosity is inversely proportional to both the temperature and the amount of gas in solution at pressures up to the bubble point where it is a minimum. Under reservoir conditions viscosity increases with pressure above the bubble point. A typical viscosity curve for a 35°API oil is shown in Figure 9-9. Viscosities of crude oils under reservoir conditions may vary from a fraction of a centipoise for high-gravity oils to several hundreds of centipoises for low-gravity asphaltic-base oils.

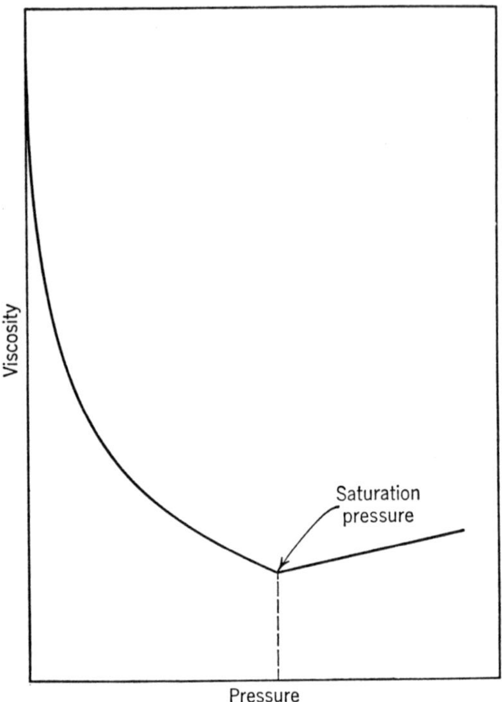

Figure 9-9. Idealized viscosity pressure diagram of a reservoir oil.

Z-Factor Curves

No natural gas under compression and/or changing temperature follows Boyle's and Charles' laws exactly. Graphs have been prepared by researchers to provide the necessary correction factors over a wide range of pressures and temperatures for the single-component hydrocarbon gases. Most of the graphs show the Z-factor plotted against the *reduced pressure* for various *reduced temperatures*. The reduced pressure is defined as the ratio of the absolute pressure of the gas to its critical pressure, and the reduced temperature is defined as the ratio of the absolute temperature of the gas to its critical temperature.

Similar graphs of Z-factors for natural gases are prepared using *pseudoreduced pressures* and *pseudoreduced temperatures*. These values are based on *pseudocritical pressures* and *pseudocritical temperatures* which are the molecular average critical pressures and molecular average critical temperatures, respectively. The method of calculating these values for a natural gas[7] is shown in Table 9-3.

Table 9-3
Computation of Pseudocritical Temperature and Pressure of a Natural Gas*

(1) Component	(2) Mole fraction	(3) Individual absolute critical temperature T_c, °R	(4) Absolute pseudocritical temperature T_c, °R (4) = (2) × (3)	(5) Individual absolute critical pressure P_c, psia	(6) Absolute pseudocritical pressure P_c, psia (6) = (2) × (5)
CH_4	0.8319	344	286	673	560
C_2H_6	0.0848	549	46.5	712	60.4
C_3H_8	0.0437	666	29.1	617	27.0
$i\text{-}C_4H_{10}$	0.0076	732	5.56	544	4.13
$n\text{-}C_4H_{10}$	0.0168	766	12.86	551	9.26
$i\text{-}C_5H_{12}$	0.0057	829	4.72	483	2.75
$n\text{-}C_5H_{12}$	0.0032	846	2.71	485	1.55
C_6H_{14}	0.0063	914	5.75	435	2.74
			393.2		667.83

* By permission of McGraw-Hill Book Company, Inc.[7]

For example, the pseudoreduced temperature T_r of the gas in Table 9-3 at a temperature of 200°F and 3000 psia is

$$T_r = \frac{T}{T_c} = \frac{460 + 200}{393.2} = 1.679$$

and the pseudoreduced pressure P_r for the same condition is

$$P_r = \frac{P}{P_c} = \frac{3000}{667.83} = 4.492$$

When the composition of the natural gas can be determined and the pseudo-reduced temperature and pressure calculated, the Z-factor graph for natural gases (Figure 9-10) may be used.

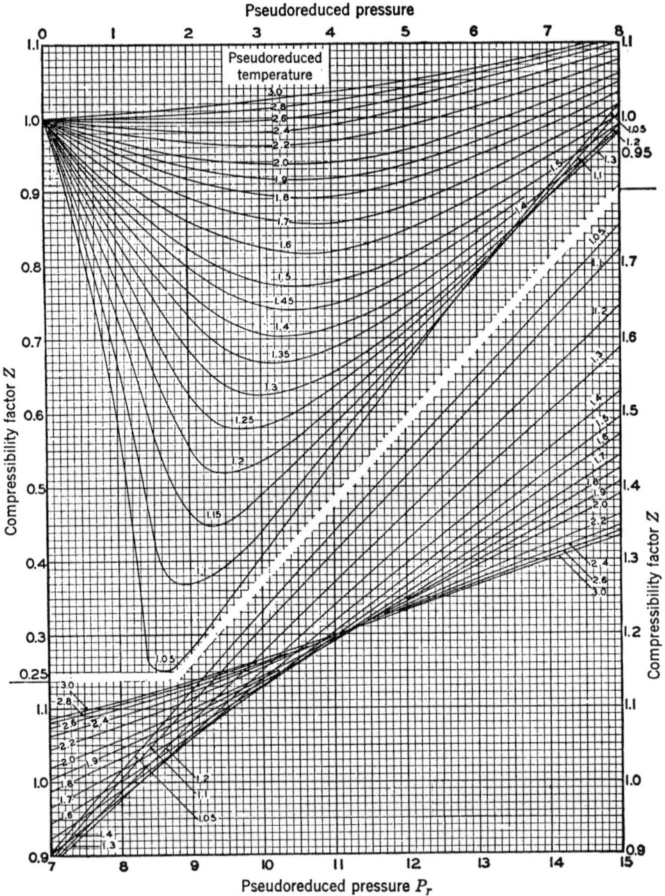

Figure 9-10. Compressibility factors for natural gases. From Brown et al.[8]

REFERENCES

1. Fred Hoyle, *Frontiers of Astronomy*, New American Library of World Literature, New York, 1957, pp. 72-74.
2. M. King Hubbert, "Entrapment of Petroleum under Hydrodynamic Conditions," *Bull. AAPG*, Vol. 37, No. 8, August 1953, pp. 1954-2026.
3. William Carruthers Gussow, "Differential Entrapment of Oil and Gas: A Fundamental Principle," *Bull. AAPG*, Vol. 38, No. 5, May 1954, pp. 816-853.
4. E. C. Lane and E. L. Garten, "'Base' of Crude Oil," *USBM Rept. Invest.*, 3279, 1935.
5. A. J. Kraemer and E. C. Lane, "Properties of Typical Crude Oils from Fields of the Eastern Hemisphere," *USBM Bull. 401*, 1937, pp. 148-149.
6. C. R. Dodson and M. B. Standing, "Pressure-Volume-Temperature and Solubility Relations for Natural-Gas-Water Mixtures," *API Drill. and Prod. Pract.*, 1944, p. 176.
7. Sylvain Pirson, *Oil Reservoir Engineering*, 2nd ed., McGraw-Hill, New York 1958, pp. 346-347.
8. G. G. Brown, D. L. Katz, G. G. Oberfell, and R. C. Allen, *Natural Gasoline and the Volatile Hydrocarbons*, Natural Gas Processors Suppliers Assn., Tulsa, Okla. 1948.

Additional References

J. W. Amyx, D. M. Bass, Jr., and R. L. Whiting, *Petroleum Reservoir Engineering–Physical Properties*, McGraw-Hill, New York, 1960.

T. E. W. Nind, *Principles of Oil Well Production*, McGraw-Hill, New York, 1964.

Frank W. Cole, *Reservoir Engineering Manual*, Gulf Publishing Company, Houston, Texas, 1961.

10 ELEMENTS OF RESERVOIR BEHAVIOR

John F. Carll described the basic principles of production of oil by dissolved gas in 1880.[1] Apparently very few in the oil industry became acquainted with his work and little attention was given to the improvement and understanding of reservoir behavior until approximately 1930. A great deal had been learned some years earlier from a practical viewpoint about decline rates and various decline curve methods of estimating recoverable reserves; yet little was known or agreed upon concerning the exact function of natural gas in oil production. Producers also had learned to rejuvenate wells and whole fields by many methods. Rather efficient techniques had been developed for pulling a vacuum on wells of low productivity and for injecting air, gas, or water in exhausted wells in order to drive oil to others still on production. Thus by 1921 there were two distinct production practices or methods of oil recovery, namely *primary recovery* and *secondary recovery*.

Primary recovery has been defined as[2]

> . . . the oil, gas, or oil and gas recovered by any method (natural flow or artificial lift) that may be employed to produce them through a single well bore; the fluid enters the well bore by the action of native reservoir energy or gravity.

In contrast with the above, secondary recovery was defined as[3]

> . . .the oil , gas, or oil and gas recovered by any method (artifical flowing or pumping) that may be employed to produce them through the joint use of two or more well bores. Secondary recovery is generally recognized as being that recovery which may be obtained by the injection of liquids or gases into the reservoir for the purpose of augmenting reservoir energy; usually, but not necessarily, this is done after the primary-recovery phase has passed.

The above definition was not intended to eliminate repressuring, pressure maintenance, and gas cyclying, operations that are usually commenced long before the native reservoir energy is exhausted. Johnson and van Wingen proposed a different definition for secondary recovery which they believed to be more in line with actual practice. Accordingly it was defined as[4]

> . . . recovery by any method (natural flow or artificial lift) of that petroleum which enters a well as a result of augmentation of the remaining native reservoir energy (as by fluid injection) after a reservoir has approached its economic production limit by primary recovery methods.

Accordingly, secondary recovery occurs when a reservoir is rejuvenated after once having reached economic production limits. Inasmuch as no definition is absolute, engineers now speak of "improving oil recovery" as an all-inclusive term, which is equally misleading since it could include gas lifting and pumping wells as secondary recovery methods.

PRIMARY RECOVERY OF OIL

Pirson referred to the various ways in which oil is moved through reservoirs during production operations as being due to *internal driving energy, external driving energy, potential energy*, and *surface energy*.[5] One, and probably more than one, of these energies is active at all times in the reservoir while reservoir fluids are in motion. Oil in itself has no inherent energy so any movement must be due to gas and water under pressure, and surface and gravity forces. There is no marked difference in the forces that are operative in both primary and secondary recovery. Primary recovery of oil, however, is said to be accomplished by the following:

1. Dissolved gas drive (internal driving energy).
2. Gas-cap drive (external driving energy).
3. Water drive (external driving energy).
4. Gravity drive (potential energy).

There is no thought that surface energy is not operational during any of the above four recovery methods. Each has its characteristics which may be identified during production operations, i.e., the characteristics of one kind of drive predominate at any one time yet one or more other drives may be contributory at the time to the movement of the oil.

Miller[6] described a method by which the type of drive in an oil reservoir could be determined in a qualitative way and more specifically when the type of drive changed. The method consists of plotting the cumulative decline in reservoir pressure against the cumulative oil recovery on log-log graph paper. Zero slope of the line connecting the points indicates a 100 per cent water drive. Slope angles of 45° show that each succeeding unit of pressure decline results in equal increments of production. Slope angles greater than 45° indicate lesser recovery per unit of pressure decline, and angles of less than 45° indicate increased recovery per unit of pressure decline due probably to effectiveness of a water drive, gravity drainage, and perhaps to a minor extent an expanding gas cap. Thus it is possible to determine when there is a change in recovery efficiency and perhaps the type of drive.

All virgin reservoirs have some dissolved gas in the reservoir liquids and under certain conditions may have in addition some reserve of gas-cap gas. Percentagewise, very little dissolved gas and associated gas were marketed before the late 1940's. Before then much of the gas produced with the oil was wasted by flaring in the field. The amount of such wastage since the birth of the industry is beyond estimation. Texas in 1948-1949 finally forced operators of oil properties where excessive waste of gas occurred to close-in their wells until gas pipeline outlets were provided, markets were found, or the produced gas was returned to the reservoir. Proportionately, gas wastage in Oklahoma and Kansas was probably equally high during the period 1915 to 1930. Many gas caps were completely wasted, particularly in Oklahoma, by allowing the discovery well to "blow" itself into an oil well. Today scarcely any gas percentagewise is wasted in our domestic oil fields.

An initial producing gas-oil ratio (GOR) of 10,000 to 100,000 scf/bbl from other than a free gas cap would be indicative of a condensate well; that is, the reservoir fluid moving into the producing well is mostly if not completely in gaseous or vapor form and hydrocarbon liquids may be recovered from the gas at the surface. A GOR of more than 100,000 scf/bbl is indicative of a gas well; that is, the reservoir fluid moving into the producing well is in a gaseous or vapor form and small quantities of liquid hydrocarbons might or might not be recoverable from the gas at the surface depending upon its original composition. Obviously no exact ratios can be stated because each kind of production grades into the other.

Natural gases from which commercial quantities of hydrocarbon liquids may be extracted are commonly called *wet gases*. Gases from which hydrocarbon liquids cannot be extracted in commercial quantities are called *dry gases*. There are no clean-cut distinctions between the two. Wet gases under reservoir conditions may grade into condensates and condensates may grade into distillates and high-gravity crude oils. Condensates and distillates may or may not be liquids under either original or subsequent reservoir conditions depending on the combination of reservoir pressure and temperature. Efficient production operations aim at maintaining high reservoir pressures by injection of produced gas from which the hydrocarbon liquids have been extracted. If produced, the liquids picked up by the dry gas are extracted and the dry gas returned to the reservoir again. Such as process of producing liquid hydrocarbons from a reservoir is called a *recycling operation*.

Dissolved Gas Drive

Most of our oil reservoirs produce by dissolved gas drive sometimes called depletion drive. Unfortunately it is the least efficient of the four basic drives. Its mechanism depends on expansion of dissolved gas. The rate of flow and the ultimate oil recovery depend primarily on the amount of gas originally dissolved in the oil, the efficiency of its utilization, and the degree of its exhaustion.

Dissolved gas drive may be associated with either a gas-cap drive or a water drive, or with both simultaneously. Usually it appears to be the sole producing mechanism as indicated by a more or less rapid decline in reservoir pressures and production rates combined with an increasing produced gas-oil ratio. At times an artificial gas cap will form at excessively high oil production rates if the reservoir dips steeply. The recovery mechanism is inherently inefficient at every rate of oil production.

Oil is recovered by dissolved gas drive as a result of progressively lower pressures within the body of the moving fluids. Undersaturated reservoirs

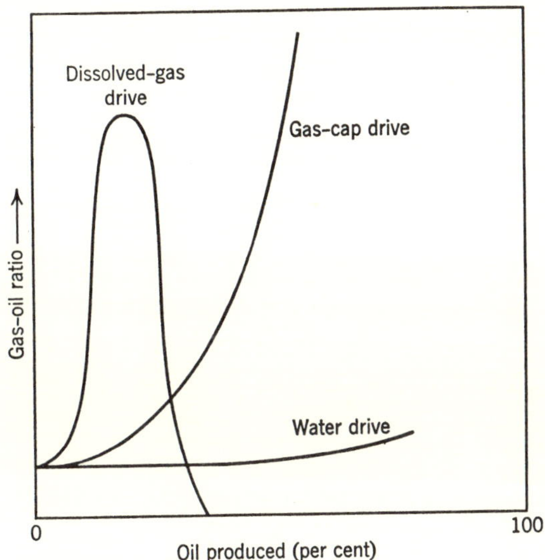

Figure 10-1. Idealized GOR-recovery relationships under the three kinds of reservoir drives.

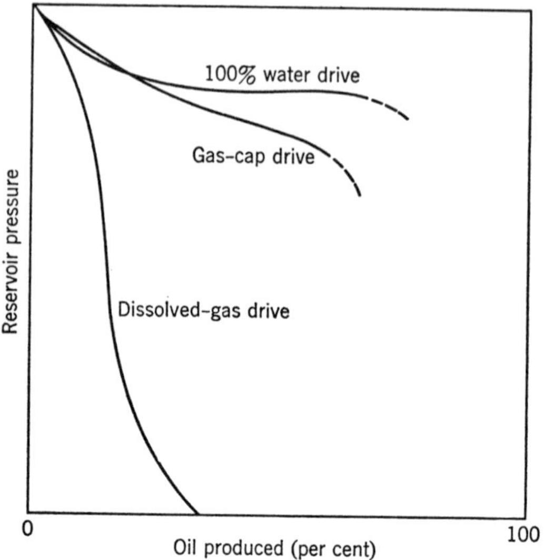

Figure 10-2. Idealized pressure recovery relationships under the three kinds of reservoir drives.

with original pressures above the bubble point may produce considerable oil by expansion of the oil (increase in B-factor) and interstitial water. Careful control of the producing gas-oil ratio is necessary.

Table 10-1

Recoveries under Dissolved Gas Drive[9]

Approximate depth range, ft	9,000				3,000			
Solubility, scf/bbl	1,200				400			
Shrinkage, per cent of stock tank oil	60				20			
Interstitial water	40		20		40		20	
Gravity, °API	26	46	26	46	26	46	26	46
Viscosity, cp	20	2	20	2	20	2	20	2
Primary recovery								
per cent of pore space	7.6	13.3	6.7	12.2	9.5	18.4	8.6	17.4
per cent of original stock tank oil	20.3	35.4	13.4	24.4	18.9	36.8	12.9	26.1

Inasmuch as maximum reduction in reservoir pressure occurs immediately about the well bore during production, a higher percentage of gas comes out of solution in this area. This increase in free gas saturation increases the relative permeability of the reservoir sand to gas and decreases it to oil. A well producing under this mechanism will show a constantly increasing GOR until such time as scarcely any oil is produced along with the gas. This condition may be reached when the free gas saturation of the reservoir reaches some 20 to 40 per cent with an average of approximately 30 per cent. The end result is an overall inefficient and nearly exhausted recovery mechanism.

A generalized recovery-GOR relationship for a dissolved gas-drive reservoir is shown in Figure 10-1. The effects of a delayed water drive are indicated on the low rate-low pressure end of the production graph. Recovery-reservoir pressure relationships are shown in Figure 10-2.

Recoveries from dissolved gas-drive fields during the early days of the industry probably averaged about 10 per cent.[7] Estimates made in 1944[8] which probably included other than dissolved gas-drive fields showed an

Table 10-2

Primary Recovery in Per Cent of Stock Tank Oil in Place for Depletion-Type Reservoirs[10]

Oil solution GOR (cf/bbl)	Oil gravity (°API)	Sand or sandstones			Limestone, dolomite, or chert		
		Maximum	Average	Minimum	Maximum	Average	Minimum
60	15	12.8	8.6	2.6	28.0	4.0	0.6
	30	21.3	15.2	8.7	32.8	9.9	2.9
	50	34.2	24.8	16.9	39.0	18.6	8.0
200	15	13.3	8.8	3.3	27.5	4.5	0.9
	30	22.2	15.2	8.4	32.3	9.8	2.6
	50	37.4	26.4	17.6	39.8	19.3	7.4
600	15	18.0	11.3	6.0	26.6	6.9	1.9
	30	24.3	15.1	8.4	30.0	9.6	(2.5)
	50	35.6	23.0	13.8	36.1	15.1	(4.3)
1000	15	...	...	...	...	...	...
	30	34.4	21.2	12.6	32.6	13.2	(4.0)
	50	33.7	20.2	11.6	31.8	12.0	(3.1)
2000	15	...	...	...	...	...	...
	30	...	...	...	...	...	...
	50	40.7	24.8	15.6	32.8	(14.5)	(5.0)

average ultimate recovery from 35 principal U. S. fields of 311 bbl/acre-ft and from 223 U. S. fields representing about one-fourth of the discovered reserves at the time an average of 187 bbl/acre-ft. Assuming average porosities of 25 per cent and average interstitial water saturation of 30 per cent, average recoveries would have been 28 and 14 per cent, respectively. Elliott and Morris presented recovery data ranging from 12.9 to 36.8 per cent for dissolved gas-taype fields as shown in Table 10-1.[9] Arps used a well-known depletion equation assuming oil gravities from 15° to 50°API, gas solubilities from 60 to 2000 scf/bbl, and six types of reservoir rocks to arrive at the recoveries shown in Table 10-2.[10]

Recoveries under dissolved gas drive can be increased to some extent by producing the wells at all times at the maximum efficient rate (MER), i.e., at lowest gas-oil ratio. Such rates may not always be economic from an income viewpoint. Recoveries may be higher when other recovery mechanisms are or become partially effective. It is largely the abused and inefficient dissolved gas-drive reservoirs that provide the best opportunities for secondary oil recovery.

Gas-Cap Drive

A free gas cap occurs in an oil reservoir only when all the gas present will not dissolve in the oil phase under existing conditions of reservoir pressure and temperature. It may be either an original or an artificial gas cap; in either case it may be maintained or enlarged by injection of produced gas.

An efficient gas-cap drive requires a rather steeply dipping reservoir with continuous beds of relatively high permeability, a high ratio of reservoir gas-cap volume to oil volume, and very careful control of oil and net gas production rates. The locations of the producing wells and their sand penetrations are equally important. The expanding gas cap may appear to be the sole recovery mechanism for a time or it may appear to be in conjunction with dissolved gas. Only rarely is an edgewater drive also indicated and then it is usually not very effective. The gas cap must expand downstructure to be efficient. Any bypassing gas wastes much of its energy available for moving oil. Ideally, expanding solution gas would raise the oil to the surface. Some downward coning of gas will occur finally at almost any economic oil production rate with the end result that every producing well near the gas-oil boundary becomes rate-sensitive in time. Generalized production gas-oil ratio and production pressure graphs for a gas-cap drive reservoir are shown in Figures 10-1 and 10-2, respectively.

Theoretically, under favorable conditions a high-ratio gas-cap-to-oil reservoir may have long-lived flowing wells with an overall high recovery efficiency. Under actual operating conditions normal reservoir recoveries range from approximately 20 to 50 per cent of the original oil in place with an average of approximately 30 to 35 per cent. Recoveries above 40 per cent usually can be attributed to return of gas and/or influence of other natural recovery mechanisms. The relatively low residual oil saturations following high recoveries by gas-cap drive combined with an unfavorable distribution of reservoir fluids usually eliminates subsequent secondary recovery operations.

Water Drive

Natural water drive as a recovery mechanism occurs in both oil and unassociated gas reservoirs. It is somewhat more complex than gas drive as an oil recovery mechanism, inefficient under unfavorable conditions, very efficient under favorable conditions, yet more easily controlled at all times. There are two kinds od drives—edgewater and bottom water. Edgewater drives are most frequent and usually the most efficient. Bottom water drives differ from edgewater drives in that they are operative only where the reservoir beds are nearly flat or of very low dip, making a widespread oil-water contact. The lower the dip, unless the reservoirs are unusually thick and uniform, the more difficult it is to deplete the reservoir efficiently because of water coning into the wells.

A natural water drive may be a complicated process. Its effectiveness may be indicated or perhaps dominate throughout the productive life of a reservoir; it may contribute somewhat passively and undetected to the maintenance of reservoir pressures throughout the productive life of the reservoir; or it may take over as the dominant recovery mechanism only late during the productive life of the reservoir. Edgewater, especially if a part of a large aquifer, may contain enough, dissolved natural gas to provide an oil recovery mechanism solely by its expansion as upstructure oil wells are produced. The principle is the same as the increase in the *B*-factor of reservoir oils as pressure is reduced by production from a high original reservoir pressure down to the bubble point.

A water drive is more efficient as a recovery mechanism than a gas drive primarily because of the higher viscosity of water under reservoir conditions compared to natural gas. Its major recovery mechanism is thought to be a "pistonlike" action. Not all the oil is removed by the water drive with the result that the relative permeability of a reservoir sand to water may be greater than to oil at relatively high oil saturations. Like gas-cap drives, a water drive tends to be more efficient in steeply dipping, high-permeability sands of uniform texture. Probably most of the so-called bypassed oil remains in sand layers of low permeability. The East Texas oil field is the all-time outstanding water drive in the United States. Recoveries by active water drives usually do not exceed 40 to 50 per cent of the reservoir oil as a whole, because reservoir pressures and recoveries depend considerably upon relative rates of water influx to reservoir fluid withdrawals. Generalized recovery graphs are shown in Figures 10-1 and 10-2.

Gravity Drive

A gravity drive is not thought to be common although at times it may be operative only partially and unnoticed. Moderate to steeply dipping continuous reservoir beds of other than low permeabilities are believed to be prerequisites. It might be contributory to oil recovery under those conditions during a gas-cap drive. Gravity drive hardly could become effective until upstructure wells had produced most of their oil by other means and it could not occur under conditions of an active water drive. The Oklahoma City field, Oklahoma, is a well-known example. The Homer field, Louisiana, is another, but injection of gas upstructure aided the downward movement of the oil. With gravity acting as a supplementary drive, recoveries in all cases from the reservoir as a whole should be higher than for dissolved gas drive alone.

Pressure Maintenance Operations

Pressure maintenance, gas condensate, and recycling operations are considered by some engineers to be a form of secondary oil recovery inasmuch as recoveries are improved. Such operations are initiated preferably during the early productive life of the reservoir and are herein considered to be primary recovery operations.

Pressure maintenance includes many related kinds of operations; however, all are based on the injection of fluids, usually gas and/or water. The maintenance of pressures by the injection of produced water in the East Texas field is expected to result in an overall economic oil recovery of approximately 90 per cent of the original oil in place. Other projects include the recovery of condensates and volatile oils by injecting and/or recycling produced gas which has been stripped of all hydrocarbon liquids carried to the surface during production. Condensate wells are of more than average depth. Reservoir pressure and temperature also are relatively high. Reservoir fluids if not originally in a vapor form will vaporize at some reduced pressure and probably condense at a still lower pressure. Scientific production operations are aimed at keeping an original vapor in that form until reaching the surface, or in evaporating any hydrocarbon liquids in the reservoir and bringing them to the surface in vapor form.

In general an attempt is made in pressure maintenance operations to keep all gas in solution, to prevent oil migration into any gas cap, to prevent edgewater encroachment, to maintain the viscosity of the reservoir oil, to keep wells flowing, to keep hydrocarbon liquids dissolved in the gas phase, and as an overall objective, to obtain maximum oil recovery. Pressure maintenance operations are expensive from an investment and operating viewpoint. Evaluations should be made of the reservoir with and without pressure maintenance operations. Some of the factors to be considered are the following:

1. Composition and properties of the reservoir fluids.
2. Volumes of reservoir fluids.
3. Productivity of wells (number needed for production).
4. Injectivity of wells (number needed for injection).
5. Permeability variations (probability of bypassing).
6. Effect of any gas cap or water drive.
7. Markets for production.
8. Field installation and operating costs.
9. Plant installation and operating costs.
10. All taxes.
11. Discounted net income at suitable interest rate with and without pressure maintenance operations.
12. If many owners, can it be unitized?

In view of the above it is obvious that no worthwhile evaluation of properties of these kinds can be made until after a more or less comprehensive reservoir engineering study has been made.

SECONDARY RECOVERY OF OIL

Some statistics show that more than one-fourth of our domestic oil production is produced by secondary recovery methods. Basically these methods include a number of production practices which are not secondary recovery methods by definition. The injection of air, gas, and water as production practices developed largely through trial and error over the period 1870 to 1950 in the Appalachian fields. These procedures were picked up in other old shallow producing areas particularly in Illinois, Oklahoma, and

North Texas. Air and gas drives were not widely practiced until about 1910. Water floods were accidental at first and then were practiced somewhat secretly beginning approximately in 1907 in the Bradford field. Pennsylvania legalized the practice of water flooding in the Bradford field and adjoining areas in 1921. The Bradford field was particularly susceptible to water flooding. Early floods consisted of nothing more than allowing surface and subsurface waters to enter the producing sands through certain wells. A study of its progress shows how practices went from those so-called dump floods to pressure flooding, then to high-pressure flooding, and from circle floods to line floods and on to pattern floods. What was called the "poor man's branch of the industry" as late as 1945 suddenly took on the mark of respectability.

The most widely used methods of secondary recovery are still water-flooding and gas injection. They are used most often separately, but are used increasingly in combination, simultaneously and/or alternately, and in conjunction with various beneficial additives. Secondary recovery processes became increasingly important following World War II and even more so as a result of the Arab Oil Embargo and industry's greater realization of the following:

1. Secondary recovery oil has no finding costs.
2. Costs of finding, developing, and producing new primary oil reserves are generally higher than for secondary recovery reserves.
3. Some two-thirds of the original oil content of reservoir sands may remain as residual oil following economic primary recovery operations.
4. Allowable crude oil prices as of the mid-seventies made secondary recovery of oil more attractive than in the past.
5. Secondary recoveries often exceed primary recoveries.
6. Secondary recovery operations have rarely been curtailed or prorated.
7. Payouts are usually faster.
8. Federal income tax laws of the mid-seventies slightly favored secondary recovery operations.

Obviously water or gas can be injected at any time during the productive life of a reservoir and industry seldom waits now for the reservoir to reach an economic limit without rejuvenation.

Trends of relative permeability curves for reservoir sands combined with the unfavorable ratio of the viscosities of water and gases to average reservoir oils show that the relative permeability to water and gases increases rapidly as the oil is produced. As a result the percentage of oil in the produced fluids goes to zero at relatively high residual oil saturations. These ratios and the residual oil saturations may be determined in the laboratory quite accurately. Nevertheless, multiple sand layers, lensing, cross-bedding and natural variations in porosity and permeability of the reservoir, both vertically and horizontally, combined with an uncertain swept area make it very difficult to predict secondary oil recoveries with any degree of accuracy. Experience in the same field or with similar reservoir conditions is invaluable.

Waterfloods usually must be discontinued for economic reasons by the time the injected water volume equals 150 to 200 per cent of the reservoir pore volume, or when the produced water-oil ratio reaches 20:1, which leaves residual oil saturations in the swept portions of the reservoir sand of approximately 15 to 45 per cent. Generally the lower the permeability and porosity of the reservoir sand the higher the residual oil saturation. Overall averages for a pattern flood are approximately 25 to 30 per cent residual oil except under very favorable reservoir conditions.

During the late 1950's when primary oil production was feeling the effects of proration, particularly in Texas, an effort was made to get the Texas Railroad Commission to prorate waterfloods also. Those with waterflood operations had always maintained that curtailment of a flood once started was very harmful to ultimate recovery. Theoretically there should be no marked differences in recoveries, but there is a wealth of production records of actual waterfloods to show that either accidental or purposeful curtailment of waterfloods is detrimental.[11, 12] Similar data have been used to show that there are no losses in ultimate recovery.[13]

Generally air or gas drives are not as efficient as waterfloods and many attempts have been made to recover additional oil by following air or gas drive operations with a waterflood. Additional oil can be so recovered in the laboratory and sometimes in the field but the economics of the additional recovery is usually questionable.

Improved Secondary Recovery Methods

Since 1945 a number of new recovery methods and modifications of the old have been developed in the research laboratory and many of them have been tried and found to be successful under favorable conditions in the field. Some of the more important methods are as follows:

- High-pressure gas injection
- Condensing gas drive
- Alternate gas and waterflooding
- LPG-slug
- Miscible-phase floods
- Micellar/Surfactant floods
- Alcohol-slug floods
- Attic-oil recovery
- Detergent flooding
- Fire flooding
- Reverse combustion
- Steam flooding

Each of these methods can be subdivided into several variations. For example, in high-pressure gas injection the gas may be natural gas or an enriched natural gas, carbon dioxide, or nitrogen; it may be a continuous high-pressure gas drive or the gas may be followed with water or a lean gas; or it might be a slug of rich gas followed by lean gas which in turn is followed by water, etc. So-called fire flooding and steam flooding are particularly adaptable to shallow reservoirs of low gravity–high-viscosity oils which cannot be produced profitably under ordinary primary recovery methods or without the application of heat within the reservoir.

The use of surfactants in water followed by a gas to form foams and/or emulsions is proving to be moderately successful in reservoirs of low oil saturations.[14-17] "Tertiary" recovery processes are being tried to recover the residual oil left after a successful secondary operation. The predicted economics of tertiary recoveries is questioned.

RECOVERY OF NATURAL GASES

Almost all natural gas is produced under conditions that would correspond to primary recovery in oil production. Associated natural gas in a gas cap may be produced before, at the same time, or after the oil ring has been produced. The production rate depends on operating procedures. The production rate of dissolved gas is a function of the rate of oil production and the state of depletion of the oil reservoir. Obviously gas production rates from such fields is curtailed when oil production rates are curtailed.

Nonassociated natural gas reserves usually are produced under conditions of depletion (volumetric) drive, although some of the smaller nonassociated gas reservoirs, especially those located on salt domes of the Gulf Coast area, may be under conditions of active water drive. Large nonassociated natural gas reserves producing by depletion drive may have very high recovery efficiencies. Low recovery efficiency is due primarily to poor well completions, low permeability and lenticularity of reservoir sands, and relatively high abandonment pressures. Recoveries may be 85 to 90 per cent of the original gas in place under favorable conditions for reservoirs under depletion drive. Water drive reservoirs, particularly if small, may have low recoveries of 35 to 50 per cent. Large water drive reservoirs may approach depletion drive reservoirs in recovery efficiency.

REFERENCES

1. John F. Carll, "The Geology of the Oil Regions of Warren, Venango, Clarion, and Butler Counties, Pennsylvania," *Sec. Geol. Survey of Pennsylvania*, 1875-1879, Vol. 3, 1880, pp. 262-263.

2. *Secondary Recovery of Oil in the United States*, American Petroleum Institute, New York, 1942, p. 255.
3. *Ibid.*
4. *Secondary Recovery of Oil in the United States*, American Petroleum Institute, New York, 1950, p. 4.
5. Sylvain J. Pirson, *Oil Reservoir Engineering*, 2nd ed., McGraw-Hill, New York, 1958, pp. 1-2.
6. H. C. Miller, "Oil Reservoir Behavior Based Upon Pressure-Production Data," *USBM Rept. Invest.*, 3634, 1942, p. 11.
7. Charles A. Ashburner, "Petroleum and Natural Gas in New York State," *AIME Trans.*, Vol. 18, 1887-1888, p. 915.
8. S. F. Shaw, "Acre-Foot Recoveries from Oil Fields," *Petr. Engr.*, February 1946, pp. 86-96; and March 1946, pp. 254-262.
9. George R. Elliot and W. L. Morris, "Oil Recovery Prediction," *Oil and Gas J.*, June 16, 1949, p. 84.
10. J. J. Arps, "Estimation of Primary Oil Reserves," *Trans. AIME*, Vol. 207, 1956, p. 184.
11. J. N. Breston and R. V. Hughes, "Relation between Pressure and Recovery in Long Core Water Floods," *Trans. AIME*, Vol. 186, 1949, pp. 100-110.
12. Bert Murphy, "Effects of Water Injection and Oil Production Practices on Water-Flood Performance," *Interstate Oil Compact Bull.*, Vol. 16, No. 1, June 1957, pp. 51-85. (Also includes discussions by R. B. Bossler, Claude R. Hocott, and J. F. Buckwalter.)
13. J. K. Jordon, W. M. McCardell, and C. R. Hocott, "Effect of Rate on Oil Recovery by Water Flooding," *Oil and Gas J.*, May 13, 1957, pp. 98-130.
14. W. B. Gogarty and H. Surkalo, "A Field Test of Micellar Solution Flooding," *Trans. AIME*, Vol. 253, 1972, pp. 1161-1169.
15. W. R. Foster, "A Low-Tension Waterflooding Process," *Trans. AIME*, Vol. 255, 1973, pp. 205-210.
16. S. C. Jones and K. D. Dreher, "Cosurfactants in Micellar Systems used for Tertiary Oil Recovery," *Trans. AIME*, Vol. 261, 1976, Pt. II, pp. 161-167.
17. R. N. Healy and R. L. Reed, "Multiphase Microemulsion Systems," *Trans. AIME*, Vol. 261, 1976, Pt. II, pp. 147-160.

Additional References

Ted A. Armstrong, "What Makes a Steam Project Succeed?," *Oil and Gas J.*, May 24, 1965, pp. 60-64.

T. R. Blevins and R. H. Billingsley, "The Ten-Pattern Steamflood, Kern River Field, California, *Trans. AIME*, Vol. 259, 1975, Pt. I, pp. 1505-1514.

R. W. Buchwald, Jr., W. C. Hardy, and G. S. Neinast, "Case Histories of Three In-Situ Combustion Projects," *Trans. AIME*, Vol. 255, 1973, pp. 784-792.

C. G. Bursell and G. M. Pittman, "Performance of Steam Displacement in Kern River Field," *Trans. AIME*, Vol. 259, 1975, Pt. I, pp. 997-1004.

A. K. Dandona and R. A. Morse, "Down Dip Waterflooding of an Oil Reservoir having a Gas Cap," *Trans. AIME*, Vol. 259, 1975, Pt. I, pp. 1005-1016.

Ralph E. Davis and Lee Hillard Meltzer, "A Method of Predicting Availability of Natural Gas Based on Average Reservoir Performance," *Trans. AIME*, Vol. 198, 1953, pp. 249-258.

William C. Goodson, "Development and Evaluation of Gas-Condensate Reservoirs," *Petr. Engr.*, April 1959 and subsequent issues.

W. C. Hardy, P. B. Fletcher, J. C. Shepard, E. W. Dittman, and D. W. Zadow, "In Situ Combustion in a Thin Reservoir containing High Gravity Oil," *Trans. AIME*, Vol. 253, 1972, pp. 199-208.

L. W. Holm and J. A. Josendal, "Mechanisms of Oil Displacement by Carbon Dioxide," *Trans. AIME*, Vol. 257, 1974, pp. 1427-1438.

T. W. Nelson and J. S. McNeil, Jr., "Oil Recovery by Thermal Methods," *Petr. Engr.*, February 1959, pp. B27-B32; and March 1959, pp. B75-B100.

F. M. Perkins, Jr., C. W. Arnold, and R. J. Blackwell, "Where We Stand on Miscible-Displacement Research," *Oil and Gas J.*, July 30, 1962, pp. 160-163; Aug. 6, 1962, pp. 118-122; Aug. 13, 1962, pp. 118-126.

Fred H. Poettman, "In Situ Combustion: A Current Appraisal," *World Oil*, April 1964, pp. 124-128; May 1964, pp. 95-98.

T. G. Roberts, "A Permeability Block Method of Calculating a Water Drive Recovery Factor," *Petr. Engr.*, September 1959, pp. B45-B48.

Charles Robert Smith, *Mechanics of Secondary Oil Recovery*, Reinhold Publishing Corporation, New York, 1966.

11 CLASSIFICATION OF OIL PROPERTIES AND RESERVES

Many classifications of oil properties with respect to status and/or quality have been proposed based largely on whether the property is drilled, producing, or commercially productive. Each classification must be based in part on other classifications; for instance, a property may be considered as commercially productive yet it may be neither drilled nor producing; or at times the property may be producing and not be commercially productive, that is, unprofitable to operate.

Oil properties and oil property interests have been shown to be a function of lands, lands owned in fee, and lands to which one has certain mineral rights. Lands may consist of areas scarcely large enough to accommodate the drilling equipment for a well up to continuous areas of thousands of acres. The owner of a property in fee is limited only as to area because he owns the minerals from the surface to unlimited depths. The owner of only the mineral rights in whatever form, however, is not only limited with respect to area but also may be limited with respect to depths. That is, he may have rights only from the surface to a given depth, rights only between given depths, or rights only below a certain depth. Or his rights may be limited to all or a part of one or more specified formations in which production may occur, that is, to all or a part of one or more particular reservoirs.

STATUS OF RESERVOIRS

A property interest may have one or more reservoirs. They may be developed reservoirs, partially developed reservoirs, undeveloped reservoirs, and what have been called prospective or untested reservoirs. Usually there is no question as to the meaning of the first three terms. The word *reservoir* must be used consistently as previously defined or the terms prospective reservoir and untested reservoir will be used ambiguously. The terms in valuation work should be limited to those formations that *are* reservoirs, but not necessarily commercially proven reservoirs. All these terms refer to status and may be used in a classification of properties somewhat as follows:

I. The property or property interest
   A. Proved reservoirs
      1. Developed
         a. Producing
            (1) Under primary recovery
               (*a*) Oil
               (*b*) Condensate
               (*c*) Gas
               (*d*) Marginal
            (2) Under secondary recovery
         b. Nonproducing
            (1) Under primary recovery
               (*a*) Oil
               (*b*) Condensate
               (*c*) Gas
               (*d*) Marginal
            (2) Under secondary recovery
      2. Undeveloped
         a. Under primary recovery
         b. Under secondary recovery
   B. Untested reservoirs
      1. Under primary recovery
      2. Under secondary recovery

Obviously a property in a single reservoir field with only one kind of production can be classified as to kind quite easily. Yet the foregoing breakdown will be found incomplete with respect to other properties. For example, a single reservoir may be productive of both oil and gas; it may be producing from one part by primary recovery; another part may be producing by secondary recovery; and one or more parts may be untested or not drilled. Many gradations are hard to classify. *Developed* refers to those drilling units on which wells have been drilled and completed for production purposes in the subject reservoir. *Undeveloped* refers to those permissible drilling units on which no wells have been drilled for production purposes into the subject reservoir. Only developed drilling units can be productive and considered as definitely proven, but they may become nonproductive at some future time. If the condition is temporary such drilling units should still be listed as productive. Other developed nonproducing drilling units may occur because of shut-in wells, that is, wells off production on account of unprofitability while awaiting abandonment, a farmout, secondary recovery operations, etc.

A valuation of oil property interests would include valuations of all the recoverable fluid contents of all reservoirs held under the subject interests. The interests might include one or more reservoirs, or portions of reservoirs, containing oil, gas, and condensates, singly or in any combination. Some reservoirs are difficult to classify with respect to contents or kinds of production. The kind of well with respect to production depends in part on the methods of well completion and operation. Usually some gas is produced with all crude oils. Generally speaking, a well with an initial gas to oil ratio of less than 10 Mscf per bbl would be an oil well; that is, the reservoir fluid moving into the producing well is partially if not entirely in liquid form and the dissolved natural gas is separated from the oil at the surface.

CLASSIFICATION OF RESERVES–OIL

Oil property valuations are concerned primarily with the quantities of crude oil, gas, and condensate, distillate, or gas liquids at a specified time remaining to be recovered from the fields, pools, and reservoirs in which designated property interests are held. These quantities of salable production may be thought of in a broad sense as the reserves of the property interests. They are what gives value to any property interest.

The word *reserves* has become a word of many meanings. Many adjectives or modifiers have been used to designate an exact meaning. Some of the more common terms used throughout industry are the following:

Original reserves	Secondary Recovery reserves
Primary reserves	Tertiary Recovery reserves
Recoverable reserves	Remaining reserves
Proven reserves	Probable reserves
Semiproven reserves	Ultimate reserves

Hypothetical reserves

Those outside the industry might be inclined to think of reserves as connoting the holding or keeping back of oil or gas for higher prices, for future marketing or use, particularly in case of an emergency. A bank must maintain certain reserves of money for emergency purposes just as we have our Army Reserves. The industry is different in that it draws upon its reserves not only during emergencies but also throughout every normal operating day. Many times the modifier has confused rather than clarifying the exact meaning intended.

Proved Reserves

The American Petroleum Institute publishes the most authoritative annual reserves data for the various producing states, the nation as a whole, and Canada. The introduction to the report for 1964 carried the following definitions and explanations with respect to our nation's oil reserves:[1]

> The reserves listed in this Report, as in all previous Annual Reports, refer solely to "proved" reserves. These are the quantities of crude oil which geological and engineering data demonstrate with reasonable certainty to be recoverable in the future from known oil reservoirs under existing economic and operating conditions. They represent strictly technical judgments, and are not knowingly influenced by attitudes of conservatism or optimism.
>
> Both drilled and undrilled acreage of proved reservoirs are considered in the estimates of the proved reserves. The proved reserves of the undrilled acreage are limited to those drilling units immediately adjacent to the developed areas, which are virtually certain of productive development, except where the geological information of the producing formations insures continuity across other undrilled acreage.
>
> Additional reserves to be obtained through the application of fluid injection or other improved recovery techniques for supplementing the natural forces and mechanisms of primary recovery are included as "proved" only after testing by a pilot project or after the operation of an installed program has confirmed that increased recovery will be achieved.
>
> The proved reserves . . . may be considered as the known and established underground working inventory of oil recoverable under prevailing conditions. These estimates are subject to future revisions, either downward or upward, even though the *presently established* "proved" reserves are considered accurate in the light of current information.
>
> The proved crude oil reserves estimates do not include:
>
> (1) Oil whose recovery is subject to reasonable doubt because of uncertainty as to the geology and reservoir characteristics, or economic factors.
>
> (2) Oil in untested prospects.
>
> (3) Oil that may become available by fluid injection or other methods from reservoirs in which such operations have not yet been applied.
>
> (4) Liquid hydrocarbons that may become available through processing of natural gas.
>
> (5) Oil that may be recovered from oil shales, coal, and other substitute sources.

These definitions and comments show that data concerning reserves, and proved, remaining, recoverable, and primary reserves of crude oil should be identical. The same is true with respect to reserve data for natural gas, natural gas liquids, and production from secondary recovery and fluid injection operations, all of which are complied separately.

Many not in the industry question the need for annual additions and revisions, sometimes called extensions and revisions, to previously compiled proved reserve data. The fact that additions and revisions for any year are necessary and may amount to several times the quantities of reserves credited to discovery wells was explained by Heald, as follows:[2]

This is not necessarily alarming. The large revisions and extensions mean that it is not possible, during the year of discovery, to appraise the importance of the discovery. It is our practice to talk only of the oil that is definitely in sight, and to let future years demonstrate that the deposit was larger and richer than we could predict with assurance. . . . [A]ll students of petroleum reserves now realize that years must elapse after discovery before the recoverable oil in a field of real importance can be accurately estimated. They also know that every important change in the price of oil, in the cost of labor, in reservoir engineering and in many minor factors, will increase or reduce the amount of oil that can be recovered from wells and fields before economic conditions force their abandonment.

The Society of Petroleum Engineers of the AIME has adopted the following defintions of proved reserves for property evaluation:[3]

Proved Reserves–The quantities of crude oil, natural gas and natural gas liquids which geological and engineering data demonstrate with reasonable certainty to be recoverable in the future from known oil and gas reservoirs under existing economic and operating conditions. They represent strictly technical judgements, and are not knowingly influenced by attitudes of conservation or optimism.

Undrilled Acreage–Both drilled and undrilled acreage of proved reservoirs are considered in the estimates of the proved reserves. The proved reserves of the undrilled acreage are limited to those drilling units immediately adjacent to the developed areas, which are virtually certain of productive development, except where the geological information on the producing formations insures continuity across other undrilled acreage.

Fluid Injection–Additional reserves to be obtained through the application of fluid injection or other improved recovery techniques for supplementing the natural forces and mechanisms of primary recovery are included as "proved" only after testing by a pilot project or after the operation of an installed program has confirmed that increased recovery will be achieved. When evaluating an individual property in an existing oil or gas field, the proved reserves within the framework of the above definition are those quantities indicated to be recoverable commercially from the subject property at current prices and costs, under existing regulatory practices, and with conventional methods and equipment. Depending on their development or producing status, these proved reserves are further subdivided:

1. Proved Developed Reserves—Proved reserves to be recovered through existing wells and with existing facilities;

a. Proved Developed Producing Reserves—Proved developed reserves to be produced from completion interval(s) open to production in existing wells;

b. Proved Developed Nonproducing Reserves—Proved developed reserves behind casing of existing wells or at minor depths below the present bottom of such wells which are expected to be produced through these wells in the predictable future. The development cost of such reserves should be relatively small compared to the cost of a new well.

2. Proved Undeveloped Reserves—Proved reserves to be recovered from new wells on undrilled acreage or from existing wells requiring a relatively major expenditure for recompletion or new facilities for fluid injection.

Probable Reserves

The specific purpose of an oil property valuation may call for estimates of reserves that are not definitely proven. Such reserves may play an important part in negotiations for sales, trades, and mergers. They may be needed in making management decisions. And they are important when settling estates. The terms probable reserves, semiproven reserves, possible reserves, and hypothetical reserves have been suggested for such use.

The term semiproven reserves may have more meaning than the others because it does indicate a considerable degree of proof. Lahee[4] limited probable primary reserves to that area which had not been drilled outside of the area of proved reserves in a pool or reservoir and which probably contains recoverable oil under conditions similar to those which control the proved oil. Arps[5] would include primary reserves behind the casing of existing wells when not conclusively proven. Both Arps and Lahee would include some secondary recovery oil under probable reserves because it could not be called proven until such recovery processes had been put into operation or their effectiveness had been fairly demonstrated. Only as a result of such calculated probable reserves could an installation for fluid injection be warranted.

Hypothetical Reserves

Lahee[6] used the term hypothetical reserves to mean those reserves which may be present in nonproducing regions where geological or other information suggests that they may be present if drilled for. Arps[7] suggested the term "possible reserves" be used when data were lacking for a higher classification. Estimates of either hypothetical or possible reserves would be only "educated guesses."

Arps' Reserves Classification

Paine[8] maintained that the reserves of a property were either proven or not proven and preferred the expression "recoverable oil" as being more meaningful. Eggleston[9] thought only proved reserves should be dignified by the term "reserves" and that such terms as probable and possible should be confined to the descriptive part of a valuation report.

In pointing out the many specific purposes of valuation reports and the occasional need for terminology somewhat less severe than the "proved definition," Arps[10] suggested a classification of reserves for valuation purposes somewhat as follows:

1. Primary reserves
   - A. Proved
     - (1) Developed
       - (*a*) Producing
       - (*b*) Nonproducing
     - (2) Undeveloped
   - B. Probable
   - C. Possible
2. Secondary reserves
   - A. Proved
     - (1) Developed
       - (*a*) Producing
       - (*b*) Nonproducing
     - (2) Undeveloped
   - B. Probable
   - C. Possible

Nearly all of our nation's proven crude oil reserves fall into the primary reserves classification. It is possible someday that oil recovered by secondary recovery methods may total a large fraction of that recovered by primary methods. On the whole primary recovery methods of oil have been quite inefficient. The greater the inefficiency in a particular reservoir the better the probability of equal or even greater efficiency under secondary recovery methods. Nevertheless, industry has learned through costly mistakes that anticipated secondary recovery reserves, in fields depleted by primary recovery methods, are rarely proven until some time after the secondary recovery method has been in full operation.

Compilations of proved reserves with respect to certain properties may also be grouped:

1. Gross reserves.
2. Controlled reserves.
3. Net reserves.

Gross reserves are the total proved reserves available for production and sale from a property irrespective of any ownership of production in part by royalty interests, co-owners, or outside interests.

Controlled reserves are the total proved reserves available for production from a property in which a company, its subsidiaries, or affiliates may have a partial ownership interest, yet all production from the property is expected to be available to the company.

Net reserves or *company net interest* in a property are the gross reserves less all other reserve interests such as co-owner, royalty, and obligatory production payments by the company.

Proved reserve data compiled by the American Petroleum Institute are gross reserves. The oil property valuation engineer must usually calculate the gross reserves of a property first and then arrive at the controlled or net reserves. Controlled reserves are of primary interest to a company for management purposes. The value of a property is based on the income to be derived from production and sales of net reserves, making them of primary interest in valuation studies with respect to sales, trades, mergers, and bankers' loans.

CLASSIFICATION OF RESERVES–GAS AND GAS LIQUIDS

Much of the preceding discussion with respect to the classification of oil reserves is applicable to a classification of reserves of natural gas and natural gas liquids. We also have our remaining reserves, recoverable reserves, original reserves, proven reserves, semiproven reserves, probable reserves, ultimate reserves, etc., with respect to natural gas and natural gas liquids. Nevertheless, there are important differences. In terms of yearly consumption our proven reserves of natural gas are much greater than for crude oil. Natural gas production and reserves are classified usually only according to the mode of occurrence in the reservoir, namely: associated gas and nonassociated gas.

Few in industry actually explore for natural gas except in the Appalachian area and to a lesser extent in West Texas and adjoining western Oklahoma. Most of our natural gas reserves are associated with oil production operations or were discovered in wildcatting for oil. Major reasons for this are related to the magnitude of proven reserves of natural gas, the strict regulation by the Federal Power Commission of prices that can be paid at the wellhead for gas marketed through interstate pipelines, and the greater problems of transporting it to centers of consumption. Generally speaking, it is more necessary to find something big in the way of a gas field than in the case of oil. A single oil well might be economic even though the production has to trucked to its market. Natural gas can be marketed only through a pipeline, a pipeline whose cost must be justified before being built.

Hydrocarbon liquids produced or extracted at the surface from wet-gas and condensate-type reservoirs may be sold separately or they may be added to marketed crude oil to reduce its viscosity, to increase its gravity, or merely to dispose of them profitably. The reservoir is usually classified according to the major economic marketable kind of production, that is, gas, oil, or condensate.

U. S. PROVEN RESERVES OF OIL AND NATURAL GAS

The American Petroleum Institute showed the following proven oil reserves as of January 1, 1977[11] and oil, natural gas liquids, and natural gas reserves as of January 1, 1966.[12] The American Gas Association[13] compiled the gas and natural gas liquids reserves as of the end of 1976.

	Estimated proved reserves at end of year	
	1965	1976
Crude Oil	31,352,391 x 10^3 bbl	30,942,166 x 10^3 bbl
Natural gas liquids	8,023,534 x 10^3 bbl	6,401,967 x 10^3 bbl
Total	39,375,925 x 10^3 bbl	37,344,133 x 10^3 bbl
Natural gas	286,468,923 x 10^6 scf	216,026,074 x 10^6 scf

Resource Base and Potential Reserves

The American Petroleum Institute has been recognized since World War II by all in the petroleum industry to have compiled the most authoritative annual proven reserve estimates for oil and natural gas. Its reports are summations of unprejudiced data provided by knowledgeable committees within the industry. Possible recoveries from oil shales and undeveloped tar sands are not included. Resource Base and Potential Reserves, or other synonymous terms used in their stead, are also estimates. But these estimates, some of which may be classified as scientific studies, are still in a "guesstimate" category in comparison API's published compilations.

One of the earliest predictions of recoverable oil reserves was made in 1908 by the U. S. Geological Survey. The ultimate yield of petroleum was said to be 10-24.5 x 10^9 bbl. As of 1976 the Nation had produced 4.5 times that maximum figure and still had an estimated 37 x 10^9 bbl of proven reserves. Thus the ultimate yield as of January 1, 1977 would have to be

138×10^9 bbl, not including as much as 5×10^9 bbl produced prior to 1918. No one in today's industry would accept that ultimate yield figure. All know there are billions of barrels of petroleum still to be discovered by added exploratory drilling and through scientifically improved exploration and production practices. However, there would be many ifs attached to all estimates. The billions of barrels of petroleum that may someday be recovered from oil shales and tar sands, if included, would have to be justified by many "ifs" also.

Landsberg, Fishman, and Fisher[14] were very critical in 1963 of reserve terminology used by the American Petroleum Institute and the industry. They were particularly critical of the so-called proved reserves, which concept was said to have very limited application and could prove positively harmful in long-range analyses. They emphasized that the size and development of proved reserves could not throw much light on the adequacy of the country's oil resources except to serve as example of how rising demand has continuously "created" the supply to meet it. The ultimate reserves concept to them did not allow for improved technology and was affected by a conservative bias. They used the term *resource base* to mean the estimated total occurrence of oil in nature. Hubbert,[15] Weeks,[16] and Zapp[17] of the U. S. Geological Survey, and many others, have made estimates of our so-called *potential oil reserves*, which would correspond in part to an oil resource base, exclusive of the oil that might be recovered from our oil shales and tar sands. These estimates vary from 200 to 600 billion bbl in the ground awaiting future recovery. Schurr, Netschert, et al.,[18] after carefully studying the reserve problem from a resource base and other viewpoints, concluded:

> The total crude oil awaiting (potentially available for) future recovery in the United States can be inferred from expert opinion to be on the order of 500 billion barrels. This includes present proved reserves, the currently unrecoverable content of known reservoirs, and the total content of undiscovered reservoirs, without regard to present or future technologic feasibility of discovery and recovery.

Thus the total oil to be discovered would amount to approximately 200 billion bbl of which approximately 80 billion bbl would be recoverable by today's methods.

Estimates of offshore potentials were quite speculative at the time. Relatively few wells had been drilled. Known offshore salt domes in the Gulf Coastal area offered the best prospects. The industry still had to learn the problems of deep-water offshore drilling and production and to rely upon their suppliers in developing safe and efficient equipment for the purpose. As of the mid-70's wells were being drilled beyond the edge of the Continental Shelf.

The same criticisms leveled at the petroleum industry over reserve terminology and lack of hypothetical estimates have been stressed likewsie with respect to natural gas. These criticisms increased at the time of the Arab Oil Embargo and more so in 1977 upon the proclamation of the President of an energy crisis. Since 1974 an effort has been made to base prospective estimates upon more factual data.

Moody[19] presented a representative company viewpoint of the world oil-reserve situation as of January 1, 1975. All oil reserves were divided into three categories, namely: proved reserves, prospective reserves, and undiscovered potential reserves. Estimates for the United States of proved plus prospective reserves were 51 x 10^9 bbl. Recoverable-undiscovered reserves were estimated to be 30 x 10^9 bbl onshore and 55 x 10^9 bbl offshore, making a grand total of only 85 x 10^9 bbl. Moody did not quote any resource base data.

The U. S. Geological Survey issued a revision[20] in 1974 of its earlier oil and gas resource estimates. The classifications used were: measured reserves, indicated reserves, inferred reserves, and undiscovered recoverable reserves. Estimated volumes as of the end of 1972 for the latter three classifications were:

Indicated–inferred reserves	
Crude oil and natural gas liquids	
Onshore.	22.0 - 38.5 x 10^9 bbl
Offshore	3.0 - 5.0 x 10^9 bbl
Total.	25.0 - 43.5 x 10^9 bbl
Natural Gas	
Onshore.	97.0 - 205.0 x 10^{12} cf
Offshore	23.0 - 45.0 x 10^{12} cf
Total.	120.0 - 250.0 x 10^{12} cf
Undiscovered recoverable reserves	
Crude oil and natural gas liquids	
Onshore.	135.0 - 270.0 x 10^9 bbl
Offshore	65.0 - 130.0 x 10^9 bbl
Total.	200.0 - 400.0 x 10^9 bbl
Natural gas	
Onshore.	605 - 1,210 x 10^{12} cf
Offshore	395 - 790 x 10^{12} cf
Total.	1,000 - 2,000 x 10^{12} cf

Moody stated in 1977 that most knowledgeable estimaters now agree that undiscovered crude oil in the U. S. is in the range of 100 - 150 x 10^9 bbl and the undiscovered gas total is put at 400 - 600 x 10^{12} cf.[21]

Resource base, potential reserves, and all such figures are in the same category as Lahee's hypothetical reserves for the nation as a whole and are very likely to be misunderstood by anyone outside the petroleum industry without detailed explanations and careful analysis. Such hypothetical data may be bandied about by economists, government officials, and within industry, but they are meaningless when it comes to meeting the everyday petroleum needs of our nation at competitive prices. Gonzalez[22] analyzed the role of economics in determining future production in view of the so-called resource base or total occurrence of oil in nature. The theme of his analysis was that "production depends upon economics–not physical existence." This is certainly true with respect to the Nation's residual oil. Notwithstanding all the new scientific methods of recovering oil, only an average of approximately 30 per cent is being recovered today. Doscher and Wise state that the Nation's residual oil presently amounts to 295.71 x 10^9 bbl, a very important fraction of any estimated resource base.[23] Trillions of barrels of oil may exist on earth just as there are trillions of dollars' worth of gold in the waters of our seas, yet all must agree that these simple facts are no clue as to how much might be produced or made available for use under a free enterprise system.

REFERENCES

1. *Proved Reserves of Crude Oil, Natural Gas Liquids and Natural Gas*, Vol. 20, Dec. 31, 1965, American Petroleum Institute, New York.
2. Kenneth C. Heald, "High Lights of Domestic Developments in 1946," *Bull. AAPG*, Vol. 31, No. 7, July 1947, p. 1125.
3. "SPE Board Adopts Committee Reports on Reserve Definitions," *J. Petr. Tech.*, July 1965, p. 815.
4. Frederick H. Lahee, "The Terminology of Petroleum Reserves," *Proc. Fourth World Petroleum Congress, Sec. 2*, Drilling–Production, 1955, pp. 561-565.
5. J. J. Arps, "Discussion of 'What Are Petroleum Reserves,'" *J. Petr. Tech.*, July 1962, p. 724.
6. Lahee, *op. cit.*
7. Arps, *op. cit.*
8. Paul Paine, *Oil Property Valuation*, Wiley, New York, 1942, pp. 44-45, 46.
9. W. S. Eggleston, "What Are Petroleum Reserves?," *J. Petr. Tech.*, July 1962, pp. 719-724.
10. Arps, *op. cit.*, pp. 724-725.
11. *News Release*, American Petroleum Institute, Washington, D. C., April 7, 1977.

12. *Proved Reserves of Crude Oil, Natural Gas Liquids and Natural Gas* (API), *op. cit.*, pp. 4-5, 17.
13. *AGANEWS*, (News release), American Gas Association, Arlington, Virginia, April 7, 1977.
14. Hans H. Landsberg, Leonard L. Fishman, and Joseph L. Fisher, *Resources in America's Future*, Johns Hopkins Press, Baltimore, Md., 1963, pp. 389-390.
15. M. King Hubbert, "Nuclear Energy and the Fossil Fuels," *API Drill. and Prod. Pract.*, 1956, pp. 7-25.
16. Lewis G. Weeks, "Fuel Reserves of the Future," *Bull. AAPG*, Vol. 42, No. 2, 1958, pp. 431-441.
17. A. D. Zapp, "Future Petroleum Producing Capacity of the United States," *USGS Bull. 1142-H*, 1962, 36 pp.
18. Sam H. Schurr, Bruce C. Netschert, et al., *Energy in the American Economy, 1850-1975*, Johns Hopkins Press, Baltimore, Md., 1960, pp. 356-359.
19. "World Crude Resource May Exceed 1,500 Billion Barrels," (Summary of paper presented before Ninth World Petroleum Congress), *World Oil*, September 1975, pp. 47-50, 56.
20. "USGS issues Revised U. S. Oil and Gas Resource Estimates," *World Oil*, June 1974, p. 96.
21. "Vast Potential seen for U. S. Oil, Gas," *Oil and Gas J.*, June 20, 1977, pp. 32-33.
22. Richard J. Gonzalez, "Production Depends on Economics–Not Physical Existence," *Oil and Gas J.*, March 30, 1964, pp. 59-64.
23. T. M. Doscher and F. A. Wise, "Enhanced Crude Oil Recovery Potential–An Estimate," *J. Petr. Tech.*, May 1976, pp. 575-585.

12 ESTIMATION OF RECOVERABLE OIL RESERVES

Even the simplest of oil property valuations as we know them today were not made before 1911 because all valuations depend on reasonably accurate estimates of (1) recoverable reserves of oil, gas, and other hydrocarbon liquids, and (2) estimated rates of recovery. The need for such estimates originated through passage of the Tariff Act of 1913, that first income tax statute under the Sixteenth Amendment. Industry and particularly the government worked on the problem of estimating reserves during the next five years and developed a number of acceptable methods. Some are still widely used today. Accuracy has been improved over the years as a result of better and more complete reservoir and production data. Newer methods have come into use which at one time depended on the development of accurate methods of measuring porosities, permeabilities, and saturations of reservoir rocks, properties of reservoir fluids, and measurement of reservoir pressures and temperatures.

That early need for accurate reserve estimates was almost eliminated for a time by passage of the 27½ per cent depletion allowance in 1926. Beginning with the early 1930's, and more so since World War II, the need has grown by leaps and bounds in order to develop, operate, and deplete the reservoirs efficiently, to consummate an ever increasing number of sales, trades, mergers, oil property loans, and to comply with the many governmental requests.

Properties for which recoverable reserves must be computed may be classified as (1) fully developed, (2) partially developed, and (3) entirely undeveloped properties. Only the last classification can be defined accurately; it consists of those properties on which no commercially productive wells have been drilled. A property may be fully developed according to an accepted or designated well-spacing pattern in one or more reservoirs, yet be only partially developed at other prospective reservoir depths. Estimates of recoverable reserves should be made, even before the first well is put on production. Once on production, future production rates and reserves become increasingly important to the operator.

Before a prospective area is tested the chances of finding commercial production must be considered carefully with respect to the ultimate value of what may be found, ultimate value necessarily being a function of recoverable reserves. Accordingly, an untested property has only a 100 per cent speculative value, despite the need to place some value upon it. Once production is found, a better, yet far from future prospective valuation can be made based upon probable productive area, productive sands, porosities, permeabilities, saturations, and fluid characteristics under reservoir conditions. That is, the property takes on a more definite value as more is learned about the reservoir and its contents, and still more during actual production operations. The newly discovered property may be either enhanced or lessened in value upon completion and production of each additional well. Estimates of reserves during the drilling-up stage of all except very large properties are limited almost entirely to comparisons with other similar properties and volumetric estimates. Such estimates are still far from reliable until after the property is drilled-up and either it or individual wells are produced for sufficient time to show declines in production rates. Reliability in any method of reserve estimations increases with cumulative production. It is very important to learn as early as possible during the productive life of the field all physical characteristics of the reservoir, its fluids, and the kind of reservoir drive. For example, early estimates of recoverable oil by primary production methods in the Citronelle field of Alabama had to be drastically reduced. After months of development drilling the original reservoir gas-oil-ratio was found to have been only 160 scf/bbl.

Crude oil reserves usually are classified into those recoverable under primary methods and those recoverable under secondary methods. Gas and other petroleum liquids usually are recovered without resorting to secondary methods, although some engineers classify recycling and pressure maintenance operations as secondary recovery.

The methods used most often in estimating reserves of oil recoverable under primary methods of recovery are:

1. Decline curve methods.
2. Cumulative production curve methods.
3. Volumetric methods.
4. Material balance methods.

Each method depends on a projection of past production trends and operating conditions into the future and includes a certain amount of "guesstimating," The oil property valuation engineer will find that each method is widely used where applicable, accepted, and equally important in his work. The choice of method depends chiefly on the availability of reliable production and related data with respect to time. The amount and reliability of the data may be a function of the age of the property. Some very old properties may have been on production long before the actual need for accurate data was recognized. Material balance methods are handicapped most often through lack of fieldwide data. The most data usually exist with respect to the more recent wells, properties, and fields developed and produced by the more progressive companies.

Production decline curves, acre-foot yields, etc., are based on averages. Averages may be the arithmetical mean, the median, or the mode. The total of all the numbers divided by the number of numbers gives and arithmetic mean. The sum of the daily productions of a well divided by the number of productive days is an example. The mode is the number occurring most frequently in a group of numbers. The median is that number which divides a group of numbers into two equal groups of numbers. The occurrence of an erratic number has the greatest effect on the arithmetic mean and the least effect on the mode. Even so, on the whole, whether dealing with one well or a group of related wells, and one property or several related properties, the arithmetic mean has proven most satisfactory for decline curve purposes. It is prudent, however, to compare its value for the period with the mode and the median for the same period in order to detect probable errors in production rates.

DECLINE CURVE METHODS

Any production decline method must be used with a degree of caution and understanding when predicting reserve estimates and ultimate recoveries. Experience combined with the history of the wells in question is a prerequisite to proper interpretation. Production records are kept in many ways, that is, in terms of barrels of oil and water, and thousands of standard cubic feet of gas, per day, per month, and per year. Monthly production records are kept most often, but electronic computer records on tape are quite often recorded on a monthly basis for only the last six or twelve months; previous monthly production data are kept summarized as annual data only. Such records often will be found too incomplete and abbreviated for diagnostic purposes. This lack of suitable production data and other basic information becomes more serious to the evaluation engineer when bottom-hole pressures, net sand thicknesses, and porosities may be incomplete to totally lacking.

A production record of an abandoned well and the known causes of changes in rate of production are shown in Figure 12-1. How to project such a production decline curve into the future can be quite puzzling. Obviously

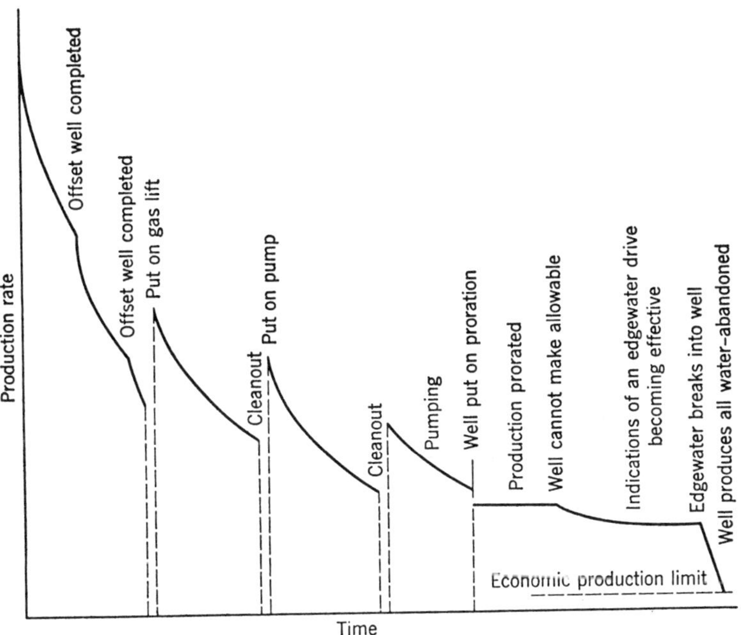

Figure 12-1. An idealized production rate-time curve of an abandoned oil well.

it cannot be projected accurately unless the well was producing at full capacity at all times or at some essentially constant percentage of full capacity. Even then one must assume that whatever controlled the rate of production in the immediate past will persist to the same extent in the future. The longer the period of regular decline, without a change in operating conditions, the better the compensation of errors and irregularities in production rates. A well on proration may or may not show a decline with time. If it does, the rate of decline is representative only as long as those conditions prevail. Any single well in a reservoir may have many irregularities in its decline rate which may appear to ruin its usefulness. Plotting of the average of many wells in the reservoir with respect to time may iron out many irregularities. A chronological decline curve for the reservoir as a whole may be more useful. Nevertheless an attempt should be made to account for the causes of all irregularities. Changes in surface conditions are fairly obvious. Except for the effects of the drilling of a new offset well, subsurface changes are sometimes difficult to account for. Decline curves are sometimes useful to detect and analyze the trouble.

The question always arises regarding the cutoff point of the curve with respect to production rate. This is mostly a matter of economics, however; at times wells may be kept on production for various reasons long after costs have exceeded income. Often this is done to hold a lease or wells on the lease until other prospective zones can be evaluated or sometimes until secondary recovery operations can be initiated. The last portion of the decline curve should be given the most weight as a general rule in predicting future performance. It is this trend with respect to costs and future income that determines the economic limit and the extent of repairs that may be made profitably.

Obviously, there is no substitute for wide experience in working with decline curves, especially when one's past predictions have been checked and analyzed against subsequent production records. Stevens and Thodos[1] have shown how to select the best kind of decline curve for a set of declining production data by plotting the three kinds of curves and analyzing the scatter of rate points by the least-squares method. The oil property valuation engineer can rarely take time to go to such length and other errors are usually more significant.

The earliest publication of a production decline curve for purposes of estimating future rates of production through its extention was in 1909.[2] Many kinds of production curves were developed during the next fifteen years whose primary purpose was to provide a means of predicting future recoveries. The first manual for the oil and gas industry under the Revenue Act of 1918 was published the next year. Acceptable methods or predicting recoveries through use of decline curves were included.

Cutler[3] in 1924 listed five kinds of production decline curves in use at that time:

Production decline curve. A curve based on the past performance of a well or wells; it shows the amount of oil produced by the well or average of a group of wells during equal successive period of time. . . .

Future production curve. A curve based on the extension of a production decline curve: it shows the estimated average future production carried to a stated economic limit for wells in the same tract, pool, or field, with relation to the amount of oil that they produced during the past year of other time unit. . . .

Appraisal curve. A curve based on the past performance of wells; it shows the estimated average ultimate production to a stated economic limit of the wells in the same tract, pool, or field, with relation to the amount of oil that they produced during the initial year. . . .

Future cruve. A curve that shows the future production carried to an economic limit of a well or average well of a property or field with references to the remaining life of the well; . . .

Rate-of-production curve. A curve auxiliary to the production decline curve; it shows the daily rate of production of wells whose yearly production is shown by the production decline curve. . . .

In trying to write formulae to use in extrapolation of decline curves, early workers noted that many of their equations were similar to those describing straight lines on semilogarithmic and log-log graph paper. Occasionally a production-time relationship is plotted on Cartesian coordinate paper to test its regularity.

Most production rate-time graphs or decline curves have been found to be of the two following kinds, basically:

1. Constant percentage decline.
2. Hyperbolic decline.

In addition an occasional well or related group of wells will have a straight-line decline when plotted on Cartesian coordinates. Cutler's early work showed that the hyperbolic-type decline curve was most frequently applicable at the time. Arps[4] called the first type "exponential" and added a previously little known third class in 1944 referred to an "harmonic decline curves." Many decline curves that resemble either the hyperbolic or harmonic types during their earlier lives, and perhaps much of their productive lives, will resemble semilogarithmic decline at low production rates. And at some production rate any kind of decline curve may lose its identity. So-called gravity drainage and delayed water drives are often given as explanations.

Several modifications of rate-time graphs and related mathematical relationships have been developed and widely used since Cutler's original presentation. It must be remembered at all times that "fits" to any curve, or portions thereof, are empirical and include all factors influencing the production rates. Some factors may be thought of as definitely known, possibly known, and many entirely unknown. Some or all having an effect on plotted portions of a curve are expected to so continue during the period of any extension of the curve.

Ramey[5,6] studied the production history of 202 leases covering 76 pools and 50 formations located in Colorado, Kansas, New Mexico, North Dakota, Oklahoma, Texas, and Wyoming. Production history for a 5-years period was analyzed and extrapolated over the next 5-years period. He concluded that hyperbolic and harmonic types of decline curves were more prevalent than previously indicated in the literature. An exponential decline curve gave the best fit in the most cases, but that the harmonic type decline was better applicable under certain extremes.

Higgins and Lechtenberg studied the application of the usual production decline curves to 90 of California's older oil fields.[7] Actually two separate but related studies were made. The usual four decline equations were tested in the first study to determine which would fit best the production data over an initial 9-year period, then extrapolating the curves so obtained over the next 6 years of production data and then beyond to determine future rates of production. The constant percentage decline method resulted in the best fit over the first 9 years. Arps' harmonic declines matched the data better for the periods 9 to 15 years and to the point where cumulative production equalled reserves. The second study was made to see if application of a least-squares technique would improve the extrapolations. The merits of two other forms of Arps' harmonic equation were tested also. It was concluded that the Arps' equation, though excellent in principle, does not represent perfectly the average performance in many fields. A reasonable assumption would be that if the 6 years' forecasts are within better than the normal expectancy of accuracy for petroleum engineering predictions, then a longer forecast would have a good background.

Recoveries and reserves prior to Culter's studies were estimated largely through comparisons with abandoned and older wells and properties produced from the same formations. Occasionally an older and shallower property lacking todays basic reservoir information is evaluated through comparisons with similar properties in various stages of exhaustion. While less true today, many properties have been so screened initially in all old shallow production areas throughout the Nation in order to estimate potential recoverable secondary-recovery reserves.

Cumulative-recovery to production-rate graphs can be plotted for all three of the usual kinds of decline curves. Only semilogarithmic declines give really obvious straight-line extensions without considerable study. Other cumulative curves not directly related to declining production are sometimes plotted. They are briefly discussed under cumulative curve methods.

The Constant Percentage Decline Curve

The constant percentage decline curve is easy to plot and to use for estimation of crude oil reserves. It also has been referred to as a *semilogarithmic decline curve*, a *die-away curve*, an *appraisal curve*, and an *organic decay curve*, the opposite of the organic growth curve (Equation 6-4) derived in Chapter 6. Using $q_1, q_2, q_3, \ldots q_n$ to represent decreasing rates of production over equal periods of time, and letting D equal the constant percentage decline in production rates per unit of time:

$$\begin{aligned} q_1 &= q_1 \\ q_2 &= q_1 - q_1D \\ &= q_1(1 - D) \\ q_3 &= q_2 - q_2D \\ &= q_1(1 - D) - q_1(1 - D)D \\ &= q_1(1 - D)^2 \end{aligned}$$

and

$$q_n = q_1(1 - D)^{n-1} \tag{12-1}$$

D being a constant percentage decline, $(1 - D)$ is also a constant percentage or ratio that may be designated as R, which makes Eq. 12-1 become:

$$q_n = q_1R^{n-1}$$

or

$$R^{n-1} = \frac{q_n}{q_1} \tag{12-2}$$

and

$$\log R = \frac{\log q_n - \log q_1}{n - 1} \tag{12-3}$$

There is rarely any need to plot production rates versus time on Cartesian coordinates. A more useful and time-saving first step is to use semilogarithmic graph paper, always plotting production rates on the logarithmic scale. Base-10 paper is to be preferred. The data may be considered to follow a constant percentage decline if a straight-line decline can be accepted as an average. Or perhaps only the last part of the graph shows a straight-line trend. On such straight-line trends any two points q_1 and q_n may be selected for use in Eq. 12-3 to solve for the value of R. These points should be considered as accurate production rates at their respective times. This means that the value of R can be solved to three and four significant figures. From this point on one would be dealing with an idealized decline curve which theoretically matches the trend of production rates.

Equation 12-2 is in the form of a geometric progression and the sum N_{p_n} of q_1 $q_2, q_3, \ldots q_n$ would be:

$$N_{p_n} = q_1 + q_1R + q_1R^2 + \cdots q_1R^{n-1} \tag{12-4}$$

Multiply by R

$$N_{p_n}R = q_1R + q_1R^2 + q_1R^3 + \cdots q_1R^n \tag{12-5}$$

Subtract Equation 12-4 from 12-5

$$N_{p_n}R - N_{p_n} = -q_1 + q_1 R^n$$

and simplifying

$$N_{p_n} = \frac{q_1(1 - R^n)}{1 - R} \tag{12-6}$$

Equation 12-6 is applicable to one or more wells where q_1 and R are based on the same units of time as n. Some inaccuracies are inherent in the use of the equation; however, it is sufficiently accurate for all valuation purposes. Equation 12-6, if representative of an average of each of W wells, n is in months and q_1 is in average barrels per day per month (30.4 days), could be rewritten and used as

$$N_{p_n} = (30.4W)\,\frac{q_1(1 - R^n)}{1 - R} \tag{12-7}$$

A constant percentage decline curve representing the average daily production per month for each of 20 wells on a property is shown on Cartesian coordinates in Figure 12-2 and on semilogarithmic coordinates in Figure 12-3.

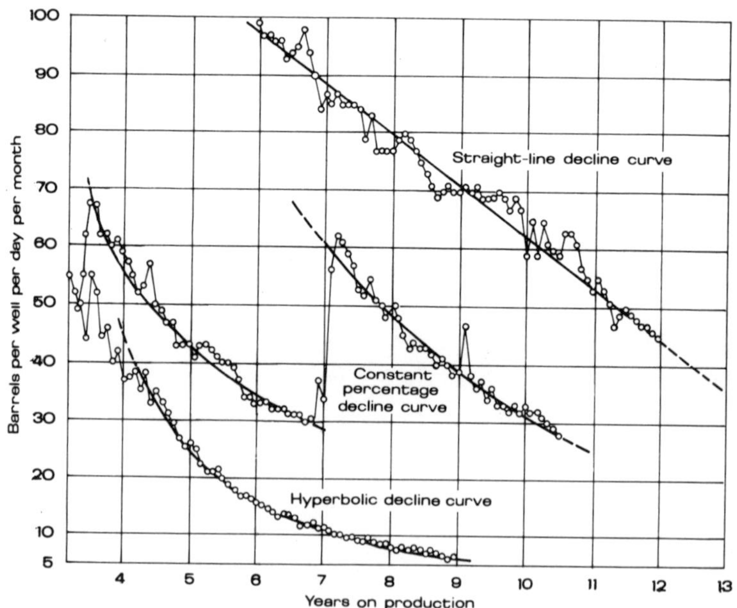

Figure 12-2. The three most common types of oil well production decline curves.

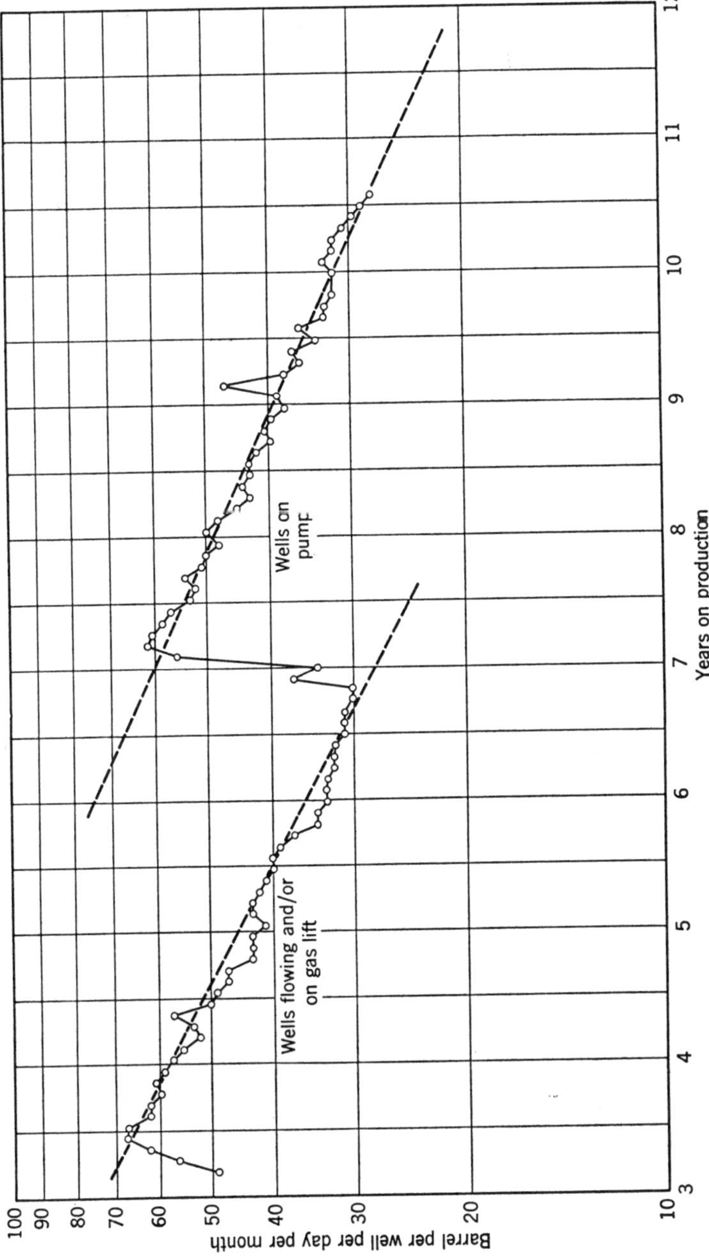

Figure 12-3. Constant percentage decline curves of oil wells plot as straight-line declines on semilogarithmic graph paper.

Another solution to the problem of estimating future production from a constant percentage decline curve assumes

q = the instantaneous production rate at any time n, and
D = the constant fractional decline rate per unit of time.

Then

$$Dq = -\frac{dq}{dn} \quad (12\text{-}8)$$

$$\sum_{q_0}^{q_n} \frac{dq}{q} = -D \int_{n=0}^{n=n} dn$$

and

$$\ln q = -Dn + C \quad (12\text{-}9)$$

When $n = 0$, q must equal q_0, the initial rate, and

$$\ln q_0 = C$$

Substituting in Equation 12-9

$$\ln q = -Dn + \ln q_0 \quad (12\text{-}10)$$

$$\ln \frac{q}{q_0} = -Dn$$

$$\frac{q}{q_0} = e^{-Dn}$$

and

$$q = q_0 e^{-Dn} \quad (12\text{-}11)$$

According to Equation 12-10

$$n = \frac{\ln q_0/q}{D} \quad (12\text{-}12)$$

and

$$D = \frac{\ln q_0/q}{n} \quad (12\text{-}13)$$

Letting N_p equal the accumulative production during the time q_0

declines to q

$$N_p = \int_{n=0}^{n=n} q dn = \int_{n=0}^{n=n} q_0 e^{-Dn} dn \tag{12-14}$$

$$= \left[-\frac{q_0}{D} e^{-Dn} \right]_{n=0}^{n=n}$$

$$= -\frac{q_0}{D} e^{-Dn} - \left[-\frac{q_0}{D} \right]$$

$$N_p = \frac{q_0}{D} - \frac{q_0 e^{-Dn}}{D} \tag{12-15}$$

Equation 12-11 shows

$$q_0 e^{-Dn} = q$$

which by substituting into Equation 12-15 gives

$$N_p = \frac{q_0 - q}{D} \tag{12-16}$$

or

$$N_{p_n} = \frac{q_0 - q_n}{D} \tag{12-17}$$

Equations in the form of 12-17 are preferred by some engineers, yet they are neither more accurate nor easier to use than Equation 12-6. The chief disadvantage to using Equation 12-17 is that today's production records are rarely kept in a form suitable for use. This is especially true with respect to tape records. For instance, if the accumulative production N_{p_n} is to be calculated for a twelve-month period and D is calculated on a periodic basis, q_0 must be on the same periodic basis as of the first day of the period n_0 and q_n must be on the same periodic basis as of the last day of the period n_n. This can be accomplished by plotting the decline data, drawing the average decline curve, and interpolating the needed values.

The Hyperbolic Decline Curve

Production decline curves plotted on Cartesian coordinates often resemble hyperbolas in general shape and relationships. Some of the earliest diagnostic work with them was done by Lewis and Beal[8,9] for the U.S. Bureau of Mines. The method was first explained in detail and favored in 1924 by Cutler[10] who thought the method was more scientifically founded and accurate than the constant percentage method.

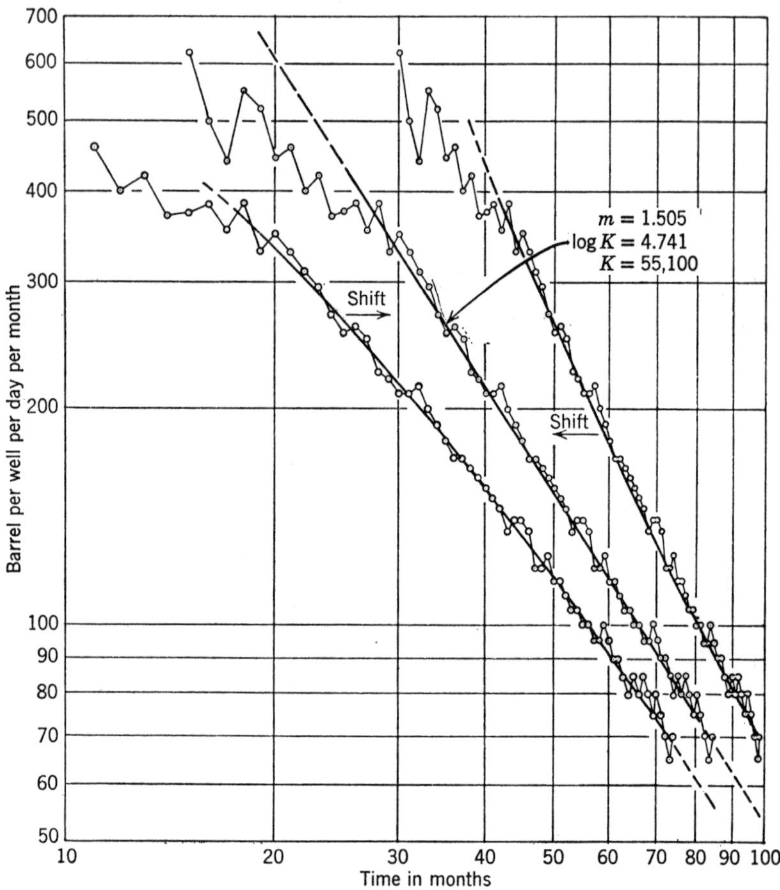

Figure 12-4. Hyperbolic-type oil well decline curves may be shifted to plot as straight lines on log-log graph paper.

The equation for a hyperbola is $y = Kx^{-m}$. The constant K is the value of the y-ordinate when $x = 1$, the slope of the curve being $-m$. Using production symbols,

$$q = Kn^{-m} \tag{12-18}$$

when n represents the units of time corresponding to production rate q and K is a constant at initial time $n = 1$. Equation 12-18 may be written as

$$\log q = \log K - m \log n \tag{12-19}$$

Any curve of this type when plotted on logarithmic (log-log) coordinates with log q plotted on the y-axis and log n plotted on the x-axis may be shifted to the right or to the left in the direction of convexity by adding a constant C. The value of C is increased or decreased $(x + C = n)$ until a position is reached where the decline curve plots as a straight line. The representative straight line so obtained can be extended to the value of log q_n, representing the logarithm of q at any time n, including the value of q at the economic production limit, and back to the value of log q_n where $n = 1$, in order to obtain the value of K. The value of $-m$ can be picked from the graph to complete the equation for the decline curve. Future production rates and accumulative production may be computed through use of these values. The equation for the curve may be recorded or, if desired, two points n_1 and n_2 with their corresponding values of q_1 and q_2.

Figure 12-4 of a hyperbolic-type decline curve shows how it had to be shifted in the direction of its convexity to get the best straight-line projection of all production values. The value of K must be calculated in this case because it falls off the graph. The value of $-m$ may be scaled off with sufficient accuracy, being equal to the ratio of the distance between any two values of q on the straight-line projection divided by the distance between the two corresponding values of n. A value of $m = -1$ would represent a very low decline rate. Where m is greater than -10 the curve loses much of its accuracy and becomes practically worthless as $m \to -20$.

The equation of the decline curve $(q = Kn^{-m})$ may be used to find the time at which any future production rate q_n occurs: Future production N_{p_n} can be calculated also.

Let

$$dN_{p_n} = q\,dn = n^{-m}K\,dn$$

$$N_{p_n} = K\int_{n_1}^{n_2} n^{-m}\,dn$$

$$N_{p_n} = \frac{K}{1-m}\left[n_2^{1-m} - n_1^{1-m}\right]$$

$$N_{p_n} = \frac{K}{m-1}\left[\frac{1}{n_1^{m-1}} - \frac{1}{n_2^{m-1}}\right] \tag{12-20}$$

Constants must be used as necessary; data plotted as barrels/day/month would change Eq. 12-20 to

$$N_{p_n} = \frac{30.4K}{m-1}\left[\frac{1}{n_1^{m-1}} - \frac{1}{n_2^{m-1}}\right] \tag{12-21}$$

The values of K and m for the hyperbolic decline curve of the average daily production in barrels per day per month for seven wells on a property in Figures 12-2 and 12-4 are based on the shifted curve of Figure 12-4.

Equation 12-19 is used. The value of m has been found by scaling the straight-line decline curve and equals 1.505. The value of K will be found to be 55,000 (rounded). To find the future production for the first year through use of Eq. 12-21, n_1 would be the eighty-fifth month and n_2 would be the ninety-seventh month. For the seven wells for the first year

$$N_{p_{n-12}} = 7\left[\frac{30.4 \cdot 55{,}000}{1.505 - 1}\right]\left[\frac{1}{85^{(1.505-1)}} - \frac{1}{97^{(1.505-1)}}\right] = 158{,}700 \text{ bbl}$$

The value of n at the economic limit of 10 bbl per day will be found through Eq. 12-19 to be 307 months, making the future life of the wells 222 months. The future production for 222 months would be

$$N_{p_{n-222}} = 7\left[\frac{30.4 \cdot 55{,}000}{1.505 - 1}\right]\left[\frac{1}{85^{(1.505-1)}} - \frac{1}{307^{(1.505-1)}}\right] = 1{,}173{,}000 \text{ bbl}$$

Other Types of Decline Curves

Occasionally as wells reach low production rates, and at times throughout their productive lives, production decline data will plot as straight lines on Cartesian coordinates. Gravity drainage and an effective water drive are sometimes offered as explanations for the unusual behavior. Such a decline curve is shown in Figure 12-2. These wells often make some water production upon completion which may or may not increase with time. If it increases, the rate of increase may be more or less the same as the decrease in the rate of oil production, that is, the overall rate of fluid production does not change. A natural water drive is indicated. Other times gravity drainage in combination with a slow, but active, water drive may be effective, where

q = the rate of oil production,
n = the units of producing time,
C = a constant,
m = the slope of the decline curve,

$$q = C - mn,$$

and the accumulative production N_p for any time period $n_n - n_0$ would be the area under the curve

$$N_{p_n} = \left[\frac{q_1 + q_n}{2}\right][n_n - n_0] \tag{12-22}$$

Proper constants must be added unless q_1 and q_2 are for the same units as n.

The Harmonic Decline Curve. Harmonic decline may occur when the drop in production per unit of time expressed as a fraction of the production rate is proportional to a fractional power between 0 and 1 of the production rate.[11,12] The nominal decline rate D is proportional to the production rate q, or

$$D = -\frac{dq/dt}{q}$$

where the subscript i represents initial conditions,

$$q_n = q_i(1 + nD_i)^{-1} \tag{12-23}$$

and

$$N_{p_n} = \frac{q_i}{D_i} \ln \frac{q_i}{q_n} \tag{12-24}$$

The harmonic type of decline, being a special type of hyperbolic decline, is not often so recognized until late during the productive life of the wells.

CUMULATIVE PRODUCTION CURVE METHODS

Cumulative curves of the production from wells, leases, and fields are very convenient for estimating ultimate recoveries graphically and for checking arithmetical calculations. They can also be used to obtain future recoveries by difference. The ease of plotting cumulative production with respect to production rate induces one to plot this data on Cartesian coordinates first rather than plotting a production decline curve to see if a straight line is clearly indicated.

Constant Percentage Decline

A straight-line relationship between cumulative production and rate when plotted on Cartesian coordinates is indicative of constant percentage decline. The trend can be extrapolated to any economic production rate limit. In conjunction with a production decline curve rates can be picked for any dates and production between these two dates can be taken from the cumulative graph.

Figure 12-5 is a cumulative production graph of the 20 wells whose average production rates plotted as a constant percentage decline curve in Figures 12-2 and 12-3. Extrapolation of the graph to an economic limit of 3 bbl per day gives an ultimate recovery of 2,400,000 bbl. The ultimate recovery can also be calculated mathematically using equations 12-3, 12-7 and 12-17. The last production rate was 28 bbl per day at the end of 126 months. Corresponding cumulative production amounted to 1,640,000 bbl.

To solve for R of Equation 12-3 take $q_1 = 60$ and $q_n = 28$ (average daily production per month) at $n_1 = 84$ and $n = 126$ (months), respectively. Solving, $R = 0.9816$.

To find the future life of the property from time when $q_1 = 28$ bbl/day to time when $q_n = 3$ bbl/day through use of Eq. 12-3

$$n - 1 = \frac{\log 3 - \log 28}{\log 0.9816}$$

Solving, $n = 121$ months.

Future production can be calculated by years and/or for the entire future life through use of Eq. 12-7. The first month's production would be at the rate q_1 of 27.5 bbl/day/well, and

$$N_{p_{12}} = 30.4 \cdot 20 \left[\frac{27.5\,(1 - 0.9816^{12})}{1 - 0.9816}\right]$$

$$= 181{,}500 \text{ bbl during the first year.}$$

For the 121 months future life

$$N_{p_{92}} = 30.4 \cdot 20 \left[\frac{27.5\,(1 - 0.9816^{121})}{1 - 0.9816}\right]$$

$$= 813{,}000 \text{ bbl}$$

making an ultimate recovery for the 20 wells of

$$1{,}640{,}000 + 813{,}000 = 2{,}453{,}000 \text{ bbl}$$

Equation 12-13 gives the value of D, the constant fractional decline rate in barrels per day per month of 0.01846. Substituting in Eq. 12-17:

$$N_{p_n} = (30.4 \cdot 20)\frac{27.5 - 3}{0.01846}$$

$$= 807{,}000 \text{ bbl}$$

Ultimate recovery would amount to 2,437,000 bbl.

Many prefer to plot all decline curves on semilogarithmic graph paper in order to easily identify rate-time changes. They also prefer to use the corresponding cumulative production graphs. If later data varies sufficiently

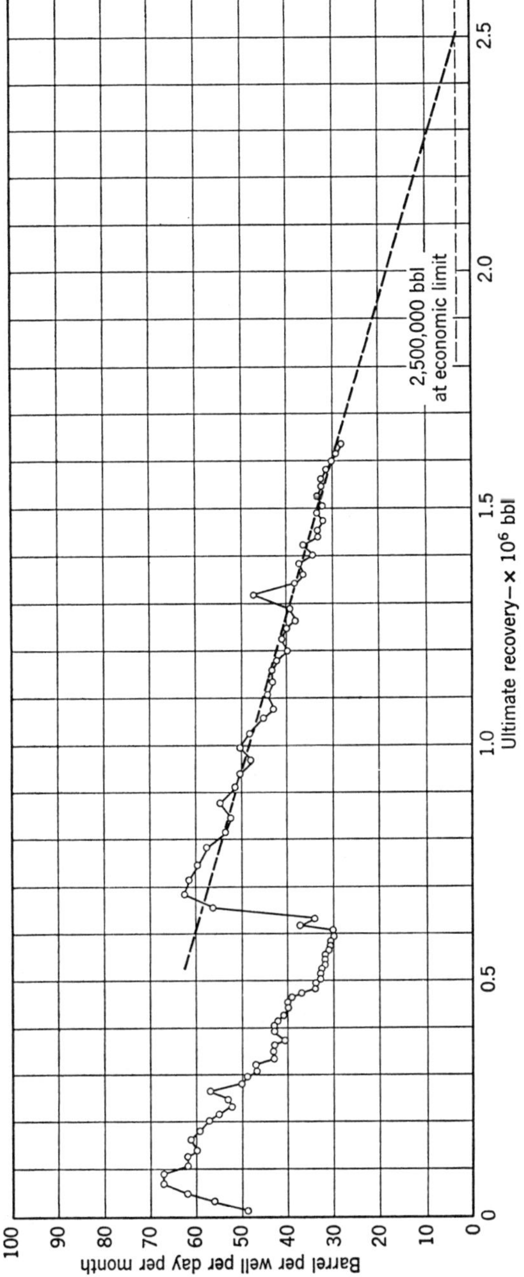

Figure 12-5. Cumulative production rate graph of constant percentage decline wells.

from previous estimates, new extensions adjusted to the latest data are used. Thus, all future estimates are said to be continuously based upon the latest identifiable production trends. Accuracy greater than three, and possibly four, significant figures is rarely needed.

Hyperbolic Decline

Arps[13, 14] has described a method whereby rate-cumulative relationships following a hyperbolic decline may be plotted on an especially prepared graph paper to get a straight line. Such a procedure is probably as time consuming as using Eqs. 12-20 or 12-21 on the assumption that the decline curve has been straightened first.

Slider[15] has described a method of decline curve analysis whereby a series of preconstructed idealized hyperbolic decline curves are plotted on transparent paper. Each curve represents a combination of a decline rate and values of m varying between 1.0 and zero. All actual production curves are plotted on semilogarithmic graph paper and each is matched to the hyperbolic curve of best fit. The values of K and m picked from the matching hyperbolic curves are used to estimate future and ultimate recoveries by the usual methods.

Some engineers hesitate to use all hyperbolic-type decline curves to estimate future and ultimate recoveries. It can be shown that values of m less than 1 would indicate infinite recoveries with infinite time. Lefkovits and Mathews[16, 17] have shown that the hyperbolic decline curves with values of m approaching or equal to 2 are useful in calculating reserves and ultimate recoveries from gravity drainage reservoirs in the stripper stage. They stress the effects of this common condition where two sands of widely differing characteristics contribute to the overall production rate.

Compaction of poorly consolidated sand reservoirs as production pressures decline also may contribute to changes in the slope of previously established decline rates as well as estimates of reserves and ultimate recoveries based upon them. Outstanding examples of reservoir sand compaction are the Post-Eocene reservoir sands of the Bachaquero field on the Bolivar Coast of Venezuela[18] and the Long Beach Harbor area of California.[19]

Harmonic Decline

If the cumulative production rate data do not plot with a straight-line trend on Cartesian coordinates, the same data may be tried on semilogarithmic coordinates with rates on the logarithmic axis. A straight-line trend of the data on semilogarithmic coordinates is diagnostic of harmonic-type decline.[20] If the decline is harmonic one has a graphical means of obtaining future and ultimate recoveries at any abandonment rate with sufficient accuracy.[21]

Other Cumulative Curves

Four other cumulative production curves are often plotted, sometimes as a check against other estimates, and other times as the only applicable method. They are:

1. Gas-oil ratios vs., cumulative oil production. Semilog paper is often the most satisfactory.
2. Log of cumulative gas production vs., log of cumulative oil production.
3. Log of the percentage of produced oil from wells making water vs., cumulative oil production.
4. Subsea elevation of an evenly advancing bottom, top, or edgewater drive, or any combination thereof, vs., cumulative oil production.

Reserves and ultimate recoveries of water-drive reservoirs are usually and sometimes better estimated by volumetric methods using a recovery factor. Very few water-drive reservoirs produce over long periods of time under only one type of drive. Most produce under combinations of driving forces such as gravity, gas dissolved in water, bounding edgewater under pressure higher than current reservoir pressure, so-called bottom water drives, up-dip drives, and dispelled water resulting from compaction of the reservoir sands. The beginnings of some kind of a water drive can often be recognized on a production decline curve upon noting a favorable change in the rate of decline not explainable through surface operations. This often may be observed long before the wells show any signs of making appreciable water. Or the production rate may stay relatively constant and sometimes even increase, in the absence of proration and/or a gas-cap drive. As a rule most water-drive reservoirs produce for a time initially under depletion-type drive providing the wells are properly completed. It is when water is recognized as the dominant driving force that regular decline curves become less and less applicable.

Oil wells are rarely abandoned because they produce no oil. They are abandoned for economic reasons. Many times the oil in the drainage area of the abandoned well may be produced by an adjoining well. Most wells are abandoned on account of overall lifting and cleanout costs, and if producing water too, treatment costs, exceeding the value of the produced oil.

VOLUMETRIC METHODS

Volumetric methods of estimating recoverable oil reserves are convenient and sufficiently accurate to use providing reliable data are available. Sometimes only volumetric methods can be used, particularly under active water-drive conditions, and often in case of any effects of a gravity drive. These methods also may provide a check against other estimates. If the reservoir properties and oil characteristics are known, the most doubtful factor in all cases is the value of the so-called recovery factor, a value that must be based on the production histories of similar reservoirs or experience. Volumetric methods often are said to be applicable only to depletion-drive reservoirs; actually by using the recovery factor it is applicable to all drives, where

N = STB of oil initially in the reservoir,
ΔN = STB of recoverable oil,
A = area of reservoir in acres,
h_n = net reservoir sand thickness in feet,
ϕ_e = average effective porosity of reservoir sand expressed as a decimal fraction,
S_w = per cent interstitial water saturation,
S_g = per cent free gas saturation upon abandonment,
B_{oi} = initial oil formation volume factor,
B_{oa} = oil formation volume factor at abandonment,
E_R = recovery efficiency factor expressed as a decimal fraction.

$$N = 7758\ Ah_n\phi_e \frac{1 - S_w}{B_{oi}} \qquad (12\text{-}25)$$

$$\Delta N = 7758\ Ah_n\phi_e \frac{(1 - S_w)E_R}{B_{oi}} \qquad (12\text{-}26)$$

At times data may be available through experience, interpretation of electric logs, core analyses, etc., to provide some concept of the free gas saturation of the reservoir sand upon abandonment. This value may be used on the assumption that the interstitial water saturation did not change during the 100 per cent depletion drive. Equation 12-5 would become:

$$\Delta N = 7758\ Ah_n\phi_e \left(\frac{1 - S_w}{B_{oi}} - \frac{1 - S_w - S_g}{B_{oa}}\right) \qquad (12\text{-}27)$$

Equation 12-26 is used most often. It is particularly useful with respect to salt-dome reservoirs under active water drive in the Gulf Coast region.

Water-drive recovery factors are usually thought to be relatively high, but for small faulted water-drive reservoirs, irrespective of dip, and for very low-dipping water-drive reservoirs with relatively thin net sands, the recovery factor may be quite low.

Recovery factors for a wide range of conditions in complete depletion-drive reservoirs may be approximated from a series of charts calculated by Wahl and others.[22] Representative ultimate recoveries, expressed as the per cent of residual oil originally in place, are given in the following table for oils having a GOR of 1300, a B-factor of 1.6, and a bubble-point pressure of 2250 psia:

Dead oil viscosity (cp)	Per cent ultimate oil recovery for interstitial water saturations of:		
	10 per cent	30 per cent	50 per cent
10	17.8	16.0	12.7
2	26.7	23.8	17.6
$\frac{1}{2}$	35.8	30.8	21.8

Craze and Buckley[23] have shown that 80 per cent of the residual gas saturations (S_g) in 27 widely scattered gas-cap and depletion-type reservoirs ranged from 20 to 40 per cent of the total pore space with an average of 30 per cent. Neglecting the B_o-factor would indicate average oil recoveries of the same magnitude, which are on the high side in comparison with those of Tables 10-1 and 10-2.

Isopachous maps may be made to obtain weighted average net sand thicknesses by planimetering individual areas for use in calculating net sand volume. Either the trapezoidal rule or Simpson's rule, as used in earth work calculations, is applicable, where

h = isopach contour interval of net sand in feet,
a_o = area in acres enclosed by the zero contour,
$a_1, a_2, a_3 \ldots a_n$ = area in acres enclosed by the first, second, third, etc., contour,
t_n = average estimated net sand thickness in feet above the nth contour.

Simpson's Rule

$$Ah_n = \tfrac{1}{3}h(a_o + 4a_1 + 2a_2 + 4a_3 + \cdots 2a_{n-2} + 4a_{n-1} + a_n) + t_n a_n \qquad (12\text{-}28)$$

(An even number of equal contour intervals n, that is, an odd number of contours, is required.)

Trapezoidal Rule

$$Ah_n = \tfrac{1}{2}h(a_o + 2a_1 + 2a_2 + 2a_3 + \cdots 2a_{n-1} + a_n) + t_n a_n \quad (12\text{-}29)$$

Average porosities of reservoir sands at well locations can be plotted, contoured, and planimetered to obtain a weighted average porosity through use of either of the above rules; or $h\phi_e$ values may be contoured to save one step in arriving at the value of $Ah_n\phi_e$ (acre-feet net pore volume) for the reservoir sand.

Volumetric estimates, and especially those made during the very early life of productions operations, may be subject to large errors. They come about largely through use of inaccurate estimates of reservoir characteristics, fluid contents, and their properties. Every item used to calculate N and ΔN in the above equations is subject to error. All errors may be accumulative and not necessarily compensating. The following percentage errors are not uncommon in volumetric estimates:

porosity	15 per cent
interstitial water or oil content	5
recovery factor	10
productive area	10
sand thickness	10

which if additive could produce a cumulative error of 50 per cent or more in a reserve estimate.

Keplinger[24] has discussed these sources of errors primarily with respect to the estimation of natural gas reserves. He shows possible magnitude of errors amounting to nearly 100 per cent. His discussion is equally applicable to volumetric estimates of oil. Extensions of decline curves, likewise during early productive life of a reservoir, are subject to considerable error. However, if production rates are declining with time, decline curves should be used as a check, if not to supersede volumetric estimates.

MATERIAL BALANCE METHODS

The application of the principles of the material balance in oil and gas reservoirs was partially developed in 1930 by Coleman, Wilde, and Moore[25] although the credit usually is given to Schilthuis[26] who showed in 1936 how it could be used to calculate reserves. Katz[27] also presented a paper on its use the same year. The principle of the material balance is nothing more than an application of the conservation of matter within a reservoir, that is, the volume of reservoir fluids produced plus the amount remaining in a reservoir must be equal to the amount present initially. The equations to express the balance can become quite complicated on account of the many factors involved. For example the oil may increase in volume as reservoir pressure is reduced through production operations until the bubble point is reached; further pressure decline reduces to B_o-factor of the oil. In addition, gas coming out of solution may occupy reservoir space and formation edge-waters may enter the former oil zone. It can be complicated still more if gas and/or water are injected to maintain reservoir pressures. The more complex adaptations of the equation and the many applications fall within the realm of the reservoir engineer who is primarily interested in calculating the initial contents of a reservoir, watching for and calculating any water encroachment, and predicting reservoir pressures and recoveries. The chief interest of the material balance equation to the evaluation engineer is that it furnishes another method of estimating original oil and gas in a reservoir to which a recovery factor may be applied for valuation purposes.

The chief drawbacks to its use are the amount, the relative accuracy of, and the complications of the required data that arise in cases of multiownership reservoirs and when the production from two or more reservoirs or multiple zones are commingled in the same well bore. The essential data required for a reservoir study are as follows:

1. Initial average reservoir pressure and successive average reservoir pressures corresponding to production times t.
2. The volumes of all reservoir fluids produced during each time interval, all measured under standard conditions.
3. The volume of any initial free gas cap and the initial volume of the oil phase.
4. The formation volume factors for the oil and the gas and the solution gas pressure relationships.
5. Knowledge of any water encroachment and estimates of the volumes at various times t, where

N = initial reservoir oil volume in cubic feet,
N_p = cumulative oil produced in cubic feet at time t and pressure p,
R_s = solution GOR at any time t,
R_p = cumulative GOR,
R_{si} = initial solution GOR,
B_o = oil formation volume factor at any time t,
B_{oi} = initial oil formation volume factor,

B_g = gas formation volume factor,
B_{gi} = initial free gas formation volume factor,
B_t = a two-phase or total formation volume factor equivalent to the volume in barrels occupied by one stock tank barrel of oil and its initial dissolved gas content under conditions of reservoir temperature t and any reservoir pressure p,

$$B_t = B_o + (R_{si} - R_s)B_g \quad \text{and} \quad B_{ti} = B_{oi}$$

G = initial reservoir free gas volume in standard cubic feet,
G_t = reservoir free gas volume at time t in standard cubic feet,
m = ratio of initial reservoir free gas volume to initial reservoir-oil volume,

$$\frac{GB_{gi}}{NB_{oi}}$$

W = initial reservoir water volume in cubic feet,
W_e = cumulative water influx in cubic feet,
W_p = cumulative produced water in cubic feet.

The change in oil volume.

Initial volume of reservoir oil $= NB_{oi}$
Oil volume at time t and pressure $p = (N - N_p)B_o$
Change in oil volume to time $t = NB_{oi} - (N - N_p)B_o$ (12-30)

The change in the free gas volume.

Initial free gas volume $= GB_{gi} = NmB_{oi}$
Initial free and dissolved gas $= (NmB_{oi}/B_{gi}) + NR_{si}$
Produced gas at time $t = N_pR_p$
Gas remaining in solution $= R_s(N - N_p)$
Free gas volume at time $t = G_i =$

$$\left[\left(\frac{NmB_{oi}}{B_{gi}} + NR_{si}\right) - N_pR_p - (N - N_p)R_s\right] B_g$$

$$\text{and } G - G_t = NmB_{oi} - \left[\left(\frac{NmB_{oi}}{B_{gi}} + NR_{si}\right) - N_pR_p - (N - N_p)R_s\right] B_g \quad (12\text{-}31)$$

The change in water volume.

Initial volume of water in reservoir $= W$
Produced water at time $t = W_p$
Reservoir volume of $W_p = B_wW_p$
Volume of water influx at time $t = W_e$
Increase in reservoir water volume $= W_e - B_wW_p$ (12-32)

Combining terms.

$$NB_{oi} - NB_0 + N_pB_o + NmB_{oi} - NmB_{oi}(B_g/B_{gi}) - NR_{si}B_g + N_pR_pB_g + NR_sB_g - N_pR_sB_g = W_e - W_pB_w \quad (12\text{-}33)$$

Adding $N_pB_gR_{si}$ to both sides of the equation, substituting B_{t_i} for B_{oi}, B_t for $(B_o + R_{si}B_g - R_sB_g)$, and simplifying and solving for N

$$N = \frac{N_p[B_t + B_g(R_p - R_{si})] - W_e + W_pB_w}{B_t - B_{ti} - mB_{ti}(1 - B_g/B_{gi})} \quad (12\text{-}34)$$

Equation 12-34 is applicable to those dissolved gas-drive reservoirs having active water drives and free gas caps. Where there is a free gas cap and no water drive, $W_e = 0$ and Eq. 12-34 becomes

$$N = \frac{N_p[B_t + B_g(R_p - R_{si})] + W_pB_w}{B_t - B_{ti} - mB_{ti}(1 - B_g/B_{gi})} \quad (12\text{-}35)$$

Where there is an active water drive, but no free gas cap, $m = 0$ and Eq. 12-34 becomes

$$N = \frac{N_p[B_t + B_g(R_p - R_{si})] - W_e + W_pB_w}{B_t - B_{ti}} \quad (12\text{-}36)$$

Where there is no active water drive, no produced water, and no free gas cap, Eq. 12-34 becomes

$$N = \frac{N_p[B_t + B_g(R_p - R_{si})]}{B_t - B_{ti}} \quad (12\text{-}37)$$

For use above the bubble-point pressure the following modification of Eq. 12-37 may be useful for many estimates despite its being subject to large errors:

$$N = \frac{N_pB_o}{B_o - B_{oi}} \quad (12\text{-}38)$$

In this equation both N and N_p may be taken as stock tank barrels, N_p and B_o being the accumulative oil production and formation volume factor, respectively, at reservoir pressure p. Efforts have been made to modify Equation 12-38 by taking into account the compressibility factors for the interstitial water and the reservoir rock, but the result is still subject to large error.

Many other corrections to, and modifications or adaptations of, the material balance equation with respect to properties of reservoir rocks and fluids will be found in the technical literature.[28, 29] Tracy[30] has described a more usable and less time-consuming form of the original Schilthuis equation with which estimates of instantaneous gas production instead of incremental oil production are used. Plots of instantaneous gas-oil ratio and average reservoir pressure vs. cumulative oil production are used to establish trends from which future ratios can be picked. Havlena and Odeh[31] also plot one variable group vs. another variable group. The group selection depends upon the production mechanism. A straight-line relationship is established. Unfortunately the method has certain limitations and can not be computerized. Rose[32] has described ways to make material balance forecasts for reservoirs with initial production data lacking. Very small changes in estimated initial reservoir pressures, gas-oil ratios, *B*-factors, etc., can produce large ultimate discrepancies. Hence, new calculations often have to be made using variations in the estimated variables in an effort to match actual performance.

Much of the technical literature dealing with the use of material-balance equations is intended to aid in the solution of special reservoir recovery mechanisms. Scaled physical modelling of producing operations were developed during the 1950's which were very useful in studying reservoir behavior under various conditions. These were largely superseded by mathematical modelling as the large electronic computors became available. Many highly sophisticated mathematical modelling studies are now being made of the many phases of reservoir behavior under all imaginable conditions. They do provide solutions and explanations to production problems difficult, if not impossible, to solve by older material-balance methods.[33,34] Many of them are of greater theoretical than practical value for evaluation purposes and their use is best left to the experienced reservoir engineer.

ESTIMATES OF GAS-CONDENSATE RESERVOIRS

Oil recoveries from gas-condensate reservoirs depend chiefly on the original composition of the reservoir fluids and the production procedure.[35-37] The liquid contents of the reservoir usually are dissolved in the gas phase and the production procedure is aimed at keeping all reservoir hydrocarbons in a single-phase state if possible until in the separator at the wellhead. Any retrograde condensation within the reservoir sand could result in lost liquid recovery.

Estimates of liquid recoveries are functions of the recoverable gas in the reservoir and should be made by experienced reservoir engineers. There may be alternative recovery procedures such as pressure depletion, pressure maintenance, and cycling by dry-gas drive, one being efficient, another less efficient from a liquids-recovery viewpoint, but net profits are not necessarily commensurate with relative recoveries. Estimates of net deferred income and profit should be a joint effort with the valuation engineers, the chief object being to adopt the most profitable procedure and timing.

Estimates of reservoir liquids are based on laboratory analyses of the reservoir fluids and usually are reported in barrels per millions of standard cubic feet of produced gas. Estimates of recoverable liquids will be some percentage of the reservoir liquids depending on a number of important factors as enumerated by Arthur:[38]

1. Field characteristics:

a. Occurence or absence of black oil.
b. Size of reserves of products.
c. Properties and composition of reservoir hydrocarbons.
d. Productivity of wells.
e. Permeability variation which controls the degree of bypassing of injected gas.
f. Degree of natural water drive existing.
g. Field development and operating costs.
h. Plant installation and operating costs.

2. The market demand for gas and liquid petroleum products.
3. Future relative value of the products.
4. Rate of interest to convert future profits to present worth.
5. Existence or absence of competitive producing conditions between operators.
6. Taxes, both ad valorem and income.

Even with the best of engineering and economic valuations large errors in profitabilities sometimes occur because of changed market conditions, unaccountable reservoir conditions, unexpected plant and operating costs, changes in procedure, plain errors, and many other reasons.

Successful operations recover from 50 to 75 per cent of the original liquids in place. Recoveries as high as 80 to 85 per cent occur under vey favorable conditions. Such high recoveries are possible under conditions of a partial water drive and at times by simultaneous injection of water and dry gas. In all cases where gas is cycled at high pressures, the gas remains in the reservoir for future recovery. Occasionally gas-condensate reservoirs are more profitable when produced only by controlled depletion methods, depending largely on the market for the gas.

ESTIMATES UNDER IMPROVED OIL RECOVERY METHODS

Improved recovery methods include secondary recovery as defined in Chapter 10 as well as those borderline methods that may be defined as either primary or secondary recovery.

Waterflooding

High-pressure peripheral waterfloods, particularly if initiated during the early life of a reservoir, should perform similarly to natural water drives. Waterflooding as a secondary recovery measure usually relies upon "pattern flooding," and none of the theoretical and analytical methods for estimating performance is entirely satisfactory. This is due primarily to the usual heterogeneous character of the majority of reservoir sands, both vertically and laterally. Calculation procedures must assume the following:

1. The individual layers or zones of permeable sands within the reservoir and their fluids contents are uniform laterally.
2. The performance of the reservoir as a whole corresponds to the summation of the performance of all its individual layers.
3. No gravity segregation of fluids occurs.
4. The displacement action is piston-like.
5. No interlayer flow occurs.

Earlougher and Guerrero[39] have reviewed the better-known procedures for estimating waterflood recoveries as presented by Stiles,[40] Dykstra and Parsons,[41], and Prats,[42], et al., and have compared the estimated recoveries using those methods with what the authors called their *empirical method*. Their method is based on experience gained from studies of data from core analyses and flood-pot tests for several thousand wells. The procedures depends chiefly on the ability to predict the following:

1. Oil and gas saturations at the start of the flood.
2. The volume of injection water required before a kick in oil production occurs. Usually 0.6 to 0.8 of fill-up is required where fill-up volume for most depleted sands amounts to 20 to 25 per cent of the pore volume.
3. The peak oil production rate that occurs when the ratio of the effective water injection rate to the oil production rate varies from 2 to 12 with a normal average of 4 to 6 for good floods.
4. The cumulative water injection at the economic limit which ranges from about 1.25 to 1.7 pore volumes with a normal average of 1.5.
5. Peak production time which ranges from 4 to 10 months for an injection rate of approximately 1 bbl/day/acre-ft of sand.
6. The average production decline after peak rate which usually ranges from 30 to 70 per cent per year for pumping production.

The rate of oil recovery from a waterflood depends largely on the rate of water injection. Earlougher stated that for flowing floods with an injection rate of 0.8 bbl per day per acre-ft the recovery rate will approximate the following:[43]

1st year	0	per cent	of	ultimate
2nd year	29	"	"	"
3rd year	22	"	"	"
4th year	16	"	"	"
5th year	11	"	"	"
6th year	8	"	"	"
7th year	5	"	"	"
8th year	4	"	"	"
9th year	3	"	"	"
10th year	2	"	"	"

Earlougher and Guerrero[44] use the following equations, where

A = area, acres,
B_o = oil formation volume factor,
B_{oi} = oil formation volume factor at discovery conditions,
D = average production decline, fraction per year,
D' = average rate of decline, fraction per year,
h_n = net sand thickness, feet,
ϕ = porosity, fraction,
Q_{oa} = abandonment oil production rate, STB/D,
Q_{op} = peak oil production rate, STB/D,
S_c = factor for converting total saturation to effective saturation—essentially a porosity correction factor, fraction of pore volume, ranges from 0 to .025,
S_o = oil saturation at beginning of flood at reservoir conditions, fraction of pore volume,
S_w = minimum interstitial water saturation, fraction of pore volume,
t_1 = time in days at which first oil production "kick" occurs,
t_2 = time in days at which peak rate is reached,
t_3 = time in days at which rate begins to decline.

$$S_o = \left(\frac{1 - S_c - S_w}{B_{oi}}\right) B_o \tag{12-39}$$

$$\text{Fill-up volume in barrels} = 7760 \cdot \phi \cdot A \cdot h_n(1.00 - S_c - S_o - S_w) \tag{12-40}$$

where

$$D' = \ln (1 - D)$$

and

$$Q_{oa} = Q_{op}\, e^{-D't}$$

$$\text{Reserves in STB of oil} = (Q_{oL}/4)(t_2 - t_1) + Q_{op}(t_3 - t_2) + 365\left(\frac{Q_{op} - Q_{oa}}{D'}\right) \quad (12\text{-}41)$$

The reserves calculated through use of Equation 12-41 correspond to the area enclosed by a semilogarithmic graph of production rate versus time. A comparison of actual recoveries with predicted recoveries for a Bartlesville sand flood and a Lower Yates sand flood, both in the midcontinent area, is given in Table 12-1. Earlougher and Guerrero's empirical method can be calculated in a fraction of the time required for the other methods. It gave the best prediction, and the Prats and others method was next best.

Table 12-1

Comparison of Recovery Predictions with Actual Recovery[45]

Method of estimating	Recovery, STB			
	Flood 1		Flood 2	
	Predicted	Actual	Predicted	Actual
Stiles	2,320,000	753,000	2,199,000	2,000,000
Modified Stiles	2,177,000	753,000	2,315,000	2,000,000
Dykstra-Parsons	1,742,000	753,000	995,000	2,000,000
Prats et al	2,410,000	753,000	2,579,000	2,000,000
Empirical	758,000	753,000	1,833,000	2,000,000

Callaway[46] presented an equation that is often useful in evaluation work, not for predicting actual recovery but for evaluating the degree of tolerance in each variable in the recovery factors under waterflooding, where

B_{oi} = initial oil formation volume factor,
B_o = reservoir volume factor during the waterflood operations,
S_{or} = residual oil saturation after waterflood, per cent,
S_w = interstitial water saturation, per cent,
R = primary recovery efficiency, per cent,
E = overall sweep efficiency, per cent reservoir volumes,
N_{wf} = fraction of original oil in place which can be recovered by waterflooding,

$$N_{wf} = \left(1 - R - \frac{B_{oi}}{B_o}\right)\left[1 - E\left(1 - \frac{S_{or}}{1 - S_w}\right)\right]$$

Variables R and E are clearly the most important providing S_w is not too high. The higher the primary recovery and the lower the sweep efficiency, the lower the waterflood oil recovery for any given interstitial water saturation.

Higgins et al.,[47] have prepared graphs for quickly estimating waterflood performance. Later Leighton and Higgins[48] presented an improved mathematical model and computorized program for frontal displacement of reservoir fluids. Mathematical and computorized calculations are invaluable for many purposes, including valuations, but they must be used with caution, especially if made before or just after starting operations. Waterflooding is rarely a cut and dried procedure. Some floods never perform according to expectations. Evaluations made during the early life of a flood should be based upon other close-by similar operations if possible. For example, experience in the Bradford Field, Pennsylvania, and surrounding areas was not entirely applicable to the shallow fields of Illinois, Kansas, Oklahoma, Ohio, and West Texas. Closely spaced wells, pattern-flooding, and high injection pressures are certainly desirable from an early payout and profitability viewpoints. Strongly folded and faulted areas in Ventura County, California, caused problems with "thief zones" on account of pressure-parting along the numerous thrust faults.[49] Lower than anticipated injection pressures had to be adopted despite prolonging of operations, less certain recovery estimates, and reductions in estimated cash flow and net incomes. Unexpected incompatibility of available injection waters and/or irregularities in the adopted sweep-out patterns may be equally detrimental.

Low-Pressure Gas Cycling

Low-pressure gas cycling is widely practiced as a secondary recovery operation in shallow fields in the Appalachian area, Kentucky, the Illinois Basin area, and throughout the older midcontinent fields. It is most successful in former depletion-drive fields where water production has never been a serious problem. Yuster and Day[50] developed methods for predicting recoveries which were based on long-core laboratory tests and field checks in the air-gas drive regions of Pennsylvania. Pressure gradients used in the laboratory were comparable to those used in Pennsylvania field operations, namely, 0.13 to 3.88 psi/ft. A gas drive would not be expected to recover oil efficiently in view of the wider differences in the viscosities of gas and oil in comparison with water and oil. Oil recoveries per unit volume of injected gas decline logarithmically with time at constant injection rates. For example, an 800-md core saturated with 50 per cent oil, 40 per cent water, and 10 per cent free gas, initially, was gas driven at a pressure gradient of 1.3 psi/ft and showed the following reduction in oil saturation:

At 1 pore volume gas injection–9 per cent reduction.
At 10 pore volumes gas injection–17 per cent reduction.
At 100 pore volumes gas injection–25 per cent reduction.
At 1000 pore volumes gas injection–33 per cent reduction.

The residual oil saturation of 17 per cent at 1000 pore volumes of injected gas is much lower than practical in field operations.

Yuster and Day[51] developed equations for calculating recoveries from low-pressure Pennsylvania gas drives where injection rates on a property are kept relatively constant over long periods of time. Their calculated recoveries for five operations checked actual recoveries quite closely. The method is probably most applicable in oil property valuation work with respect to operating projects.

Other Improved Recovery Methods

It is almost impossible to predict with any degree of accuracy the probable recoveries under most of the newer recovery methods. For example, sometime after miscible phase and other high-pressure gas flooding had been literally perfected in the laboratory, recoveries of 100 per cent of the residual oil in place after primary recovery operations were predicted for the swept portions of the reservoir. Some years later recoveries in the field under the various methods of miscible phase recovery were quite disappointing. It was thought that more oil percentagewise would be recovered than by waterflooding, but profitabilities were seriously questioned with respect to many operations. Horner stated:[52]

> . . . Miscible fluids sweep less of the reservoir than does water in the waterflooding process. Fifty to 70 per cent of the reservoir is likely to be swept by a miscible flood; 70 to 90 per cent is likely to be swept by a waterflood.
>
> On the other hand, miscible displacement will recover 90 to 100 per cent of the oil in the portion it sweeps. Waterflooding will recover 40 to 80 per cent of the oil in that portion of the reservoir which it sweeps.
>
> . . . One which sweeps only 50 per cent of the reservoir might be quite uneconomical. One that sweeps 70 per cent or close to it probably would be quite attractive.

These data would indicate an average recovery of 50 per cent under waterflooding and 73 per cent under miscible displacement. Others believe that miscible displacement projects will recover only 10 to 15 per cent more oil than waterflooding, making top recoveries range from perhaps 55 to 65 per cent.[53] Costs of even the simplest miscible project are much higher than for a waterflood and to obtain the above average recoveries a follow-up waterflood would often be necessary. On the other hand some companies plan to follow their less successful waterfloods with miscible flooding. On the whole the economics of the average miscible displacement project is doubtful.

For a time the so-called "fire floods" held great promise of increased recoveries from depleted primary fields and espcially from low-gravity crude oil reservoirs of shallow depth. Some companies still have great faith in them under certain favorable conditions. "Steam floods" have stolen the limelight, and perhaps rightfully so, because the process was developed in the field instead of in the research laboratory. The process offers the most promise in relatively shallow low-gravity crude reservoirs such as occur in California where other recovery methods have not been profitable.

T. R. Blevins and R. H. Billingsley have described a 10 inverted seven-spot injection pattern with 32 producing wells on a 61 acre project in the Kern River Field, California.[54] The 70-foot sand section at a depth of 700 feet has a permeability of 7,600 md, 35 per cent porosity, and 52 per cent initial oil saturation. The oil has a viscosity of 2,710 cp at 85°F and 4 cp at 350°F. Steam flooding showed a sweep efficiency of 60 per cent.

Bursell and Pittman also have described steam flooding in the Kern River Field and stated that recoveries were 50 to 70 per cent of the original oil in place.[55] They concluded that the oil recovery was definitely a function of reservoir oil viscosity. Fairfield has prepared a simplified method of estimating steam drive recoveries which takes into account operating costs.[56] Over 13 per cent of the current oil production in California was ascribed to the 29 fields utilizing steam soak methods in 1967.[57] Several other experimental projects were in operation in Wyoming and Venezuela at that time.[58] The U. S. Bureau of Mines believed that recent developments in thermal methods of oil recovery could double the proved reserves of oil in the United States.[59] API estimates of proved oil reserves at that time were approximately 30 x 10^9 bbl. Yet in 1969 thermal recovery was referred to as "A Troublesome Neophyte."[60]

During the mid-1970's many laboratory, computor, and field research projects were in progress dealing with "microemulsion" and "micellar" systems secondary and tertiary oil recovery. Most of them were outgrowths of miscible-phase recovery methods. Improved types of water-soluble and oil-soluble surfactants offered considerable promise.[61,62] It was hoped that the higher allowable prices for secondary and tertiary recovery oils might offset earlier discouragement with surfactants. The high costs of all such products at the time combined with very high absorption rates on the sands were very important factors.

It is what shows up in the stock tank that determines the real value of any prospective producing oil property, not the staggering amounts believed to be, nor even known to be, in the reservoir sands. Actually not enough oil has been produced from a sufficient number of reservoirs by these newer methods of recovery and there is so much to learn about field operations and procedures, particularly with respect to the overall economics, that the valuation engineer must be overly cautious in dealing with them.

REFERENCES

1. W. F. Stevens and G. Thodos, "Estimating Oil Reserves from Production Data," *Petr. Engr.*, February 1961, pp. B-46–B-51.
2. David T. Day, "The Petroleum Resources of the United States," *USGS Bull. 394*, 1909, pp. 44 and Pl.V.
3. Willard W. Cutler, Jr., "Estimation of Underground Oil Reserves by Oil-Well Production Curves," *USBM Bull. 228*, 1924, p. 6.
4. J. J. Arps, "Analysis of Decline Curves," *Oil and Gas Property Evaluation and Reserve Estimates*, Petr. Trans. Reprint Series, No. 3, Soc. Petr. Eng'rs, AIME, 1970, pp. 93-102.
5. H. J. Ramey, Jr., "The Ability of Rate-Time Decline Curves to Predict Future Production Rates," *Masters Thesis*, Dept. Petr. Eng., Uni. of Tulsa, Okla. (Includes an excellent bibliography).
6. H. J. Ramey Jr., and E. T. Guerrero, "The Ability of Rate-Time Decline Curves to Predict Production Rates," *Oil and Gas Property Evaluation and Reserve Estimates*, Petr. Trans. Reprint Series, No. 3, Soc. Petr. Eng'rs., AIME, 1970, pp. 103-105.
7. R. V. Higgins and H. J. Lechtenberg, "Production Decline Curves Using Data from California Oilfields," U. S. Bureau of Mines, *Rept. Investigations No. 7547*, 1971, 28 pp.
8. J. O. Lewis and Carl H. Beal, "Some New Methods of Estimating the Future Production of Oil Wells," *AIME Bull. 134*, February 1918, pp. 477-504.
9. Carl H. Beal, "The Decline and Ultimate Production of Oil Wells," *USBM Bull. 177*, 1919, p. 27.
10. Culter, *op. cit.*, pp. 22-35.
11. J. J. Arps, *op. cit.*, pp. 102-103.
12. J. J. Arps, "Estimation of Primary Oil Reserves," *Oil and Gas Property Evaluation and Reserve Estimates*, Petr. Trans. Reprint Series, No. 3, Soc. Petr. Eng'rs., AIME, 1970, pp. 30-33.
13. J. J. Arps, "Analysis of . . . ." *op. cit.*
14. J. J. Arps, "Estimation of . . . ." *op. cit.*
15. H. C. Slider, "A Simplified Method of Hyperbolic Decline Curve Analysis," *J. Petr. Tech.*, March 1968, pp. 235-236.
16. C. S. Mathews and H. C. Lefkovits, "Gravity-Drainage Performance of Depletion Type Reservoirs," *Trans. AIME*, Vol. 207, 1956, p. 265.
17. H. C. Lefkovits and C. S. Mathews, "Application of Decline Curves to Gravity-Drainage Reservoirs in the Stripper Stage," *Oil and Gas Property Evaluation and Reservoir Estimates*, Petr. Trans. Reprint Series, No. 3, Soc. Petr. Eng'rs., AIME, 1960, pp. 136-141.

18. H. A. Merle, C. J. P. Kentie, G. H. C. van Opstal, and G. M. G. Schneider, "The Bachaquero Study–A Composite Analysis of the Behavior of a Compaction Drive/ Solution Gas Drive Reservoir," *J. Petr. Tech.*, September 1976, pp. 1107-1115.
19. James Gilluly and U. S. Grant, "Subsidence in the Long Beach Harbor Area, California," *Bull. Geol. Soc. Amer.*, Vol. 60, (March 1949), pp. 461-529.
20. J. J. Arps, "Estimation of . . ." *op. cit.*
21. J. J. Arps, "Analysis of . . ." *op. cit.*
22. W. L. Wahl, L. D. Mullins, and E. B. Elfrink, "Estimation of Ultimate Recovery from Solution Gas-Drive Reservoirs," *Trans. AIME*, Vol. 213, 1958, pp. 132-138.
23. R. C. Craze and S. E. Buckley, "A Factual Analysis of the Effect of Well Spacing on Oil Recovery," *API Drill. and Prod. Pract.*, 1945, pp. 144-159.
24. H. F. Keplinger, "Reserve Calculation Review Shows Why Engineering Estimates Differ," *Oil and Gas J.*, January 17, 1977, pp. 60-70.
25. Stewart Coleman, H. D. Wilde, Jr., and T. W. Moore, "Quantitative Effect of Gas-Oil Ratios on Decline of Average Rock Pressure," *Trans. AIME*, Vol. 86, 1930, pp. 174-184.
26. Ralph J. Schilthuis, "Active Oil and Reservoir Energy," *Trans. AIME*, Vol. 118, 1936, pp. 33-52.
27. D. L. Katz, "A Method of Estimating Oil and Gas Reserves," *Trans. AIME*, Vol. 118, 1936, pp. 18-32.
28. Wahl, Mullins, and Elfrink, *loc. cit.*
29. William Hurst, "The Simplification of the Material Balance Formulas by the LaPlace Transformation," *Trans. AIME*, Vol. 213, 1958, pp. 292-303.
30. G. W. Tracy, "Simplified Form of Material Balance Equation," *Oil and Gas Property Evaluation and Reserve Estimates*, Petr. Trans. Reprint Series, No. 3, Soc. Petr. Eng'rs., AIME., 1960, pp. 25-35.
31. D. Havlena and A. S. Odeh, "The Material Balance as an Equation of a Straight Line," *Oil and Gas Property Evaluation and Reserve Estimates*, Petr. Trans. Reprint Series, No. 3, Soc. Petr. Eng'rs., AIME., 1970, pp. 66-70.
32. Walter Rose, "Material Balance Forecasts with Initial Production Data Lacking," *Producers Monthly*, September 1960, pp. 24-29.
33. A. S. Odeh, "Reservoir Simulation–What is It?" *Oil and Gas Property Evaluation and Reserve Estimates*, Petr. Trans. Reprint Series, No. 3, Soc. Petr. Eng'rs., AIME., 1970, pp. 106-111.
34. L. Kent Thomas and L. J. Hellums, "A Nonlinear Automatic History Matching Technique for Reservoir Simulation Models." *Trans. AIME.*, Vol. 253, 1972, Pt. II, pp. 508-514.
35. R. H. Jacoby and V. J. Berry, Jr., "A Method for Predicting Pressure Maintenance Performance for Reservoirs Producing Volatile Crude Oil," *Trans. AIME*, Vol. 213, 1958, pp. 59-64.
36. J. K. Rogers, N. H. Harrison, and S. Regier, "Comparison Between the Predicted and Actual Production History of a Condensate Reservoir," *J. Petr. Tech.*, June 1958, pp. 127-131.
37. William C. Goodson, "Gas Condensate Reservoirs," *The Petroleum Engineer*, (Reprint of an 8-chapter analysis originally in The Petroleum Engineer magazine) 1959, 27 pp., 90 references.
38. M. G. Arthur, "Economics of Cycling," *Trans. AIME*, Vol. 146, 1942, pp. 221-230.
39. R. C. Earlougher and E. T. Guerrero, "Analysis and Comparison of Five Methods Used to Predict Water Flood Reserves and Performance," *Mines Magazine*, January 1961, pp. 11-20.

40. W. E. Stiles, "Use of Permeability Distribution in Water Flood Calculations," *Trans AIME*, Vol. 186, 1949, pp. 9-13.
41. H. Dykstra and R. L. Parsons, "The Prediction of Oil Recovery by Water Flood," *Secondary Recovery of Oil in the United States*, American Petroleum Institute, New York, 1950, pp. 160-174.
42. M. Prats, C. S. Matthews, R. L. Jewett, and J. D. Baker, "Prediction of Injection Rate and Production History for Multifluid Five-Spot Floods," *Trans. AIME*, Vol. 216, 1959, pp. 98-105.
43. R. C. Earlougher, "The Valuation of Oil Properties for Secondary Recovery," paper presented at 75th Anniversary of the Colorado School of Mines, Sept. 30–Oct. 1, 1940.
44. Earlougher and Guerrero, *op. cit.*, p. 14.
45. *Ibid.*, p. 18.
46. F. H. Callaway, "Evaluation of Waterflood Prospects," *J. Petr. Tech.*, October 1959, pp. 11-16.
47. R. V. Higgins, D. W. Boley, and A. J. Leighton, "Graphs Quickly Predict Waterflood Performance," *World Oil*, Pt. 1, September 1968, pp. 89-96, Pt. 2, October 1968, pp. 152-163.
48. A. J. Leighton and R. V. Higgins, "Improved Method to Predict Multiphase Waterflood Performance for Constant Rates or Pressures," *Rept. Investigations, No. 8055*, U. S. Bureau of Mines, 1975, 66 pp.
49. M. Felsenthal and F. J. Gangle, "A Case Study of Thief Zones in a California Waterflood," *Trans. AIME*, Vol. 259, 1975, Pt. I, pp. 1385-1391.
50. S. T. Yuster and R. J. Day, "Gas Requirements in Gas-Drive Recovery," *Producers Monthly*, October 1946, pp. 21-28.
51. *Ibid.*, p. 26.
52. William L. Horner, "The Future of Miscible Displacement as a Method of Secondary Recovery," *IPAA Monthly*, October 1959, pp. 34-36, 58-59.
53. Robert J. Enright, "Are Miscible Floods Worth the Cost?," *Oil and Gas J.*, Dec. 11, 1961, pp. 43-46.
54. T. R. Blevins and R. H. Billingsley, "The Ten-Pattern Steamflood, Kern River Field, California," *Trans. AIME*, Vol. 259, 1975, Pt. I, pp. 1505-1514.
55. C. G. Bursell and G. M. Pittman, "Performance of Steam Displacement in the Kern River Field," *Trans. AIME*, Vol. 259, 1975, Pt. I, pp. 997-1004.
56. William H. Fairchild, "How to Quickly Estimate Steam Drive Oil Recovery," *World Oil*, March 1968, pp. 74-78.
57. James Burns, "A Review of Steam Soak Operations in California," *J. Petr. Tech.*, January 1969, pp. 25-34.
58. *J. Petr. Tech.*, January 1969, (a series of articles dealing with oil recoveries using steam).
59. "U. S. on the Threshold of Reserves Revolution," *Oil and Gas J.*, May 10, 1965, pp. 103-104.
60. H. J. Ramey, Jr., (an interview with), "Thermal Recovery–A Troublesome Neophyte," *J. Petr. Tech.*, January 1969, p. 7.
61. Robert N. Healy, Ronald L. Reed, and Clarence W. Carpenter, Jr., "A Laboratory Study of Microemulsion Flooding," *Trans. AIME*, Vol. 259, 1975, Pt. I, pp. 87-100, (and discussions, pp. 100-103.
62. S. C. Jones and K. D. Dreher, "Cosurfactants in Micellar Systems Used for Tertiary Oil Recovery," *Trans. AIME*, Vol. 261, 1976, Pt. II, pp. 161-167.

Additional References

P. E. Baker, "Effect of Pressure and Rate on Steam Zone Development in Steamflooding." *Trans. AIME*. Vol. 255, Pt. II, pp. 274-284.

John M. Campbell, *Petroleum Reservoir Property Evaluation*, John M. Campbell (Campbell Petroleum Series), Norman, Oklahoma, 1973.

Albert B. Cook, F. Sam Johnson, George B. Spencer, and Abdo F. Bayazeed, "The Role of Vaporization in High Percentage Oil Recovery by Pressure Maintenance," *Trans. AIME*, Vol. 240, 1967, Pt. I, pp. 245-250.

A. K. Dandona and R. A. Morse, "Down-Dip Waterflooding of an Oil Reservoir Having a Gas Cap," *Trans. AIME*, Vol. 259, 1975, Pt. I, pp. 1005-1016.

C. F. Gates and I. Sklar, "Combustion as a Primary Recovery Process–Midway Sunset Field," *Trans. AIME*, Vol. 251, 1971, Pt. I, pp. 981-986.

J. Geertsma, "Land Subsidence Above Compacting Oil and Gas Reservoirs," *Trans. AIME*, Vol. 255, 1973, Pt. I, pp. 734-744.

W. C. Hardy, P. B. Fletcher, J. C. Shepard, E. W. Dittman, and D. W. Zadow, "In-Situ Combustion in a Thin Reservoir Containing High-Gravity Oil," *Trans. AIME*, Vol. 253, 1972, Pt. I, pp. 199-208.

R. N. Healy, R. L. Reed, and D. G. Stenmark, "Multiphase Microemulsion Systems," *Trans. AIME*, Vol. 261, 1976, Pt. II, pp. 147-160.

L. W. Holm, "Use of Soluble Oils for Oil Recovery," *Trans. AIME*, Vol. 251, 1971, Pt. I, pp. 1475-1483.

H. A. Merle, C. J. P. Kentie, G. H. C. van Opstal, and G. M. G. Schneider, "The Bachaquero Study–A Composite Analysis of the Behavior of a Compaction Drive/Solution Gas Drive Reservoir," *Trans. AIME*, Vol. 261, 1976, Pt. I pp. 1107-1115.

T. E. W. Nind, *Principles of Oil Well Production*, McGraw Hill Book Company, New York, N. Y., 364 pp., 1964.

J. G. Richardson and R. J. Blackwell, "Use of Simple Mathematical Models for Predicting Reservoir Behavior," *Trans. AIME*, Vol. 251, 1971, Pt. I, pp. 1145-1154.

John O. Robertson, Jr., and George V. Chilingar, *Secondary Recovery and Carbonate Reservoirs*, American Elsevier Publishing Company, Inc., 1972, 304 pp.

Leo A. Schrider and Richard E. Cerullo, "A Decline Curve Pitfall Using Least Squares Solution," *J. Petr. Tech.*, April 1970, p. 141.

Charles R. Smith, *Mechanics of Secondary Oil Recovery*, Reinhold Publishing Corporation, New York, N. Y., 1966, 504 pp.

M. R. Todd and W. J. Longstaff, "The Development, Testing, and Application of a Numerical Simulator for Predicting Miscible Flood Performance," *Trans. AIME*, Vol. 253, 1972, Pt. I, pp. 874-882.

James A. Wasson and Leo A. Schrider, "Combination Method for Predicting Waterflood Performance for Five-Spot Patterns in Stratified Reservoirs," *Trans AIME*, Vol. 243, 1968, Pt. I. pp. 1195-1203.

D. A. Wayhan, R. A. Albrecht, D. W. Andrea, and W. R. Lancaster, "Estimating Waterflood Recovery in Sandstone Reservoirs," *API, Drill, and Prod. Practice*, 1970, pp. 252-257.

S. D. Yoelin, "The TM Sand Steam Stimulation Project," *Trans. AIME*, Vol. 251, 1971, Pt. I, pp. 987-994.

13

ESTIMATING RECOVERABLE NATURAL GAS RESERVES

The most important question regarding natural gas wells before and at the time of the passage of the Revenue Act of 1913 was the so-called "rate of deliverability." Some simple methods and "rule-of-thumb" procedures were in vogue relating declining deliverability rates to total recovery, but none proved to be wholly acceptable for income tax purposes.

DEVELOPMENT OF PRESENT METHODS

In 1916 Shaw described the common method of measuring the flow capacity of a natural gas well as follows:[1]

> The well is opened and allowed to blow off freely for 3 to 24 hours. Then a Pitot tube or spring gage is used in measuring the forward or momentum pressure of the flowing gas. The opening of a pipe connected with the gage is turned against the stream of gas. The pressure is measured by a column of water or mercury or a spring gage, according to the force of the gas. The pressure reading is then expressed in cubic feet per day by means of a set of tables, the barometric pressure, the size of the pipe being taken into account.

Shaw also discussed a volumetric method of calculating reservoir gas volumes based on the application of Boyle's law and stressed the difficulties in using it, particularly with respect to estimating porosities of reservoir sands.

Matson[2] in a related discussion described in detail the method of estimating natural gas reserves through use of Boyle's law. He presented authoritative graphs showing hourly pressure decline of individual wells on open-flow tests, or with time, and how well pressures decline with increased producing wells in a field. Johnson and Huntley[3] also were firm believers in the applicability of Boyle's law for estimation of gas reserves.

The first authoritative information on the subject of natural gas reserves was included in the *Manual for the Oil and Gas Industry* originally published by the Internal Revenue Bureau through the Treasury Department in 1919. The following methods of estimating "gas depletion" were enumerated:[4]

1. Decline in open-flow capacity.
2. Comparison with life history of similar wells or properties, particularly those now exhausted or nearing exhaustion.
3. Size of reservoir and pressure of gas, or the pore-space method.
4. Other indications of depletion.

The latter included observations of changes in pipeline and compressor station pressures, appearance of oil and/or water in a gas well, or in neighboring gas wells, and observations of *minute pressures*. Minute pressures observations had been in use for many years. The method consisted of a series of observations of pressure buildup in one minute upon closing-in wells that were flowing wide open into a pipeline. Possibilities of gas migrating away from a subject well or property were clearly understood.

The "closed pressure method" based on Boyle's law was recommended in the *Manual*.

In 1922 Ruedemann and Gardescu[5] reviewed the three methods of testing gas wells in general use at the time, namely: the open-flow capacity, the closed-in pressure, and the minute pressure above the line methods. They disapproved of the open-flow method. A "constant production to the pound decline" method was recommended instead. It was based on this principle:

> . . . the areas beneath the segments corresponding to closed pressure decline [of wells or reservoirs plotted against time] are proportionate to the amount of gas produced.

Volumes were not believed to be exactly proportional to the decline in closed-in pressure.

Johnson and Morgan,[6] in 1926, discussed the use of Boyle's law and pointed out that they knew of no commercial pool where the yield per pound loss remained constant throughout the life of the pool. Nevertheless, they believed that

> . . . the best method of estimating future production of gas wells from pressure data is to make a curve of the yield per pound loss by successive time units and to extrapolate this line directly, ...

Davis[7] reviewed the various methods in use in 1927 and emphasized the application of Boyle's law, but pointed out that the method might be very misleading where a water drive tended to maintain reservoir pressures. He advised the use of all applicable methods of reserve estimation in order to check one estimate against another.

In 1935 Stephenson[8] briefly reviewed the need for accurate estimates of natural gas reserves and stated that the most commonly used methods at the time were the porosity pressure method and the rock-pressure production-decline method. He stressed the need for accurate porosity values, sand thicknesses, production boundaries, and accurate average closed-in pressures. He advocated using a method believed to have been used first by Davis to obtain an average weighted reservoir pressure. The same procedure is still widely used with respect to both oil and gas reservoirs. It is based on the use of what is now called an *isobaric* map on which, according to Stephenson,[9]

> . . . all points of equal reservoir pressures are connected by contours; areas of equal pressure are then measured with a planimeter and the area in acres multiplied by the reservoir pressure in pounds to give a series of products. If the products thus obtained are added together and divided by the total area of the field, a weighted average pressure results from which the decline in pressure is readily calculated.

Today a company must know its natural gas reserves and rates of deliverability with much greater accuracy than in the past. Basically the information is needed for the same reasons it is needed for crude oil, but since World War II a more pressing reason has developed with respect to all gas sold through interstate pipelines. The Federal Power Commission requires comprehensive and authoritative information concerning the daily flow capacity and reserves of all connected reservoirs. Gruy and Crichton have stressed the following additional factors that are particularly contributory to today's needs for reliable estimates:[10]

1. To determine which fields contain sufficient reserves to justify the construction of pipeline outlets to serve particular markets.
2. To design pipelines necessary to serve those fields adequately.
3. To determine the location of industrial and chemical plants.
4. To finance the development of gas properties, and the construction of gas pipelines.
5. To determine fair and adequate depletion and depreciation rates.
6. To justify applications for gas pipelines before various regulatory bodies.
7. To determine the number of wells required to exploit the reserves most economically.
8. To aid in establishing values for the purchase or sale of properties, and for purposes of inheritance taxes.
9. To determine equities under unitized operations.
10. To provide a basis for calculating the economics of cycling operations.

The methods advocated by Stephenson are still widely used. Industry has learned how to measure the deviations of natural gas (the *Z*-factor), measure reservoir temperatures accurately, obtain porosities and sand thicknesses from cores, or more recently from electric well logs, and to gage gas production more accurately. Calculations of original and subsequent gas reserves in place can be made sufficiently accurate today, but industry cannot always estimate with equal accuracy the percentage that will be recovered. Here an experience factor becomes important, especially so as abandonment pressure approaches atmospheric pressure.

Large natural gas reserves are seldom left entirely shut-in or undeveloped. Small reserves may be left shut-in for long periods waiting for an outlet. Gas-cap gas may remain shut-in until most of the oil phase has been produced. It seldom takes long to develop any gas reserve because of the wide well spacing normally used. As a result a nonassociated gas field usually has only a short undeveloped or partially developed stage within the United States. Offshore gas and nonassociated gas in other nations may be undeveloped or not produced for a lack of pipelines or markets.

Reserves, purchases, and sales of natural gas are measured in standard cubic feet (scf), thousands of standard cubic feet (Mscf), and millions of standard cubic feet (MMscf). The absolute pressure and temperature at which natural gas must be measured varies from state to state. Temperatures normally are taken at 60°F. Absolute pressures range from 14.65 psia in Texas, Oklahoma, Kansas, and Missouri to 14.73 in California and 15.025 psia in the Rocky Mountain region. British Columbia also uses the 15.025 psia base, whereas Alberta and Saskatchewan use 14.40 psia. A base pressure of 14.65 psia commonly is used for intracompany purposes.

ESTIMATION OF NONASSOCIATED GAS RESERVES

Present-day estimates of nonassociated natural gas reserves usually are made through use of one or more modifications of the old *volumetric* method and the *pressure decline curve* method. Decline in pressure can be plotted with respect to either deliverability or accumulative recovery. Use of the latter method is justified through use of what is now called the *material balance concept* which shows that estimates of reserves can be obtained accurately from no more than accurate pressure and volume data. Usual sources of important errors, their probable magnitude, and suggested means of partially correcting them are considered under the heading "Common Sources of Errors in Estimates." Nevertheless, a check when possible of the material balance method against the volumetric method is desirable.

Volumetric Methods

The accuracy of the volumetric method depends on the completeness of development operations and the adequateness of accurate data with respect to reservoir area, sand thickness, porosities, interstitial water saturation, temperature, pressure, and the deviation factor of the gas,

where

G = scf of gas initially in the reservoir,
G_p = scf of gas produced, or recoverable
G_R = scf of gas remaining in reservoir at pressure p,
43,560 = gross volume in cf of 1 acre-ft of reservoir sand,
A = area of reservoir in acres,
h_n = net thickness reservoir sand in feet,
ϕ_e = effective porosity of reservoir sand, decimal fraction,
S_w = interstitial water saturation, decimal fraction,
p_i = average initial reservoir pressure in psia,
p = any subsequent average reservoir pressure in psia,
p_{sc} = base pressure in psia,
T_R = reservoir temperature in °F,
T_{sc} = base temperature in °F,
Z_p = deviation factor at pressure p,
Z_i = deviation factor at initial pressure p_i.

$$G = 43{,}560\ Ah_n\phi_e(1 - S_w)\frac{p_i}{p_{sc}}\left(\frac{T_{sc} + 460}{T_R + 460}\right)\frac{1}{Z_i} \tag{13-1}$$

and

$$G_R = 43{,}560\ Ah_n\phi_e(1 - S_w)\frac{p}{p_{sc}}\left(\frac{T_{sc} + 460}{T_R + 460}\right)\frac{1}{Z_p} \tag{13-2}$$

Subtracting

$$G_P = G - G_R = 43{,}560\ Ah_n\phi_e(1 - S_w)\frac{1}{p_{sc}}\left(\frac{T_{sc} + 460}{T_R + 460}\right)\left(\frac{p_i}{Z_i} - \frac{p}{Z_p}\right) \tag{13-3}$$

Equation 13-3 would give the scf of gas recoverable in depleting the reservoir from the initial pressure p_i down to any subsequent pressure p. In practice pressure p may be the pressure of the transmission line in which case the recoverable reserves would be reduced considerably in comparison with the reserves recoverable down to reservoir pressures approaching atmospheric pressure. Abandonment pressures can never be less than some tens of psig depending on line pressures, reservoir conditions, well depths and spacing, and overall production costs.

Another volumetric method of calculating natural gas reserves is based on the ideal gas law

$$pV = ZnRt$$

where

n = number of pound-moles,
p = absolute pressure,
R = gas law constant 10.73,
t = absolute temperature $T + 460$,
V = volume in cubic feet,
Z = compressibility factor.

$$\frac{V}{n} = \frac{10.73Z(T + 460)}{p} \tag{13-4}$$

Let $v_M = \frac{V}{n}$ = cubic feet of gas per pound-mole at reservoir conditions

then

$$v_M = Z\frac{10.73\ (T + 460)}{p} \tag{13-5}$$

and at standard conditions of temperature and pressure,

$$v_{M_{sc}} = 379.4 \text{ scf}$$

where v_{M_i} and v_{M_p} are the cubic feet of gas per pound-mole at initial conditions i and at any subsequent reservoir pressure p, respectively:

$$G = 43{,}560\ Ah_n\phi_e\ (1 - S_w)\left(\frac{379.4}{v_{M_i}}\right) \tag{13-6}$$

and

$$G_R = 43{,}560\ Ah_n\phi_e\ (1 - S_w)\left(\frac{379.4}{v_{M_p}}\right) \tag{13-7}$$

or

$$G_p = G - G_R = [43{,}560\ Ah_n\phi_e\ (1 - S_w)]\left[379.4\left(\frac{1}{v_{M_i}} - \frac{1}{v_{M_p}}\right)\right] \tag{13-8}$$

The Material Balance Method

Natural gas within a volumetric-type reservoir is free to move easily and quickly in any direction toward a pressure sink with the result that such low-pressure areas are quickly equalized upon shutting-in the producing wells. The reserves of a gas reservoir can be calculated by the material balance method when the aerial extent, sand thickness, porosity, and interstitial water content are not known. One must assume in using the material balance method that the composition of the gas does not change appreciably with time, although actual determination of the Z-factors over the productive pressure range will greatly improve the accuracy of the estimate. The method is based on the principle of the conservation of mass, i.e.:

(mass of gas produced) = (mass of gas initially in reservoir) – (mass of gas remaining in reservoir)

Where n_p, n_i, and n_r are the moles of gas produced, initially present, and remaining, respectively,

$$n_p = n_i - n_r$$

The principle of the method would be correct for a single component gas but may be used with sufficient accuracy for any multicomponent gas estimate,

where

V = the net pore volume of the reservoir in cubic feet,
V_i = the initial pore volume of gas present in the reservoir in cubic feet,
V_p = the pore volume of gas produced from the reservoir in cubic feet,
V_r = the pore volume of gas remaining in the reservoir in cubic feet,
Z = volume correction factor for natural gas,
T = temperature in °F,
p = pressure in pounds per square inch absolute,
G = standard cubic feet of gas initially in the reservoir,
G_p = standard cubic feet of gas produced, and p, sc, i, and r are subscripts representing produced, standard conditions, initially, and remaining, respectively.

$$V_p = V_i - V_r$$

Under a 100 per cent water drive V_p would be replaced by encroaching water as fast as the gas was produced and the initial pressure p_i would be maintained. Or if water encroachment was totally lacking and the reservoir produced volumetrically, the net pore volume V of the reservoir filled with gas would remain constant and the various volumes of gas expressed in standard cubic feet would be:

$$\begin{pmatrix}\text{scf of gas produced}\\ \text{from the reservoir}\end{pmatrix} = \begin{pmatrix}\text{scf of gas initially}\\ \text{in the reservoir}\end{pmatrix} - \begin{pmatrix}\text{scf of gas re-}\\ \text{maining in the}\\ \text{reservoir}\end{pmatrix}$$

or

$$\frac{V_p p_{sc}}{Z_{sc}(T_{sc} + 460)} = \frac{V_i p_i}{Z_i(T_i + 460)} - \frac{V p_r}{Z_r(T_i + 460)} \qquad (13\text{-}9)$$

$$\frac{V_p p_{sc}}{Z_{sc}(T_{sc} + 460)} = G_p, \quad \text{and} \quad \frac{V_i p_i}{Z_i(T_i + 460)} = G \qquad (13\text{-}10)$$

but

$$G_p = G - G_r$$

and

$$G_p = G - \left[\frac{V}{(T_i + 460)}\right]\frac{p_r}{Z_r} \qquad (13\text{-}11)$$

Equation 13-11 is in the form of $y = b - mx$, the equation of a straight line on rectangular coordinate paper. Thus the ultimate recovery of a volumetrically controlled gas reservoir can be solved graphically by plotting the reservoir pressure in psia divided by the corresponding *Z-factor on the* y-axis, the scf of accumulative gas produced down to each plotted pressure on the x-axis, and extending the graph to the anticipated abandonment pressure in psia, as illustrated in Figure 13-1. The pressure-volume graph in this example indicates a recovery of 22MMMscf of gas at an abandonment pressure of 200 psia.

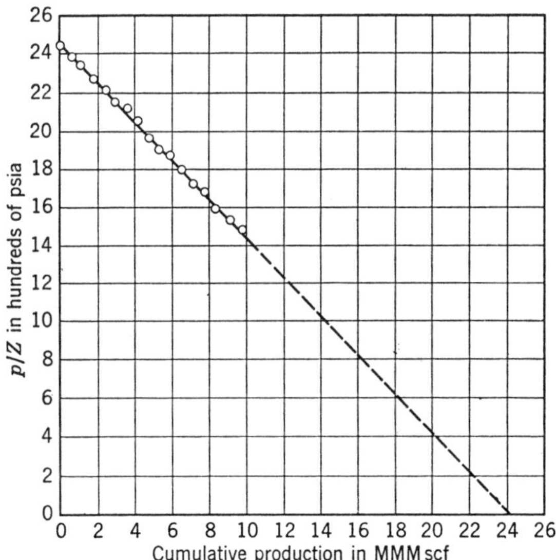

Figure 13-1. Graphical method of estimating ultimate recoveries of nonassociated natural gas.

Water-Drive Gas Reservoirs

A volumetric estimate of a natural gas reservoir made early during its productive life may have to be reduced by 50 per cent or more upon definite indications of an active water drive. An important factor is whether the reservoir pressures will be wholly or partially maintained by water influx at desired gas withdrawal rates. Another factor is that the higher the operating pressure of the reservoir the lower the recovery efficiency, other factors being equal. One might think that the smaller the areal extent of the reservoir the higher the efficiency of an active water drive, but such is not often the case. Actually, the larger the areal extent of the reservoir the more often the recovery efficiency will be higher. Several reasons appear to be obvious;

1. Gas wells are easily "drowned-out."
2. Gas wells near the edgewater line, if produced at excessively high rates, will induce vertical coning of "bottom water" and lateral cusping of advancing edgewater into the well.
3. If the areal extent of the reservoir is several hundred acres or more and/or there are sufficient upstructure producing wells, the shutting-in of edge wells, as they go to water, may not reduce either field rate of production or the total recovery appreciably.
4. Usually, the smaller the reservoir the more difficult it is to prevent an irregular water advance.
5. No wells may be located at or near the highest part structurally on the reservoir, a prevalent condition on Gulf Coast salt domes where relatively small sealed fault blocks are gas productive.

Gas reserves under strong water drives may be estimated through use of the volumetric equation and a recovery factor that is based on experience. Sand thicknesses from electric logs can be picked quite accurately. The accuracies of porosities and interstitial water contents may be subject to considerable error. In fact either one or both often may be surmised with sufficient accuracy from a general knowledge of those reservoir sands in that general area because a percentage recovery factor must be used which is either based on experience or guessed at. Recovery factors for large reservoirs may vary from 50 to 85 per cent of the initial gas in place depending on the uniformity of the reservoir section with respect to thickness, dip, porosity, permeability, and the effectiveness of the water drive. The weaker the water drive–that is, the nearer depletion drive is approached–the better the recovery, everything else being equal. Recovery factors for small water-drive gas reservoirs may vary from 30 to 60 per cent of the initial gas in place depending in part on the same uniformities as large reservoirs, and especially on the location of the wells highest structurally and the rate of gas withdrawal. In all cases the pressure of the gas at the time the advancing water invades the drainage area is a major factor in recoveries because the residual gas saturations are comparable to residual oil saturations in water-drive oil fields on a percentage of pore-space basis.[11]

The "probable" proven reserves G_P in an effective water-drive non-associated gas reservoir may be expressed by adding a recovery factor E_R to Eq. 13-1, so that

$$G_p = E_R(43{,}560\ Ah_n\phi_e)(1 - S_w)\frac{p_i}{p_{sc}}\left(\frac{T_{sc} + 460}{T_R + 460}\right)\frac{1}{Z_i} \qquad (13\text{-}12)$$

The recovery factor E_R should not be used during the initial productive life of the reservoir unless some sort of water drive is positively known. However, a water drive may be suspected and watched for at any time during the productive life of the reservoir. An unusually flat relationship between P/Z and the cumulative production may be so interpretted. Evidence of a water drive depends somewhat upon whether production rates have been radically adjusted at times. Some investigators during the early 1950's and through the 1960's were not convinced that the P/Z to cumulative production relationship was unreliable, except perhaps over the very early production period and providing all data was correct.[12, 13]

The Reservoir Limit Test

A method for obtaining an approximate estimate of the size of a continuous gas reservoir, nonassociated or associated, when only one or two wells have been drilled into it was first described by Park Jones.[14] The test was based on a transient flow method on the well by holding the rate of production constant during the test period while accurately recording the bottom-hole flowing pressure. The rate of the pressure transient through the reservoir could be determined by noting the arrival time of a pressure decrease in any closed-in nearby wells, or in cases of only one well in the reservoir, rates similarly obtained in comparable reservoirs may be used. The presence of and distance to the boundary of the reservoir, an edgewater contact, sealing and nonsealing faults, and permeability barriers were said to be determinable with fair accuracy under favorable conditions.

Cole[15] pointed out in 1958 that good geological reservoir data were needed to estimate the limits of a gas accumulation for which the gas volume had been calculated through use of material balance methods. Studies of the application of reservoir limit tests for gas accumulations have increased since

1970. Chierici et al[16] were convinced that gas reservoirs under partial water drive conditions could not be uniquely calculated if based upon past performance. This was especially true where small fluctuations in rates had occurred and pressure data was of just normal accuracy. Earlougher[17] stated that long-term pressure drawback tests as used for oil reservoirs were equally applicable to gas reservoirs. The shape of the reservoir and the location of the test well in the system were determinable. Only slope and intercept data from plots of short-term drawdown tests and long-term reservoir tests were needed. Data concerning the formation, viscosities, well-bore radius, skin factor, and initial reservoir pressure were unnecessary. Brewer[18] stated that a plot of log $(P_i/Z_i - P/Z)$ vs. log cumulative production may indicate water influx. A straight-line relationship at an angle of less than 45° indicates active pressure support.

Strobel et al[19] reported their use of reservoir limit tests in naturally factured formations. A dry gas reservoir with 3 wells located 2 and 8 miles apart over a 10 mile stretch along the crest of a structure were studied. The test data were matched with type curves for estimating the drainage area which proved to be 18 miles long by 3 miles wide. Only very high precision pressure measurements were made. Porosity and permeability also were calculated from the precision curves. The authors believed their results compared well with published data for natural fracture systems. Rossen[20] has discussed the usual methods of estimating natural gas reserves using the P/Z curve relationship. He thought the method to be particularly inadequate for gas reservoirs under abnormal pressure and substantial aquifer support. A procedure was described and recommended using a nonlinear regression analysis to observed pressures and production history.

Deliverability of Gas Wells

At times the deliverability rates of natural gas wells are as important in a valuation study as the recoverable reserves. Reservoir pressure is a measure of both deliverability and reserves, but some reservoirs are of such low permeability, wholly or in part, that wells must be spaced uneconomically close to provide an allowable and adequate deliverability rate for the reservoir as a whole. As a result the value of some natural gas reservoirs may be as much if not more a function of deliverability than a function of the volumes of reserves.

Rawlins and Schellhardt[21] of the U. S. Bureau of Mines prepared a comprehensive manual in 1936 on the measurement and use of flowing pressures in gas wells to predict deliverability. They did not approve of the older and commonly used "open-flow" test methods and showed that a series of tests at high flowing pressures were not only more accurate, but also less wasteful of gas. They used a well-known flow equation for gas wells which may be stated as

$$q_g = C(p_{ws}^2 - p_{wf}^2)^n \qquad (13\text{-}13)$$

where

q_g = rate of flow, Mscf per 24 hours,
C = a coefficient,
p_{ws} = shut-in bottom-hole pressure in psia (wellhead pressure plus pressure due to weight of gas column in well),
p_{wf} = flowing bottom-hole pressure in psia (wellhead pressure plus pressure due to weight of gas column in well),
n = exponent corresponding to the slope of the straight-line relationship between q_g and $(p_{ws}^2 - p_{wf}^2)$ when plotted on logarithmic coordinate paper, or the tangent of the angle with the y-axis.

Usually only three or four rates need be run, or just enough to give a straight line on the logarithmic plot. Such a plot is shown in Figure 13-2. The angle between the y-axis and the plotted straight line may be measured with a protractor and the corresponding tangent n of the angle found in suitable tables, or it may be calculated by picking two sets of x and y values and substituting them in the following equation:

$$n = \frac{\log x_1 - \log x_2}{\log y_1 - \log y_2} \qquad (13\text{-}14)$$

The value of n may be substituted in the Eq. 13-13, all other values being known except the constant C, and it may be calculated now through use of the following equation:

$$\log C = \log q_g - n \log (p_{ws}^2 - p_{wf}^2) \qquad (13\text{-}15)$$

The theoretical open-flow rate when p_{wf} would be equal to atmospheric pressure may be calculated, knowing C and n, or it too may be scaled from the graph. Corrections that may be applied to compensate for deviations from Boyle's law and methods of calculating the weight of the flow column in the well may be found in Rawlins and Schellhardt's publication.[22]

Northcutt[23] also has contributed a graphical solution to the open-flow potential method of testing gas wells. Completion practices often used today and modern gas-reservoir engineering have shown that the test procedures of Rawlins and Schellhardt and of Northcutt may not always be entirely reliable. Complete drawdowns, much greater care and accuracy, and sophisticated interpretations have become increasingly important.

Many gas wells are completed in fractured carbonate reservoirs. Many

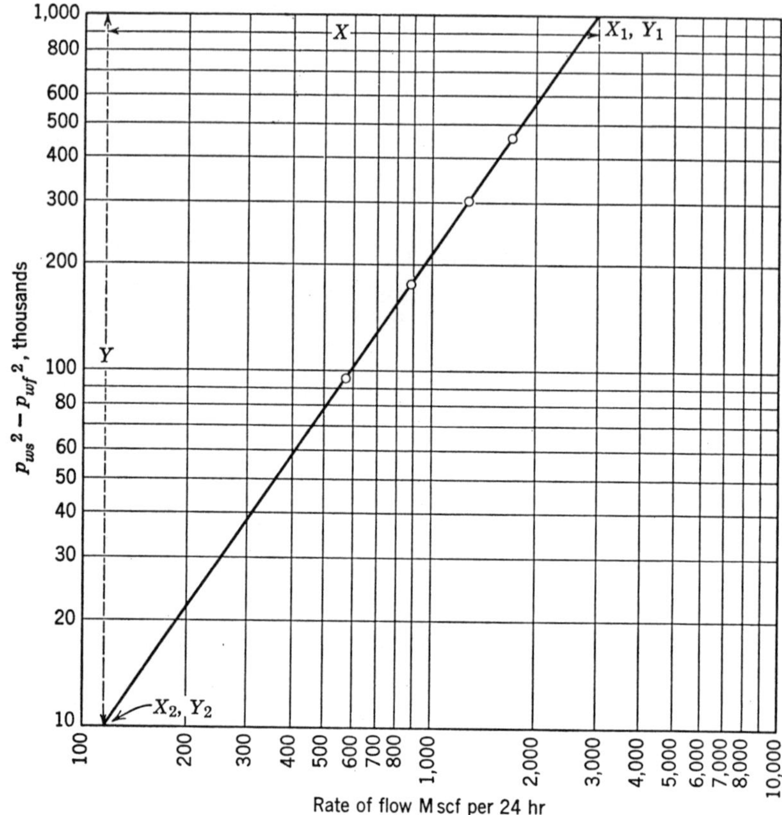

Figure 13-2. Graphical method for finding open-flow rate of natural gas wells.

other gas wells are hydraulically fractured. Adams et al[24] have discussed the meaning of the two straight-line portions of some pressure build-up curves and the use of the change in slopes to estimate distance to a fracture or fracture system. The first straight-line slope furnishes data on permeability and skin effect and the second provides means of calculating static pressure. The transition period between the two straight-line periods of flow in hydraulically fractured reservoirs may be quite long.[25,26] As a result, too-short a flow test may be misinterpretted. Turbulence also can complicate interpretations.

Odeh and Al-Hussainy[27] have described an improved method of determining the static pressure of a well from buildup data. The method requires an accurate value of pressure at time zero. Less other basic data are needed than in previous methods. Usual methods of obtaining gas well characterization normally required a series of drawdown-buildup tests. Odeh, Moreland, and Schueler[28] have provided a method whereby only one drawdown and one buildup sequence are needed to calculate an equivalent characterization. Still simpler procedures also have been published by Hossein Kazemi.[29]

Compaction of reservoir matrix as a result of production and effects upon rates and recoveries is discussed under *Common Sources of Errors*.

ESTIMATION OF ASSOCIATED GAS RESERVES

Methods to be used in estimating associated natural gas reserves depend on whether the gas occurs as dissolved gas in an oil reservoir or as a free gas cap. Obviously there is dissolved gas where there is a free gas cap.

Estimation of Free Gas-Cap Gas

Gas-cap gas may be estimated through use of any of the methods applicable to nonassociated gas reservoirs. The volumetric method generally is preferred if there are sufficient reservoir data. The initial gas reserve is not difficult to estimate unless too much of the underlying oil is produced before determining the original gas-oil contact. The tops of the gas-bearing section as well as the gas-oil contacts usually can be picked quite accurately from the electric logs. A lack of wells in the gas-cap area may present a problem because either cores or electric logs are needed to determine porosity, interstitial water saturation, and, if desired, permeability. Pressures and Z-factors are easily obtainable if any of the wells are open to the gas zone.

Gas-cap gas poses some difficult production problems for obtaining maximum recoveries of each reservoir fluid. This is particularly true when reservoir pressures are reduced excessively while producing one or the other reservoir fluid. Excessive free-gas withdrawal either with respect to total volumes or rates may cause oil to be lost to the free-gas zone. Much of this oil may never be recovered except by some later secondary recovery method. Excessive oil withdrawals either with respect to rates or total volumes may cause free gas to migrate into the oil zone. This gas can be recovered eventually, but overall oil recovery efficiency may be low unless withdrawal rates are carefully controlled.

Estimation of Dissolved Gas Reserves

Dissolved gas reserves in an oil reservoir are essentially a function of the volume of the oil in place at the reservoir temperature and pressure. The standard cubic feet of dissolved gas in a barrel of stock tank oil at any time during the productive life of a well or reservoir may be determined accurately from a bottom-hole sample of the reservoir fluid.

The original volume of dissolved gas G in a dissolved gas type of oil reservoir is the product of the estimate of the initial volume of stock tank oil N in place multiplied by the initial reservoir solution-gas-oil ratio R_{si}, which may be expressed:

$$G = NR_{si}$$

The difference between the initial volume G and the volume G_p already produced is a measure of the remaining gas reserve. However, an estimate of the economic reserve is in part a function of its rate of availability and that in turn is related chiefly to the method and rate of oil recovery. Rate of oil recovery may be so restricted by market demand or the relative permeability of the reservoir sand to gas at prevailing oil saturations that a gas pipeline outlet for the field might be uneconomical.

Usually most of the dissolved gas in an oil field is produced before the oil production rate reaches an economic limit. The volume of dissolved gas produced with each barrel of oil can be controlled in part, but is always more than the amount dissolved in that barrel of oil under reservoir conditions. Excessively high oil production rates do not affect the amount of gas originally recoverable from the reservoir, but they do result in higher gas availability rates during the early productive life of the reservoir with consequent early exhaustion of the gas and low overall oil recovery efficiency. Fields tests can be made to determine the most efficient rate to flow oil wells with respect to produced gas-oil ratios. During extended periods of proration and restricted allowable production rates, gas-oil ratios may remain relatively constant for months to years. Nevertheless, gas-oil ratios always tend to increase with time as oil is produced from dissolved gas drive fields regardless of oil production rates until most of the recoverable reserves have been produced as was illustrated in Figure 10-1.

COMMON SOURCES OF ERRORS IN ESTIMATES

For several years after World War II, estimations of natural gas reserves and deliverability rates were thought to be rather easily made, yet sufficiently accurate. As demands rapidly increased and reserve ratios declined, more attention was given to sources of errors.

Gruy and Crichton[30] have noted the following common sources of errors in making reserve estimates of natural gas:

1. Construction of a pressure curve for an early well in a field and application of the indicated reserve to later wells on a per well or per acre basis.
2. Construction of pressure-decline curves from arithmetic average pressures when well spacing is not uniform and the number of wells is not constant.

3. Construction of pressure-deline curves using pressures obtained from isobaric maps, all of which do not cover the same area.
4. Failure to check decline-curve estimates by volumetric calculations to determine whether the area, thickness, and porosity required to contain the estimated reserve conflict with the available data.
5. Failure to consider the effect of deviation from Boyle's law.
6. Assuming that neglect of corrections for temperature and compressibility are compensating.

Not only should decline-curve estimates be checked by volumetric estimates, but volumetric estimates should be checked by decline-curve estimates. Initial volumetric estimates are subject to the maximum errors, as is the case with initial oil recoveries. Keplinger[31] has reviewed the major sources of errors in a hypothetical gas reservoir. As with an oil reservoir, some errors are additive and some are subtractive. There is always the chance that all will be in the same direction. Keplinger's example assumes that condition. After completing the first well, the initial volumetric estimate was increased 93 per cent due to small increases in estimated porosity, area, and pay thickness, and a decrease in formation water content. His error would have amounted to only 27 per cent during mid-life of the reservoir and 12 per cent after the reservoir was on a declining rate. Keelan and Pugh[32] have concluded that recovery factors for carbonate reservoirs are particularly subject to error. Characteristics of the reservoir rock and wetting-phase imbibition appear to be the major factors. Wetting phases might be either oil or water, depending upon the type of accumulation. The reservoir rocks were divided into three types according to the "Archie" system. Type I is compact, crystalline, hard and dense, with no visible pore space under a hand lens. Type II is dull, earthy, or chalky, with barely visible pores. Type III is granular to sucrosic, or oölitic, with clearly visible pores. Trapped gas saturations varied from 23 to 68 per cent for samples having 80 per cent initial gas saturation. Trapped gas varied with initial gas in place, and varied linearly with initial gas saturation in Type I rocks. Porosity was not a factor in Type I and Type II. Trapped gas varied inversely with porosity in Type III. And there was little difference in trapped gas for Type II and Type III when initial gas saturation was greater than 70 per cent.

Compaction of gas reservoirs during production operations, as with oil reservoirs, is not uncommon. Variogs *et al*[33] concluded as a result of mathematical and laboratory studies that stress on low permeability rocks could result in lower production rates than might be anticipated. They also showed that pressure-drop about the flowing well-bore may be serveral times the average reservoir pressure drop. Geertsma[34] found notable subsidence above producing oil and gas fields to be the exception rather than the rule. He explained a simple procedure to single out the exceptional but real problem areas.

REFERENCES

1. E. W. Shaw, G. C. Matson, and C. H. Wegemann, "Natural Gas Resources of Parts of North Texas," (Gas in Area North and West of Fort Worth, by E. W. Shaw), *USGS Bull. 629*, 1916, pp. 23-41.
2. *Ibid.*, p. 92 (Gas Prospects South and Southeast of Dallas, by G. C. Matson).
3. R. H. Johnson and L. G. Huntley, *Oil and Gas Production*, Wiley, New York, 1916, pp. 164-167.
4. *Manual for the Oil and Gas Industry under the Revenue Act of 1918*, U. S. Government Printing Office, Washington, D. C., 1919, pp. 30-31.
5. P. Ruedemann and I. Gardescu, "Estimation of Reserves of Natural Gas Wells by Relationship of Production to Closed Pressures," *Bull. AAPG*, September-October 1922, pp. 444-463.
6. R. H. Johnson and L. C. Morgan, "A Critical Examination of the Equal Pound Loss Method of Estimating Gas Reserves," *Bull. AAPG*, September 1926, pp. 901-904.
7. Ralph E. Davis, "Estimation of Natural Gas Reserves," *Oil and Gas J.*, Vol. 26, No. 4, June 16, 1927, pp. G125-G126.
8. E. A. Stephenson, "Valuation of Natural Gas Properties," *Geology of Natural Gas*, American Association of Petroleum Geologists, Tulsa, Okla., 1935, pp. 1011-1033.
9. *Ibid.*, p. 1023.
10. Henry J. Gruy and Jack A Crichton, "A Critical Review of Methods Used in Estimation of Natural Gas Reserves," *Trans. AIME*, Vol. 179, 1949, p. 249.
11. T. M. Geffen, D. R. Parrish, G. W. Haynes, and R. A. Morse, "Efficiency of Gas Displacement from Porous Media by Liquid Flooding," *Trans. AIME*, Vol. 195, 1952, pp. 29-38.
12. E. R. Brownscombe and F. Collins, "Estimation of Reserves and Water Drive from Pressure and Production History," *Trans. AIME*, Vol. 186, 1949, pp. 92-99.
13. R. M. Hubbard and J. R. Elenbaas, "Determining Gas-Filled Pore Volume in a Water-Drive Gas-Storage Reservoir," *J. Petr. Tech.*, April 1964, pp. 383-388.
14. Park Jones, "Reservoir Limit Test on Gas Wells," *J. Petr. Tech.*, June 1962, pp. 613-619.
15. Frank W. Cole, "Gas Reservoir Material Balance can be Helpful," *World Oil*, October 1958, p. 182.
16. G. L. Chierici, G. Pizzi, and G. M. Ciucci, "Water Drive Gas Reservoirs: Uncertainty in Reserves Evaluation from Past History," *Trans. AIME*, Vol. 240, 1967, Pt. I, pp. 237-244.
17. R. C. Earlougher, Jr., "Estimating Drainage Shapes from Reservoir Limit Tests," *Trans. AIME*, Vol. 251, 1971, Pt. I, pp. 1266-1268.
18. C. L. Brewer, "Useful Plot helps Analyze Gas Reservoir Production," *World Oil*, July 1974, pp. 186-188.
19. C. J. Strobel, M. S. Gulati, and H. J. Ramey, "Reservoir Limit Tests in a Naturally Fractured Reservoir–A Field Case Study Using Type Curves," *Trans. AIME*, Vol. 261, 1976, Pt. I, pp. 1097-1106.
20. R. H. Rossen, "A Regression Approach to Estimating Gas in Place for Gas Fields," *Trans. AIME*, Vol. 259, 1975, Pt. I, pp. 1283-1289.
21. E. L. Rawlins and M. A. Schellhardt, "Back Pressure Data on Natural Gas Wells and Their Application to Production Practices," *USBM Monograph 7*, 1936, pp. 8, 19-26.

22. *Ibid.*, pp. 166-174.
23. R. D. Northcutt, "How Long Will Your Gas Last?" *Oil and Gas J.*, March 11, 1957, pp. 173-177.
24. A. R. Adams, H. J. Ramey, Jr., and R. J. Burgess, "Gas Well Testing in a Fractured Carbonate Reservoir," *Trans. AIME*, Vol. 243, 1968, Pt. I, pp. 1187-1194.
25. Keith K. Millheim and Leo Cichowicz, "Testing and Analyzing Low-Permeability Fractured Gas Wells," *Trans. AIME*, Vol. 243, 1968, Pt. I, pp. 193-198.
26. Robert A. Wattenbarger and Henry J. Ramey, Jr., "Well Test Interpretation of Vertically Fractured Gas Wells," *Trans. AIME*, Vol. 246, 1969, Pt. I, pp. 625-632.
27. A. S. Odeh and R. Al-Hussainy, "A Method for Determining Static Well Pressure from Buildup Data," *Trans. AIME*, Vol. 251, 1971, Pt. I, pp. 621-624.
28. A. S. Odeh, E. E. Moreland, and S. Schueler, "Characterization of a Gas Well from One Flow-Test Sequence," *Trans. AIME*, Vol. 259, 1975, Pt. I, pp. 1500-1504.
29. Hossein Kazemi, "Determining Average Reservoir Pressure from Pressure Buildup Tests," *Trans. AIME*, Vol. 257, 1974, Pt. II, pp. 55-62.
30. Gruy and Crichton, *op. cit.*, p. 261.
31. H. F. Keplinger, "Reserve Calculation Review shows Why Engineering Estimates Differ," *Oil and Gas J.*, January 17, 1977, pp. 60-70.
32. Dare K. Keelan and Virgil J. Pugh, "Trapped–Gas Saturations in Carbonate Formations," *Trans. AIME*, Vol. 259, 1975, Pt. II, pp. 149-160.
33. Juris Vairogs, C. L. Hearn, Donald W. Dareing, and V. W. Rhoades, "Effect of Rock Stress on Gas Production from Low–Permeability Reservoirs," *Trans. AIME*, Vol. 251, 1971, Pt. I, pp. 1161-1167.
34. J. Geertsma, "Land Subsidence Above Compacting Oil and Gas Reservoirs," *Trans. AIME*, Vol. 255, 1973, Pt. I, pp. 734-744.

Additional References

G. V. Chilingar, R. W. Mannon, and H. H. Rieke III, *Oil and Gas Production from Carbonate Rocks*, American Elsevier Publishing Company, Inc., New York, N. Y., 1972, 409 pp.

R. E. Davis and L. H. Meltzer, "A Method of Predicting Availability of Natural Gas, Based on Average Reservoir Performance," *Oil and Gas Property Evaluation and Reserve Estimates, Petr. Trans. Reprint Series, No. 3*, AIME, 1960 ed., pp. 142-151.

A. E. Essis B. and G. W. Thomas, "The Use of Open Flow Potential Test Data in Determining Formation Capacity and Skin Factor," *Trans. AIME*, Vol. 251, 1971, Pt. I, pp. 879-887.

Eugene D. Glass, "Static Pressure Volumes in Gas Wells," *The Petroleum Engineer*, September 1975, p. 90.

John D. Huppler, "Scheduling Gas Field Production for Maximum Profit," *Trans. AIME*, Vol. 257, 1974, Pt. II, pp. 279-294.

J. M. Ryan, "Limitations of Statistical Methods of Predicting Petroleum and Natural Gas Reserves and Availability," *Oil and Gas Property Evaluation and Reserve Estimates*, Petr. Trans. Reprint Series, No. 3, 1970, AIME, pp. 227-230, discussion 230-235.

E. T. Zana and G. W. Thomas, "Some Effects of Contaminants on Real Gas Flow," *Trans. AIME*, Vol. 249, 1970, Pt. I, pp. 1157-1168.

14 SOME ECONOMIC FACTORS THAT INFLUENCE PROPERTY VALUES

Several papers published in 1921 explained how to determine the values of oil properties.[1] This was about the same time our nation was supposedly running out of oil because so much had been used during World War I and finding rates were discouraging. These authors discussed the many factors to be considered in valuation work and stressed the effects of market demand for crude oil on values. Market demand always has been the major item in the overall problem of supply and demand. The industry has never had much control over demand inasmuch as prices help to determine demand. On the whole the supply was ample and at times quite excessive until the mid-70's. Any attempts at price controls had quickly failed. The industry as a whole, since its very beginning and until the mid-70's, prospered as a result of the free enterprise system. Excesses of production often occurred which were very wasteful of a limited national resource. Needless and wasteful production practices were largely corrected by 1935 upon passage of the "Connally" or "Hot Oil Act" and the "Interstate Compact to Conserve Oil and Gas." The Interstate Compact Commission was granted powers to control the latter. Despite much nation-wide, and even Congressional criticism, that control of production was indirect control of prices, these measures over time worked admirably to the advantage of the Nation. On the whole operators have not been allowed to produce wells since that time at rates over their "MER.s" (maximum efficient rates above which reservoir energy is wasted).

The U. S. industry dominated the worldwide supply and demand of oil and refined products in 1921. U. S. companies had very little in the way of foreign production operations with the exception of those in Mexico, but they did have worldwide marketing organizations which were dominated essentially by a very few companies. Any domestic domination for a few years was limited essentially to the refining and marketing of kerosene and gasoline.

Modern trends within the domestic industry show an overall tendency for the larger and/or better financed companies to purchase or otherwise acquire the smaller independent producing companies. Nevertheless the number of oil and natural gas operators in the production branch of the industry within the United States is estimated to have increased approximately 60 per cent over its total of 10,000 in 1966. The eight major companies only had an estimated 40 per cent of the Nation's production in 1976. The largest company only had approximately 9 per cent. These percentages hardly point to monopolization within the industry in comparison concentrations within the steel, automotive, and aluminum industries. The larger companies do tend to grow larger as the smaller companies tend to get smaller. Exclusive of foreign operations growth is more a matter of acquisition with some companies than growth from within. Almost all such acquisitions are a part of company policies to increase the ratios of properly proven reserves to production rates and to improve marketing coverage. Most of the larger companies find it much easier and cheaper to purchase proven domestic reserves than to explore for them.

Expanding companies usually must become self-sufficient and balanced in all branches of the industry, namely exploration and drilling, production, transportation, refining and manufacturing, and marketing. Those engaged in all branches to some extent on a national basis are usually referred to as the "majors." Basically those engaged in only exploration, drilling, and production are the "independents." An independint does not necessarily lose its favored status by expanding into all branches as long as it does not go "national." A few independents have expanded operations successfully into Canada and overseas.

At one time low profits in one branch of the industry might be offset to some extent in another branch. Today the profit margin is low in each branch by older standards. Even independent companies were attracted into foreign operations, expecially during the late 1950's and the 1960's, because of quicker anticipated payouts. Particularly since the earlier 1970's, the economics of foreign operations have changed radically. Beginning in 1975, a company has had to be almost desperate for oil production to do so. Before the 1930's most of the capital needs for growth were generated from within. Since that time an ever increasing percentage of the industry's capital needs has been supplied through loans from banks and insurance companies. As a result companies growing on borrowed money can take neither as many risks nor as much risk in comparison with pre-World War II.

Although oil property valuations have to do chiefly with production, those engaged in making such valuations will find it beneficial to learn about the other branches, particularly exploration, drilling, and marketing. The economics of each branch and of the industry as a whole are very important. Some knowledge of foreign operations will be helpful. The successes, failures, and costs in all branches of the industry, domestic and foreign, have an ultimate effect on the values of all production properties.

ESTIMATES OF FUTURE DEMAND

Estimates of the Nation's production of petroleum liquids during 1976 were 2.825 x 10^9 bbl of crude oil[2] and 0.701 x 10^9 bbl of natural gas liquids,[3] making a total of 3.526 x 10^9 bbl. Imports of crude oil and products were estimated to be 2.65 x 10^9 bbl or 43 per cent of the estimated total demand of 6.176 x 10^9 bbl. Imports amounted to only 20.4 per cent of the total demand of 4.075 x 10^9 bbl in 1965.

Use of equation 6-4 shows an increase in demand over the 11-year period of 3.86 per cent annually. Continuation at the same rate of annual increase would indicate a demand of 8.861 x 10^9 bbl of petroleum liquids in 1985, equal to an average daily production of 23.78 x 10^6 bbl. An annual future growth rate of 2.5 per cent would decrease the annual demand in 1985 to 7.713 x 10^9 bbl.

An estimated 19.719 x 10^{12} cf of natural gas were produced in the United States in 1976[4], down from a high of 22.61 x 10^{12} cf in 1973.[5] Production in 1976 was only 21 per cent higher than in 1965. It is quite probable that future production will neither exceed the 1973 high nor the 1976 rate, until perhaps for a relatively short period of time after the remainder of the 26 x 10^{12} cf of natural gas in the Prudhoe Bay area of Alaska is made available. Thus, maintenance of the 1976 rate of production until 1985 is very likely to prove to be an impossibility. According to equation 6-4, on the assumption that use equalled demand at all times between the years 1965 and 1974, increased use was at a rate of 4.215 per cent annually. Demand for natural gas in 1985, if use continued to increase at the same annual rate, would amount to 37.11 x 10^{12} cf. Should no more gas be found and increase in use continue at 4.215 per cent annually, the reserves at the beginning of 1977 would last only 8 more years. Declining production since 1973 is actually quite alarming.

MEETING FUTURE DEMANDS

The American petroleum industry in wartime and peace has met all demands for high quality products at very reasonable prices. Its going into foreign operations, particularly during the 50's and 60's was an important contributing factor until forced to give up such operations. There has always been many in and out of government who have been very critical of the industry, and particularly with respect to the claim of reasonable prices. These critics have conveniently overlooked the benefits to the Nation through prorationing production, the setting of well allowables, the size of the gasoline tax, to whom it goes, that it was higher in many states prior to 1974 than the wholesale price received by the refiner, and that it has long amounted to 5 to 8 times industry's profit per gallon.

The question arises within the industry more and more often as to how it can continue its record of adequate supply at reasonable prices in view of many recent outside counter actions. The larger companies have lost the depletion allowance; interstate gas sales have always been priced too low; federal crude oil pricing policies and many recent ambiguous regulations in force or contemplated are very counter productive. In addition, the industry is faced with likely possibilities that the government will get into one or more phases of the industry soon and that Congress may pass some kind of a divestiture bill. Either or both, like with the railroads, would soon create chaotic conditions within the industry which could eventually lead to taking over parts if not all the industry.

Many "authorities" have had our nation running out of oil periodically since 1920. Other authorities on the subject go to the other extreme, take a "Pollyanna" viewpoint, and rely on wishful thinking. Others remembering that oil is a wasting asset are hesitant to pessimistic with respect to the adequacy of future productive capacity. Another 50×10^9 bbl or more of recoverable domestic oil may be found, but on the whole it has become increasingly hard to find and that found is at an alarmingly increased cost.

The ability of the industry to meet future demands for domestic production depends on the volume of proven reserves and our ability to make them available. Both are influenced by the price of crude oil. No doubt imports of oil tended to keep oil prices down until the Arab Oil Embargo took effect and prorating of production as in Texas and some other states tended to keep prices up at least for a time. "Excessive" imports have been blamed since 1950 for most of the economic ills of the independent producer. Imports of petroleum liquids literally became a "must" during the mid-1950's and even more so after the mid-1960's. The proportions of imports in 1976 to domestic demand and to domestic production have increased to 41 per cent and to 70 per cent, respectively, as shown in Table 14-1. Only very small volumes of natural gas, partly in liquid form, were being imported as of 1976, high costs being a negative factor.

Table 14-1

Proportions of Liquid Petroleum Imports to

Year	Domestic Demand	Domestic Production
1946	7 per cent	7 per cent
1956	15	18
1966	21	27
1976	41	70

Undoubtedly imports of petroleum liquids and natural gas will become increasingly necessary in the future, probably not percentagewise with respect to total demand for several years, but certainly with respect to volume.

Reserve Ratios–1976

Published ratios of oil reserves to production vary slightly depending upon sources of data, whether the data is for crude oil alone or for crude oil plus natural gas liquids, and if the reserves were taken at the beginning or at the end of the production year.

Estimates by the American Petroleum Institute of crude oil reserves as of December 31, 1976, and production during the year[6] indicate a reserves ratio of 10.95.

Reserve ratios for natural gas and for natural gas liquids based upon the reserves and production report of the committee on natural gas reserves of the American Gas Association amount to 11.05 and 9.15, respectively, as of December 31, 1976.[7]

The reserves ratio of crude oil plus natural gas liquids is not often used. It would have amounted to 10.59 as of December 31, 1976. The reserve ratios for crude oil and natural gas were 13.0 and 33.0, respectively, as of December 31, 1946.

Estimates of Domestic Productive Capacity

Estimates of domestic crude oil production capacity will vary with the estimated volume of recoverable reserves and the ratio to production used.

The National Petroleum Council estimated domestic productive capacity of crude oil to be 11.59×10^6 bbl per day as of January 1, 1964.[8] The reserve ratio would have been 7.3 for crude oil and 8.6 for all hydrocarbon liquids using current reserve estimates of the American Petroleum Institute. Reserve ratios of 14 and 7.75 were indicated for natural gas using the Council's and the Institute's reserve data, respectively. The Department of Interior estimated the domestic productive capacity of all petroleum liquids to be 11.5×10^6 bbl per day, also as of January 1, 1964.[9] The reserve ratio would have been 8.2 using the same reserve estimate of the American Petroleum Institute. The Department made it clear that such a high level of production was contingent upon "continued strong emphasis upon both primary and secondary recovery effort," and that it could be maintained only six to twelve months.

Neither the domestic productive capacity for crude oil nor for natural gas are absolute. Both are matters of scientific and/or economic judgment. The industry has tried since the mid-30's not to exceed the MER (maximum efficient rate of production) for any field. Crude oil production has been close to the MER since 1972 which means that domestic productive capacity for 1976 was essentially 2.825×10^9 bbl, or 7.74×10^6 bbl per day. Domestic productive capacity for crude oil has been limited since 1970 to actual production, largely on account of a lack of sufficient transportation facilities. Full utilization of the Alaska pipe line after 1977 will narrow the difference between the MER and actual production for a time.

The Nation's productive capacity for natural gas has been slightly higher than actual production for several years. Necessary time lags in providing production and transportation facilities have been a factor. Federally regulated prices for interstate gas have not compensated production and delivery costs sufficiently. Thus from an economic viewpoint, and one that is practical, the productive capacity for natural gas in 1977 was 19.54×10^{12} cf.

Volumes of Future Reserves Needed

Estimates of the volumes of future reserves needed to meet the various estimates of demand depend to a great extent on an acceptable ratio of reserves to yearly production. Production rates with respect to reserves depend on well completion practices and spacing of wells primarily, exclusive of any production prorationing.

Before World War II oil wells usually were spaced on relatively small drilling units which seldom exceeded 10 surface acres. Since then industry has concluded that thousands of unnecessary production wells were drilled every year. Material shortages during World War II and ever increasing productive depths, however, forced industry to go to wider spacing. Spacing has gone to essentially 40 reservoir-acres per well through much of the midcontinent, Rocky Mountain, and Texas areas, exclusive of the Gulf Coast and North Texas. Permissible multiple completions have helped to reduce well spacing per surface acre. Coastal salt domes fields are still closely spaced on account of structural complexities, but offshore salt dome wells must be more widely spaced for economic reasons.

Advances in well-logging techniques and interpretations, selective jet-perforating, selective acidizing, selective fracturing, and particularly advances in the science of reservoir mechanics have resulted in recoveries of ever higher percentages of the original oil in place, whether primary or secondary. Industry on the whole has become convinced that closely spaced wells are not needed for efficient reservoir drainage in most reservoirs—in fact many times very closely spaced wells reduce recovery efficiency—and that the productive capacities of wells with respect to both rate and ultimate recovery have been greatly increased. Also, many smaller independent producers have been short of undrilled production well locations and money to develop deep reservoirs, and have been unable to operate profitably under some stringent state proration practices.

The Reserves to Production Ratio. The question as to when the nation's productive capacity is no longer adequate depends on a large number of factors of which the ratios of reserves to production and imports to production and any state of national emergency are the most important.

During the early months of World War II the Production Division of the Petroleum Administration for War (PAW) adopted a maximum efficient rate (MER) of annual crude oil production equal to one-twelfth of the proven reserves at the beginning of the year. Petroleum engineers had found that neither the maximum nor the minimum rates at which a reservoir might be produced was necessarily the most efficient. Ideally the MER for reservoirs producing under primary recovery methods would be that rate which utilized the available reservoir energy most efficiently in terms of ultimate recovery. The MER of a dissolved gas drive field could be determined readily by field tests to find rates at which gas-oil ratios were lowest. It is more difficult to determine for reservoirs producing under conditions of pressure maintenance, water drive, and secondary and tertiary recovery methods. The MER rate may not be the most profitable rate despite the best recovery factor, but it would be the ideal rate from a conservation viewpoint.

All estimates of domestic productive capacity are based on various compilations of the estimated MERs of the nation's oil fields. There is no certainty as to the MER of any widely spaced well being higher than that for an identically completed closely spaced well in the same reservoir. A well on 40-acre spacing might recover under favorable conditions three to four times as much oil as a well completed on 10-acre spacing in the same reservoir. But from a theoretical viewpoint the rate of production for the 40-acre well cannot exceed that of the 10-acre well for the same pressure drawdown unless the effective radius of the 40-acre well is two or more times that of the 10-acre well.

Relative production rates can be improved in those areas where wells respond satisfactorily to heavy fracturing. A 40-acre well would need an effective radius by fracturing of 100 ft in order to have the same production rate as four 10-acre wells with 0.25-ft effective radii on the same drilling unit if all produced at the same pressure drawdown.

In view of the facts that a large percentage of this nation's proven reserves of crude oil are located in the Gulf Coast area and that a larger percentage of those wells cannot be fractured successfully, at least heavily, and that wells elsewhere tend to be widely spaced, estimates involving productive capacities greater than one tenth of the proven reserves in primary recovery reservoirs are to be questioned unless the higher rate of production is understood to be for a very short period of time only and not on any sustained basis.

Secondary and tertiary recovery projects vary widely in size, depths, well-spacing, and methods of operations. Reserves to production ratios should be kept low for economic reasons. The same is true with respect to very deep and offshore wells. Future national average ratios probably will not be appreciably higher than the 1976 ratio of 8, excluding the Alaskan reserves. Actually a sustained overall future ratio of 10 - 11 would be preferable from a national security viewpoint. The previously recommended national ratio of 12 now appears to be no longer necessary in view of the increased percentage of crude production coming from enhanced recovery projects.

Reserve Requirements in 1985

Estimates of reserve requirements in 1985 amount to little more than educated guesses. There are too many variables of unknown trends. Future domestic supply is primarily a function of prices. Prices are normally a function of demands. And demands can be met only by a continuation of large volumes of imports. Federal policies will influence total demand, domestic production, and imports. Presidential policies were proposed in 1972 and in 1977 that would reduce both future consumption and imports. On the assumption that increases in domestic demand after 1977 can be kept down to 2.5 per cent per year and that the 1977 percentage of imports should not be increased, domestic production in 1985 would amount to 3.53×10^9 bbl of crude oil and 0.875×10^9 bbl of natural gas liquids. A continued import limit of 40 per cent of total demand would amount to 2.35×10^9 bbl of crude oil and 0.58×10^9 bbl of natural gas liquids. Total supply in 1985 would be 5.88×10^9 bbl of crude oil and 1.46×10^9 bbl of natural gas liquids, or a total of 7.34×10^9 bbl of petroleum liquids.

A crude reserves to production ratio of 10 would indicate a need for 35.3×10^9 bbl of crude oil reserves in 1985, or an approximate increase of 5×10^9 bbl over crude reserves in 1976.

The volume of natural gas reserves needed in 1985 is more difficult to estimate than for crude oil. Despite efforts to reduce consumption, reserves in 1985 should at least equal those in 1976, or namely 220×10^{12} cf.

Exploration and Drilling

Volumes of proven reserves of crude oil and natural gas with its natural gas liquids are found and increased only by exploration and drilling. Exploration and drilling may be thought of as including geological, geophysical, and other field exploratory work, wildcat drilling, and production drilling. These activities accounting for the major share of all production expense naturally suffer during times of excessive productive capacity.

Exploratory geological work reached its peak in the late 1920's and has been superseded largely by geophysical work, particularly with the seismograph since World War II. Seismograph exploration had dropped by 1970 to less than one-third the 1952-53 rate. Activity in 1976 was approximately one-third higher than in 1970. Almost all the United States thought to be favorable for the accumulation of oil and gas has been explored one or more times, at least in a preliminary way. Seismic work on shore now consists largely in reworking older areas and checking and exploring at greater depths. Offshore activity during the mid-70's was largely concentrated in the Gulf of Mexico and the Atlantic Continental Shelf area.

Exploratory drilling also has declined since the mid-50's. The peak was reached in 1956 with 16,207 holes of which only 3,089 were productive. The rate then declined 55 per cent to a low of less than 6,000 holes in 1971. By 1976 exploratory drilling had increased by approximately 10 per cent over 1971.

Wildcat success ratios as of 1960 were quite low, especially in comparison with the ratios in 1976. The exploration experience of 1960 showed the following:[10]

1. Only one out of nine wildcat wells finds and produces any oil.
2. Only one out of 49 wildcat wells finds a field or reservoir that will produce as much as one million barrels of oil.
3. Only one out of 213 wildcat wells finds a field or reservoir that will produce as much as ten million barrels or more of oil.
4. One out every four development wells is dry.
5. The average producing well drilled during the late fifties would produce little more than 100,000 bbl during the first twenty years of its producing life.

A few of these wildcats cost up to one or more millions of dollars to drill and test. The great majority were relatively shallow and brought the overall average down to nearly $150,000 each. Costs of finding and developing each barrel of new reserves has always been very difficult to compute. However, at the time crude replacement costs in the Mid-continent area were thought be be from 10 to 15 per cent higher than posted prices.

Estimated total of approximately \$9.0 x 10^9 and \$19.0 x 10^9 were spent on exploration, development, and production of oil and gas in the years 1970 and 1974, respectively.[11] The estimated total had increased in 1976 another estimated 32 per cent and costs were expected to increase another 10 per cent in 1977.[12]

Inadequate allowable prices for interstate sales of natural gas have been a serious deterrent to wildcat drilling for natural gas. Consensus for years within the industry has been that proven reserves will continue to decline as long as the Federal interstate pricing controls remain in effect. Suggested control of intrastate sales prices would be an added deterrent, especially in Texas. Nearly one-half of all the 530 gas-productive wildcat wells drilled in 1976 were located in that State.[13] A total of 130 new-field wildcat wells were drilled in the United States during 1976 to depths of more than 15,000 ft. The finding of natural gas was the primary object.[14] Three new-field oil discoveries and 30 new-field gas discoveries resulted, of which 17 of the gas discoveries were in Texas at average depths of 18,121 ft.[15]

Need for Increased Drilling Activity

Exploratory and development drilling in the United States dropped nearly one-third between 1956 and 1964 for obvious reasons. Excess productive capacity, scientifically controlled state proration practices, and too-low wellhead prices for both crude oil and natural gas were major factors. Ever increasing drilling, developing, completion, and production costs in comparison with those of foreign operations were very important also. These conditions essentiallys continued until the taking over of foreign operations and the Arab Oil Embargo.

Probably none of the major industries have surpassed the oil industry in better meeting all product demands at very reasonable prices since the birth of the automobile industry. Yet, on the whole, neither Congress, the public press, nor the consuming public have ever given the industry proper credit for having benefitted the Nation through foreign operations. Imports over the years until the Arab Oil Embargo were produced from concessions essentially owned 100 per cent by major United States companies. The Nation benefitted also from the net incomes and taxes so generated. And had the major companies not developed all their foreign productive capacity, a more drastic domestic supply situation would have occurred years earlier than 1972. It is doubtful that any amount of domestic drilling could have met all demands for products in 1977.

Independent producers have always been very critical of "excessive" imports. Halbouty stated in 1965[16] in a widely accepted estimate that a minimum of 60,000 wells per year should be drilled during the next ten years if future demands were to be met for the next twenty years–and that at least 25,000 of those wells each year should be exploratory tests. Actually, only approximately 350,000 wells had been drilled during the 11-year period 1966-1976,[17] of which approximately 70,000 were wildcats.[18] The industry would have been hard pressed to have recommended 600,000 drilling locations during the period 1966-75. Added reserves per hole drilled during the period averaged approximately 92,000 bbl and 67,000 bbl, including and excluding Prudhoe Bay field, respectively. Reserves added per wildcat well were actually lower because total reserves added during the year included those due to field revisions and extensions. Added reserves per well in 1976, including field extensions and revisions, amounted to approximately 171,000 bbl. Continuation of that higher rate of success would indicate a need for an average of 18,520 wildcat wells per year for the period 1977-1985, or 22,530 wildcats in 1985, in order to maintain a reserve ratio of 10.

The "Independents" were said to have drilled 83 per cent of all well completions in 1976.[19] The 22 largest companies were not included as independents. Apparently the Federal government only looks at numbers of wells drilled instead of barrels of oil added to reserve in setting up its regulations. Changes in the depletion allowance, pricing policies for old and new oil and natural gas, and threats of divestiture are far more serious obstacles to the larger and integrated companies in meeting future demands than to the small independent.

Large volumes of both associated and non-associated domestic natural gas reserves will be added by future drilling. Estimates of natural gas awaiting discovery and exploitation range up to $1{,}100 \times 10^{12}$ cf. Most of such potential reserves will be uneconomic to develop and to produce except at multiples of allowable prices in 1977. More conservative estimates taking projected economics into consideration estimate potential future reserves in the range of $300 - 400 \times 10^{12}$ cf. Nevertheless, prospects of substantially increasing proven reserves through drilling to meet demands through 1985 are far from encouraging. Nor are prospects very encouraging with respect to maintaining present reserve ratios.

Completion of the proposed gas pipe line from the Prudhoe Bay field to the lower states will increase availability substantially for a time, but it will not improve the reserve ratio; they are already included in the reserve totals. Hundreds of exploratory and production wells will be drilled in 1985. The numbers depend in part upon research progress aimed at supplying both natural and syngas from new reserve sources. Any estimates of numbers of wells to be drilled could be very misleading. Individual wells and even individual fields, unless of so-called giant size, like with crude oil, only meet demands for the equivalents of minutes to a few days. Possibly sufficient gas in liquid state as well as pipe-line gas from Mexico, the Canadian Arctic[20] and Hudson Bay areas may be imported in the future to offset domestic shortages. Price in all cases would be an obstacle. Other sources under study to help meet future demands center largely around gasification of coals,[21] oil shales and tar sands, all as an adjunct to syncrude production.

The American Gas Association estimated that the industry itself would spend \$90 x 10^9 on supply research by 1985. Approximately \$1.2 x 10^9 would be for syngas from coal. Estimates of potential natural gas reserves from low permeability formations seem to offer the most promise for increasing supplies. Economic means of recovering gas from such formations is of major importance. Potential reserves of this class could amount to another 600 x 10^{12} cf. These formations lie chiefly in the Green River, Piceance, and Uinta basins of the Rocky Mountains and in the Devonian shales of Appalachian Illinois, Indiana, Kentucky, and Michigan.[22-25] Recovery would probably have to be through drilling of hundreds of wells subjected to nuclear stimulation or massive hydraulic fracturing beyond that practiced in the mid-70's. Another source of natural gas which might be added to the "gas resource base" category of reserves is that occurring in deep-seated brines under abnormally high pressures in the Gulf Coast area. Estimates of 50-100 cf of dissolved and/or associated gas per barrel of brine indicate staggering volumes of possible reserves. The economics of its recovery and treatment for sale appeared unfavorable as of the mid-70's.

Heavy Oil, Oil Shales, and Tar Sands Development

There is a potential supply of at least one trillion barrels of known oil reserves awaiting economic means of development and production in so-called heavy oil sands and tar sands and that which may be distilled from oil shales. Crude has been produced from the tar sands in California for more than 75 years. Earliest wells were actually hand-dug to depths of approximately 100 ft and less. As of the mid-70's it was being produced commercially from conventional wells. It has been known that crude oil could be distilled from oil shales since World War I, but commercial production can hardly be called a reality as of 1977.

Heavy, Viscous Crudes. The U. S. Bureau of mines has estimated that approximately 150 billion bbl of low-gravity, high viscous crude oils occurred in sands at relatively shallow depths largely in California, North Texas, Missouri, Kentucky and eastern Utah.[26] More than 10 x 10^9 barrels have been recovered and an additional 30 to 60 billion bbl may now be economically recoverable through use of thermal, solvent, and/or other viscosity-reducing methods.

Thermal methods were developed primarily in California for use in shallow reservoirs containing asphaltic-base oils of very low gravities and high viscosities. Most of the early testing was done in wells that had essentially reached economic limits following what was mostly a slow gravity-drive recovery and some air-gas drive projects. Early applications of heat were largely considered to be an additional recovery method, namely "tertiary recovery." Early attempts to increase recoveries through injection of hot air and hot natural gas were hardly successful. More oil was recovered by injecting hot water. Later, use of steam injection and especially of superheated steam injection have proven to be the most satisfactory.

One of the earliest successful steam drives was a 9-spot pilot test of 5.6 acres in Missouri. Approximately 7,000 bbl of otherwise unrecoverable 1,000 cp, 18° API gravity crude was produced.[27] Yoelin[28] has described another of the early steam stimulation projects which was located in the Huntington Beach Offshore field, California. Only two per cent of the original oil in place had been produced. Estimated added recovery by steam injection amounting to 5 x 10^6 bbl from the 150-acre reservoir, or 700 bbl per acre-foot of sand was indicated. The semi-depleted Brea Field of California also was particularly suitable for the application of thermal energy through steam injection.[29] Remarkable success had been obtained through steam-displacement in the Kern River field, California.[30, 31] From 50 to 70 per cent of the original oil in place has been recovered from wells mostly drilled before 1915. An estimated 74 per cent of the 55,000 bbl/d production rate in 1973 was attributable to steam displacement.

Variable success has been obtained with in-situ combustion projects. It was widely tried during the 1950's-1960's. Research and field experience with the method was analyzed by Dietz in 1970.[32] One of the first real successes was in the Midway Sunset field, California[33] where only 160,000 bbl of crude had been recovered from the 150-acre field, previously. In-situ combustion had recovered 9 x 10^6 bbl of crude oil or 24 per cent of the initial oil in place after ten years of operation. The method was also found successful in recovering high gravity oils. Hardy et al[34] described its research application in the May-Libby reservoir of the Delhi field, Louisiana, where as much as 62.8 and 82.7 per cent of the 40° API gravity oil in place was recovered from the two wells in an 231 acre-ft project.

Tar Sands. The crude oil in oil-impregnated sands may be so viscous as to resemble cold tar. Extensive domestic deposits of these so-called tar sands occur in Kentucky and eastern Utah. The Athabaska tar sands located in northern Alberta Province, Canada, are the largest and best known deposit in North America where approximately 400 billion bbl of extractable oil are known to be in place of which probably 60 to 100 billion bbl could be extracted economically at present crude oil prices. The Alberta government at one time was reluctant to permit development of these deposits with the thought that they might discourage regular oil field development and exploitation indefinitely. However, a 30,000 bbl/day plant was planned to be in operation during the late 1960's. Operations since the beginning have not been completely successful. Increases in crude prices since the Arab Oil Embargo have greatly increased the incentive to develop large scale production operations. As of mid-1977 a $69 million in situ combustion research project was to be initiated which included facilities to produce one million barrels of oil by 1984.

Oil Shales. The potential supplies of oil from oil shales staggers the imagination, not only within the United States but also on a worldwide basis.[35] This potential is not in the form of a crude oil in shale; it is an organic sedimentary material called *kerogen*. It is thought to be derived from algae, minute aquatic organisms, spores, and pollens which were deposited during Eocene times in a lacrustine environment along with very fine-grained materials which have been compacted into laminated shale-like sediments. The largest and richest deposits in the United States are in northwestern Colorado, southwestern Wyoming, and northeastern Utah. The oil can be extracted from the shale only by destructive distillation of the kerogen at temperatures between 800 and 900°F. There are some 1600 sq mi of these shales in northwestern Colorado which are particularly thick and rich in kerogen. Estimates run as high as one trillion bbl of retortable oil reserves in this one area of which approximately 400 billion bbl are in the richer sections. Retortable oil content varies from less than 15 gal per ton of shale to over 30 gal. Thus on an acre-foot basis, the retortable oil content is considerably higher than the recoverable oil content of an oil reservoir sand. Not all of the shale will be economically retortable and about one third must be left behind in mining.

The U. S. Bureau of Mines worked diligently on the "shale oil problem" from 1944 to 1955 near Rifle, Colorado, and developed advanced mining and retorting techniques suitable to the area. The Union Oil Company took up the problem in the late 1950's, extracted over 50,000 bbl of oil in a retort of its own advanced design, test-marketed products refined from it, and reluctantly concluded that shale oils were noncompetitive with crude oil at the time. The Colorado School of Mines took over the Bureau's plant in 1964 determined to arrive at exact current costs in producing shale oil. Many of the major oil companies hold shale oil properties in the area and two of the companies financed the School's research. Other oil companies are working on the problems independently, both from a mining-retorting basis and an in situ combustion basis.

Interest in the development of oil shales has waxed and waned since 1918. Unprecedented interest developed during the 1960's. It was so intense that the concept of an immediate wide-scale industry needing thousands of workers and hundreds of millions of dollars investment was widely predicted for the rather sparsely populated area. Private industry would have commercial plants in full scale operation by 1970.

One plant was in operation but not on a commercial basis by 1977. And only one quite small plant was really capable of commercial operation. Many factors have worked against the budding shale oil industry. Prior to the Arab Oil Embargo imports were a very important deterrent. An over zealous press was also blamed because of its wide-spread and continuous publicity. Criticism of the oil industry and the Federal Government grew on the parts senators, and state and local citizens. The "give-a-way" by the Department of Interior, water shortages in the area, and very unacceptable environmental conditions are still common criticisms. Room and pillar methods of mining and surface retorting with its large volumes of spent shale would have created tremendous disposal problems. There would have been approximately 1.3 - 1.5 cubic yards of spent shale per barrel of production in retorting the richer zones. Strip mining presented the same drawbacks. These problems of waste disposal, air pollution, water use, and of protecting and restoring the environment have led to a gradual shift in plans toward underground retorting despite its many drawbacks too.

Some authorities were convinced as early as 1958 that shale oils could be produced at that time for considerably less than it costs industry to find and produce crude oil.[36, 37] Others by 1965 were equally certain without having actually mined or retorted any.[38] The one company that has been ready to operate on a relatively small scale for serveral years now estimates that shale oil would have to sell for $20 per barrel to be commercial at production costs in 1977. Thus inflation has become an added important deterrent to a commercial shale oil industry. Those oil companies who took up large expensive leases during the earlier 70's anticipated only small commercial plant operations by 1985. Creation of underground retorts filled with broken shale resulting from the blast may be the answer.

Price of Crude Oil and Natural Gas

Assumed future market prices for crude oil and natural gas naturally have a very important bearing on the valuation of a property. Predictions have been off considerably in the past. New discoveries were common factors. Imported oils became another. Independent producers have always thought prices were too low and usually blamed the majors. For years the majors were accused of manipulating prices to their ultimate benefit. The independents insisted upon protective import duties to keep prices up as the majors seemed to concentrate upon expanding foreign operations and imported more and more crude. The independents got some but relatively little governmental support. On the whole the consuming public certainly benefitted. The price of gasoline at the refineries scarsely increased more than 15 per cent between the years 1920 and 1970. Quality greatly increased. Total sales taxes on gasoline nearly equalled and sometimes exceeded the refinery price during the latter part of the period.

Development of vast foreign oil productive capacity and reserves certainly tended to keep domestic crude prices down to some extent. Offsetting that, domestic reserves that would have been produced and perhaps exported were saved. Those reserves so-saved are far more valuable to the Nation today. Many now looking back realize that the "free enterprise system" really worked too well for the good of the public and the Nation, and to a certain extent for the companies also. Nevertheless, once one company found it advantageous to go into foreign operations, all that were able were economically forced to follow. Anything resembling an Arab Oil Embargo was literally inconceivable. Had those companies not gone into foreign operations, the Nation would have found itself with real shortages and higher prices years sooner.

Critics of the prices of crude oil, its products, and natural gas could have given neither comprehensive nor unbiased thought to the economics of the petroluem industry. As pointed out in Chapter 7, insinuations through use of the word "profits" in reporting net incomes of major oil companies, has had what appears to be a public desire to discredit the industry. Public opinion has been so influenced that passage of such measures as the Tax Reform Axt of 1969, elimination of the depletion allowance for all larger companies, and price fixing of crude oils were all past due.

Prices paid producers for natural gas entering interstate sales are controlled by the Federal Power Commission. The consuming public is being protected against profiteering by the producers. Small volumes of natural gas have been imported from Canada and Mexico. There are hopes for larger import volumes in the future. Until quite recently it seemed that the Commission still looked at gas produced from oil wells as a by-product for which little more than handling costs were permitted. Drillers and producers of gas wells have always been very unhappy with the Commission's policies. The severe gas shortages of the winter 1976-1977 were blamed upon the producers. Accusations of gas being held back awaiting higher prices was essentially unfounded. Accumulations found and not being produced had to be of less than sufficient quantity to justify making them available to the market at prevailing prices. Thus local markets in Texas might be paying three of four times allowable interstate rates. As of mid-1977 the Federal Power Commission had hopes of controlling prices of those intrastate sales also. The consuming public on interstate lines after more than 20 years of low gas prices is essentially 100 per cent behind the Commission. As a result of too-low allowable prices some onshore gas discoveries and distant offshore discoveries have been looked upon as liabilities instead of as assets. However, prices paid the producer have been allowed to gradually increase over the past 20 years, in part as a result of inflation, but not enough to encourage sufficient exploratory drilling. Congress in 1977 neither saw fit to eliminate price controls of natural gas nor to increase prices sufficiently. Two earlier Presidents had vetoed decontrol measures.

Future prices for crude oils must be sufficient to encourage the finding of new oil as well as to develop to the maximum sources previously discussed. Secondary recovery production (tertiary probably included) as of 1975 amounted to over 47 per cent of total production. The percentage of "enhanced oil recovery," defined as that produced by pressure maintenance, secondary, thermal, and tertiary recovery operations, will increase to over 60 per cent by 1980 according to a study by Carrales and Funk.[39] Allowable crude oil prices as of mid-1977 were intended to provide an attractive price inducement for increasing production by increased drilling of new wells and through enhanced recovery operations. Even so, Carrales and Funk estimated total production in 1980 to be only 67 x 10^6 bbl more than the 2,927 x 10^6 bbl produced in 1975.

Increases in prices of Arabian light crude posted as of January 1973, f.o.b. Ras Tanura in U. S. dollars, amounted to only $0.79 per bbl above a steady price of $1.80 per bbl over the period 1965-70. Arabian light crude posted prices were essentially doubled as of October 1973 and more than doubled again to $11.65 per bbl as of January 1974. Export prices for crude oils were quickly adjusted upward comparable amounts by all petroleum exporting nations, now referred to as the "OPEC," the Organization of Petroleum Exporting Countries.

Weighted OPEC crude oil prices for 1977 were expected to average $13.57, $12.72, and $12.27 for light, medium, and heavy grades, respectively. There were variations in price ranging from -$0.68 to +$1.93 as of July 1977 depending upon differences in transportation costs, gravities, and quantities and qualities desired by the refiners.

Allowable domestic prices set by the Federal Energy Authority (FEA) in January 1976 to become effective next February 1 were $5.25 per bbl for old crude oil and $11.28 per bbl for new and stripper crude. The FEA was attempting to establish a weighted average price of $7.66 per bbl for all domestic crude production as ordered by Congressional action to become effective February 1, 1976. Old crude oil was to be that average amount produced and sold from a property during 1975. Its price was that prevailing in the field on May 15, 1973 plus $1.35, which on a national basis averaged

\$5.25 per bbl. Maximum yearly adjustments of 7 per cent for inflation and 3 per cent as a production inducement were provided. Amendments of August 15, 1976 redefined stripper crude and its allowable price and substituted a 10 per cent escalation factor for the composite price of crude in place of the previous inflation and inducement adjustments. The effective date was to be September 1, 1976. Stripper crude was defined as that produced and sold from a property whose maximum daily average per well did not exceed 10 bbl during any consecutive 12-month period after December 31, 1972.[41] Its price was allowed to advance effective September 1, 1976, under new provisions of the Energy Conservation and Production Act (ECPA) to meet the delivered cost of imported crude of like quality.

Price controls as of January 1, 1976 were based upon a calculation that 40 per cent of the domestic production was upper-tier crude and 60 per cent was lower tier. It was realized late during the year that as old crude production declined and upper-tier production increased, the average price paid would exceed the allowable. That was happening. As a result the price of upper-tier crude was reduced \$0.20 per bbl effective January 1, 1977, and reduced again in March 1977 another \$0.45 per bbl. Merklein[42] has pointed out that these price reductions were contrary to avowed Federal policies aimed at increasing production. The FEA had also been criticized for not providing sufficient inducements for increasing production and recoveries from secondary and tertiary recovery operations. Both are expensive to initiate and to operate.[43]

Field prices paid for natural gas sold in interstate commerce have been under control of the Federal Power Commission (FPC) since 1953 as per a decision of the U. S. Supreme Court. Allowable prices paid in comparison with liquid fuels have always been much too low on a BTU basis. Initial rates were set at a time when many producers were happy to receive most any small payment for what the FPC seemed to regard as a valueless by-product of crude oil producing operations. However, prices were gradually increased over the years after lengthy hearings and delays. A brief review of the more recent pricing rulings follows:

The FPC set a national rate on December 31, 1975 of \$0.235 per mcf for old natural gas, defined as that which had been flowing in interstate commerce since prior to January 1, 1973. This allowable price was increased to \$0.295 per mcf on July 1, 1976. Certain small adjustments were permitted with respect to quality, gathering costs, and location of production. In addition, *small* producers, those whose annual production did not exceed 10×10^{12} scf per year, were permitted to receive \$0.35 per mcf for 1973-74 vintage gas sold interstate. The standard pressure base for measuring all gas was changed from 14.65 psia to 14.73 psia.

The FPC first set a national base rate of \$0.42 per mcf for *new* natural gas, from wells commenced during the 1973-74 period, and being sold interstate. On December 4, 1974 the rate was increased to $0.50 per mcf. A 1-cent escalation was to be allowed on each following January 1. This rate was later increased to $1.01 per mcf for gas from wells commenced during the 1973-74 period and then decreased to $0.93 per mcf as of November 1976. A 1-cent per year escalation was to be added beginning January 1, 1977. The allowable rate for post-January 1, 1975, gas was set at $1.42 per mcf with a 1-cent per quarter escalation.

The petroleum industry appears to have become one of the most useful "whipping boys" of Congress from a political viewpoint. As a result, adequate prices to really encourage sufficient exploration for and development of necessary added productive capacity of both crude oil and natural gas, probably will never be granted. Present attitudes and policies can eventually force the petroleum industry into the same position as the present-day railroads. That would literally prove disastrous for the nation. Assuming that inflation adjustments of 5 per cent per annum are allowed, and that the price of stripper oil after 1977 is permitted to rise at the same rate, allowable prices for it would be approximately $17.00 per bbl and $22.75 per bbl in 1980 and 1985, respectively. Probably 80 per cent of the nation's crude production will be stripper oil by 1985. Present pricing policies can only encourage escalation of foreign oil prices.

The going prices for oil and gas in the vicinity of the property on the specified date of the valuation are usually used in evaluating oil and gas properties. In view of inflationary trends since World War II it would seem logical to add an inflation factor to the market value of oil and gas production in calculating future income. This could be misleading unless inflation factors are included in all costs. There are so many possibilities of wrong assumptions that the valuation engineer should eliminate all that he can. By using the going price of crude oil at the time of the valuation to calculate future income he is hoping that any increases in crude oil prices will offset the more likely increases in costs. The same is true with respect to natural gas property valuations. However, use of modern computers permits the many variations if so desired.

Values set on properties and prices paid are usually based on relatively early payouts which do not allow much time for either price or cost increases. The most satisfactory way to try to be on the safe side is to provide for a slow payout by using a slightly larger discount factor which takes care of a lot of overestimations with respect to future income. Inasmuch as valuation studies intended for loan purposes usually cover relatively short periods, adjustments of future costs and prices are rarely considered.

REFERENCES

1. Earl Oliver, "Appraisal of Oil Properties," *Trans. AIME*, Vol. 65, 1921, pp. 353-364.
2. *News Release*, American Petroleum Institute, Washington, D.C., April 1977.
3. *AGA NEWS*, American Gas Association, Arlington, Va., April 1977.
4. *Ibid.*, Am. Gas Assoc.
5. "U. S. Natural Gas Supply and Production: 1956-1975," *The Oil Producing Industry in Your State*, Independent Petroleum Association of America, 1976 ed., p. 101.
6. *News Release, op. cit.*
7. *AGA NEWS, op. cit.*
8. *Proved Discoveries and Productive Capacity of Crude Oil, Natural Gas, and Natural Gas Liquids in the United States*, March 25, 1965, The National Petroleum Council, Washington, D. C.
9. *An Appraisal of the Petroleum Industry of the United States*, U. S. Department of Interior, Washington, D. C., January 1965, p. 6.
10. Douglas H. Powell, "Evaluation of Oil Properties in After-Tax Dollars," *J. Petr. Tech.*, January 1960, pp. 28-36.
11. "Drilling Costs to Rise Moderately in 1977," *World Oil*, February 15, 1977, pp. 86-87.
12. *Ibid.*, p. 87.
13. "More Wildcats Forecast," *World Oil*, February 15, 1977, pp. 77-79.
14. *Ibid.*, p. 79.
15. *Ibid.*, p. 79.
16. Michel T. Halbouty, "Can We Meet U. S. Demands of the Next Twenty Years?," *Drilling*, February 1965, pp. 39-42.
17. "U. S. Drilling: Another Good Year," *World Oil*, February 15, 1977, p. 74.
18. "More Wildcats Forecast," *op. cit.*, p. 79.
19. "Independents drilled 83% of '76 Completions," *World Oil*, February 15, 1977, p. 90.
20. Robert E. King, "Canadian Arctic Promises Future Gas and Oil Supply," *World Oil*, December 1975, pp. 53-59.
21. P. Raimondi, P. L. Terwilliger, and L. A. Wilson, "A Field Test of Underground Combustion of Coal," *J. Petr. Tech.*, January 1975, pp. 35-41.
22. R. F. Lemon and H. J. Patel, "The Effect of Nuclear Stimulation on Formation Permeability and Gas Recovery at Project Gasbuggy," *J. Petr. Tech.*, October 1972, pp. 1199-1206.
23. Charles R. Boardman, Gregory W. Hammack, Walter F. Fertl, and Charles A. Atkinson, "Evaluation of Low Permeability Gas-Bearing Formations in Rio Blanco County, Colorado," *J. Petr. Tech.*, October 1973, p. 1125.
24. Philip L. Randolph, "Massive Stimulation Effort May Double U. S. Gas Supply," *World Oil*, August 1974, pp. 33-36.
25. John C. Foster, "A 'New' Gas Supply–Devonian Shales," *Oil and Gas J.*, January 3, 1977, pp. 100-101.
26. W. D. Dietzman, M. Carrales, Jr., and C. J. Jirik, "Heavy Crude Oil Reservoirs in the United States: A Survey," *USBM Inf. Circ.* 8263, 1965, p. 3.
27. V. V. Valleroy, B. T. Willman, J. B. Campbell, and L. W. Powers, "Deerfield Pilot Test of Oil Recovery by Steam Drive," *J. Petr. Tech.*, July 1967, pp. 956-964.

28. Sherwin D. Yoelin, "The TM Sand Steam Stimulation Project," *J. Petr. Tech.*, August 1971, pp. 987-994.
29. C. W. Volek and J. A. Pryor, "Steam Distillation Drive–Brea Field, California," *J. Petr. Tech.*, August 1972, pp. 899-906.
30. C. G. Bursell and G. M. Pittman, "Performance of Steam Displacement in the Kern River Field," *Trans. AIME*, Vol. 259, 1975, Pt. I, pp. 997-1004.
31. T. R. Blevins and R. H. Billingsley, "The Ten Pattern Steam Flood: Kern River Field, California," *Trans. AIME*, Vol. 259, 1975, Pt. I, pp. 1505-1514.
32. D. N. Dietz, "Wet Underground Combustion, State of the Art," *Trans. AIME*, Vol. 249, 1970, Pt. I, pp. 605-617.
33. C. F. Gates and I. Sklar, "Combustion as a Primary Recovery Process–Midway Sunset Field," *Trans. AIME*, Vol. 251, 1971, Pt. I, pp. 981-986.
34. W. C. Hardy, P. B. Fletcher, J. C. Shepard, E. W. Dittman, and D. W. Zadow, "In-Situ Combustion in a Thin Reservoir Containing High-Gravity Oil," *Trans. AIME*, Vol. 253, 1972, Pt, I, pp. 199-208.
35. H. M. Thorne, K. E. Stanfield, G. U. Dinneen, and W. I. R. Murphy, "Oil-Shale Technology: A Review," *USBM Inf. Circ.* 8216, 1964, p. 4.
36. Ernest P. Miller, "Shale Oil Nears Competitive Level with Domestic Production," *J. Petr. Tech.*, August 1958, pp. 25-27.
37. "Shale Oil Is Commercial Today," *Petr. Week*, July 18, 1958, pp. 72-76.
38. Henry Steele, "The Prospects for the Development of a Shale Oil Industry," *Western Economics J.*, Fall Issue, 1963, pp. 60-65.
39. M. Carrales, Jr., and Velton T. Funk, "Liquid Hydrocarbon Production in the United States, 1946-75 and 1980 Projected, Highlighting Enhanced Recovery," *USBM Information Circular 8734*, 1977, pp. 6-8.
40. "FEA spells out new two-tier price regs," *Oil and Gas J.*, January 12, 1976, p. 43.
41. "What You Should Know about Stripper Wells and Pricing," *World Oil*, December 1976, pp. 52-54.
42. H. A. Merklein, "U. S. Oil Pricing policy: Incentive for Inefficiency," *World Oil*, July 1977, pp. 141-144.
43. H. A. Merklein, "U. S. Energy Policy: It's Primarily Based on Politics," *World Oil*, June 1977, pp. 53-57.
44. "FPC Issues Opinion on Nationwide Ceiling Rates for New Interstate Gas Sales," *News Release*, Federal Power Commission, Washington, D. C., November 5, 1976.

Additional References

Lawrence K. Barker, "Producing SNG by Hydrogasifying in Situ Crude Oil Shale," *USBM, R. I. 8011*, 1975, 37 pp.

W. B. Cashion, "Geology and Fuel Resources of the Green River Formation Southwestern Uinta Basin Utah and Colorado," *USGS, P. P. 548*, 1967, 48 pp.

"The Oil Crisis: In Perspective," (18 authors), *Daedalus*, Vol. 104, No. 4, Fall Issue 1975, 293 pp.

Hollis M. Dole, "The Colony Story," *The Mines Magazine*, (Colorado School of Mines), October 1975, pp. 6-15.

J. Hagoort,, A. Leijnse, and F. van Poelgeest, "Steam-Strip Drive: A Potential Tertiary Recovery Process," *J. Petr. Tech.*, December 1976, pp. 1409-1418.

A. E. Harak, L. Dockter, A. Long, and H. W. Sohns, "Oil Shale Retorting in a 150–Ton Batch–Type Pilot Plant," *USBM, R. I. 7995*, 31 pp.

D. R. Knop and J. F. Roorda, "Economic Restraints on U. S. Energy Supply and Demand," *J. Petr. Tech.*, July 1975, p. 803.

H. J. Lechtenberg, G. L. Gates, W. H. Caraway, and O. C. Baptist, "Field Study of Viscous Oil Production by Solvent Stimulation, Wilmington, California," USBM, Technical Progress Report 62, November 1972.

D. B. Lombard and H. C. Carpenter, "Recovering Oil by Retorting a Nuclear Chimney in Oil Shale," *J. Petr. Tech.*, June 1967, pp. 727-734.

J. S. Miller, C. J. Walker and J. L. Eakin, "Fracturing Oil Shales with Explosives for In Situ Oil Recovery," *USBM, R. I. 7874*, 1974, 100 pp.

H. Niko and P. J. P. M. Troost, "Experimental Investigation of Steam Soaking in a Depletion Type Reservoir," *Trans. AIME.*, Vol. 251, 1971, Pt. I, pp. 1006-1014.

David R. Parrish, Charles B. Pollock, and F. F. Craig, Jr., "Evaluation of COFCAW as a Tertiary Recovery Method Sloss Field, Nebraska," *Trans. AIME.*, Vol. 257, 1974, Pt. I, pp. 676-686.

E. B. Rogers, Jr., "Successful Well Stimulation Program Has Revitalized a California Oil Field," *J. Petr. Tech.*, December 1976, pp. 1420-1426.

J. R. Ruark, H. W. Sohns, and H. C. Carpenter, "Gas Combustion Retorting of Oil Shale Under Anvil Points Lease Agreement: Stage II, *USBM, R. I. 7540*, 1971, 74 pp.

H. E. Thomas, H. C. Carpenter, and T. E. Sterner, "Hydraulic Fracturing of Wyoming Green River Oil Shales: Field Experiments, Phase I, *USBM, R. I. 7596*, 1972, 18 pp.

H. M. Thorne, K. E. Stanfield, G. U. Dinneen, and W. I. R. Murphy, "Oil Shale Technology: A Review," *USBM, Information Circular 8216*, 1964, 24 pp.

15 FINDING THE PRESENT-WORTH VALUE OF AN OIL PROPERTY

The oil property valuation engineer must always keep in mind that the calculated present-worth value of a property is not necessarily the "value" of the property. As pointed out in Chapter 4 the word *value* has many meanings. Each participant in an oil property transaction will arrive at *his* value of the property. The value found might be thought of as an exchange value. The calculations leading to the present worth of a property must reflect those factors that were considered essential in the determination of a present-worth value which can be used as a basis in negotiations.

The first valuation of an oil property through use of the present-worth method is believed to have been made in 1911 by W. W. Orcutt of the Union Oil Company of California.[1] Five years later Hager began an article on valuations by asking—"How many oil men ever take the time to consider the earning capacity of an oil lease or to analyze production conditions?"[2] Requa in 1918 outlined the basic steps to be taken as proposed by the Appraisement Committee of the Independent Oil Producers Agency in California.[3] Later, Oliver[4] and Beal[5] discussed and amplified the steps to be taken in greater detail. The overall principles have not changed since that time; only the basic information has been improved.

Requa stated that the following facts were to be considered:[6]

1. That from each property there will ultimately be produced a certain total quantity of oil.
2. That in the production of that oil a certain total quantity of money will be expended.
3. That a certain total amount of money will be received from the oil.
4. That the total net receipts will be the total gross receipts less the cost of development and production.
5. That the present value of the net receipts must be such an amount that when invested by a purchaser it will be returned with 8 percent interest additional during the life of the property.

Many modifications are made to that general procedure depending upon the specific purpose of the valuation. However, it is still applicable in an overall way with respect to any producing property, large or small, oil or gas. Probably no company would use a discount factor as low as 8 per cent; 12 per cent would be a more likely minimum. More care has to be given today to all the numerous taxes and particularly to the application of the depletion allowance. Unless a small producer's operating costs are considerably more than those of a major company, the small producer might be able to offer more than the major in buying a small property.

AN EVALUATION PROCEDURE

One general evaluation procedure that closely follows Requa's steps may be illustrated by an example of calculations leading to the sale of the Douglass Rankin producing oil property with 40 pumping wells. The independent owner of the property was forced to sell it and was asking $1.50 per bbl of net remaining recoverable primary reserves, plus $300,000 for the lease equipment. Future recovery was estimated by years through use of the constant percentage decline curve method and amounted to 2,158,000 bbl, gross, at an economic production limit of 5 bbl per well per day. Accordingly, the asking cash price of the property would amount to $3,132,375. It is often cheaper for a producer to buy producing reserves rather than risk present-day costs of finding and developing them. It could also be to a purchaser's advantage to buy reserves because his depletion allowance would be on a fixed cost basis instead of on the declining percentage basis. The majors and all larger producing companies not entitled to a depletion allowance find it still more advantageous.

Purchase of the property was considered by a relatively small incorporated well-servicing company. Its management thought the asking price per bbl of reserves sounded reasonable; the Federal government certainly would have to increase allowable prices for old crude oil in time; and that prices of crude would have to increase faster than production costs. The company decided to make a present worth calculation using present allowable price of crude, namely $5.25 per bbl, throughout the productive life of the property and known operating and related costs experienced on an adjacent property. All the associated gas production was under a long-term contract with an interstate pipeline company at an allowable price of $0.25 per mcf.

Estimates of Gross Income

Royalty on both the oil and the gas was one-eighth on all marketed production. Table 15-1 shows the estimates of annual gross and net production of oil and gas, annual gross income, and a cumulative gross income of $10,463,645 (column 6). No proration of oil production was practiced in this state. Rates of oil production in those states practicing oil proration would have to be adjusted accordingly, although wells of this size would be exempt in most states. Future gas production was based on future oil production estimates at ratios then existing in the field and extrapolated on the basis of experience in the area.

Production Costs, State and Local Taxes

Production costs, all taxes exclusive of federal income tax, and calculation of net income before federal income tax are accounted for in Table 15-2. The lifting cost for the crude oil including overhead, power, operating costs and repairs was estimated as $0.50 per barrel for each of the first three years with an increase of $0.20 per barrel for each of the three following three-year periods, and $0.40 per barrel for each of the last two years. Lease-handling costs for the natural gas production was estimated at $0.02 per Mscf. Both the lifting cost for the oil and the handling cost for the gas are chargeable to the gross production of each. An item of $25,000 per year is included for all except the last two years to cover intangible well expenses, cleanouts, and pump and well repairs. This item could be either over or underestimated; it was based on experience with wells of the same approximate age and producing depth in nearby areas. The purchaser assumed payment of the severance tax of $0.01 and gathering tax of $0.005 per mcf on the gross production of the natural gas, and the $0.08, $0.10, and $0.04 per bbl conservation, severance, and gathering taxes, respectively, on the gross production of crude oil.

Net Income after Federal Income Tax

Tables 15-3 and 15-4 account for deductions for depreciation, depletion, federal income tax, and the calculated net income, by years and total.

A straight line depreciation rate of $20,000 per year was used leaving a salvage value of $20,000 upon completion of primary recovery operations. State income and property taxes were estimated to be on the average 15 per cent of gross income after depreciation. Assuming the property would be operated in 1977, allowable percentage depletion rates would be: 22 per cent for 1977-80, inclusive, 20, 18, and 16 per cent for 1981, 1982, and 1983, respectively; and 15 per cent thereafter.

The larger independents and all major oil companies have many expense items during the year for such major items as geological and geophysical

Table 15-1

Gross Income from Property

Year	Production			Gross income to operator		
	(1) Gross oil production bbl	(2) Net oil production bbl $(1) \cdot \frac{7}{8}$	(3) Net gas production after royalty, Mscf	(4) Crude oil (2) · ($5.25)	(5) Gas (3) · ($0.25)	(6) Total (4) + (5)
1	356,000	311,500	311,500	$1,635,375	$77,875	$1,713,250
2	284,000	248,500	248,500	1,304,625	62,125	1,366,750
3	226,000	197,750	197,750	1,038,188	49,438	1,087,625
4	194,000	169,750	203,700	891,188	50,925	942,113
5	170,000	148,750	178,500	780,938	44,625	825,563
6	150,000	131,250	157,500	689,063	39,375	728,437
7	134,000	117,250	164,150	615,563	41,038	656,600
8	120,000	105,000	147,000	551,250	36,750	588,000
9	108,000	94,500	132,300	496,125	33,075	529,200
10	98,000	85,750	111,475	450,188	27,868	478,056
11	90,000	78,750	102,375	413,437	25,594	439,031
12	82,000	71,750	93,275	376,687	23,319	400,006
13	76,000	66,500	79,800	349,125	19,950	369,075
14	70,000	61,250	73,500	321,562	18,375	339,937
Totals	2,158,000	1,888,250	2,201,325	$9,913,313	$550,332	$10,463,645

exploration, and for wildcat drilling, none of which is chargeable to any one production operation, but to overall operations. Major companies with foreign operations must pay income and other taxes to the host nation, all of which tend to reduce their overall U.S. income taxes. Thus a company may appear to have a very large income and yet have a very small federal income tax. For these reasons the proportionate tax on any one producing property may be considerably less than the maximum percentage. This does not always permit the major company to bid more for a small property because income tax savings may be offset by loss of the depletion allowance. Some clients request that the valuation stop with the estimate of net income before federal income tax rather than disclose their tax brackets. Others request that the maximum be used supposedly to give a minimum present worth. Companies that are very active in exploration and wildcat drilling may not pay more than 10 to 15 per cent of the net income in income taxes. Occasionally a company pays no income tax; it's all a matter of policy based on when profits are to be taken. An average tax rate for the industry is probably between 20 and 30 per cent. The asking price of $1.50 per bbl was used throughout for depletion purposes in arriving at the taxable income. Irrespective of any other oil production by a purchaser, the maximum percentage depletion could have been as low as $0.7875 per bbl and never higher than $1.155 per bbl. Table 15-4 (column 22) shows yearly and total net incomes after using 48 per cent federal income tax rate. The total net income would amount to $5,245,854 after adding back in the depreciation charges.

Present-Worth Value of the Property

The present-worth value of any property is not a definite dollar value to all parties concerned. It is merely a measure of the future net income from production operations discounted at an arbitrary rate which may vary from a low of 7 or 8 per cent to a high of perhaps 25 per cent. The rate depends on a number of factors which vary widely among companies and in this sense may be said to correspond to a desired rate of "earnings of undertaking" and depends to some extent on how one interprets the term "fair market value," if that is a part of the calculations. It is sometimes a fixed percentage if certain other approximations based on it are used to determine loan values. In addition differences in income tax brackets may influence it. Irrespective of the discount factor used, its importance is of significance only during the first few years of the estimate.

Column 24 (Table 4) shows yearly and cumulative present worth values based upon 15 per cent discount factors for oil property valuations as given in Table A-1 in the Appendix. A discount factor of 15 per cent after federal

Table 15-2

Production Costs, State and Local Taxes, Excluding Depreciation

Year	(7) Intangibles	(8) Lifting costs (1) · (see text)	(9) Gas handling costs $(3) \cdot \frac{8}{7} \cdot$ ($0.02)	(10) State and local taxes on crude (see text)	(11) State and local taxes on gas (see text)	(12) Total production costs and taxes (7) + (8) + (9) + (10) + (11)	(13) Gross income (6) – (12)
1	$25,000	$178,000	$7,120	$78,320	$5,340	$293,780	$1,419,470
2	25,000	142,000	5,680	62,480	4,260	239,420	1,127,330
3	25,000	113,000	4,520	49,720	3,390	195,630	891,995
4	25,000	135,800	4,656	42,680	3,492	211,628	730,485
5	25,000	119,000	4,080	37,400	3,060	188,540	637,023
6	25,000	105,000	3,600	33,000	2,700	169,300	559,137
7	25,000	120,600	3,752	29,480	2,814	181,646	474,954
8	25,000	108,000	3,360	26,400	2,520	165,280	422,720
9	25,000	97,200	3,024	23,760	2,268	151,252	377,948
10	25,000	107,800	2,548	21,560	1,911	158,819	319,237
11	25,000	99,000	2,340	19,800	1,755	147,895	291,136
12	25,000	90,200	2,132	18,040	1,599	136,971	263,035
13	00	114,000	1,824	16,720	1,368	133,912	235,163
14	00	105,000	1,680	15,400	1,260	123,340	216,597
Totals	$300,000	$1,634,600	$50,316	$474,760	$37,737	$2,497,413	$7,966,232

Table 15-3
Net Income before Federal Income Tax

Year	(14) Depreciation	(15) Gross income after depreciation (13) – (14)	(16) Property and state income taxes (15) · ($0.15)	(17) Net income before federal income tax (15) – (16)	(18) Percentage depletion (6) · (allowable)	(19) Cost depletion (2) · ($1.50/bbl)	(20) Taxable income 17 – (allowable)
1	$20,000	$1,399,470	$209,920	$1,189,550	$376,915	$467,250	$722,300
2	20,000	1,107,330	166,100	941,230	300,685	372,750	568,480
3	20,000	871,995	130,799	741,196	239,278	296,625	444,571
4	20,000	710,485	106,573	603,912	207,265	254,625	349,287
5	20,000	617,023	92,553	524,470	165,113	223,125	301,345
6	20,000	539,137	80,870	458,267	131,119	196,875	261,392
7	20,000	454,954	68,243	386,711	105,056	175,875	210,836
8	20,000	402,720	60,408	342,312	88,200	157,500	184,812
9	20,000	357,948	53,692	304,256	79,380	141,750	162,506
10	20,000	299,237	44,886	254,351	71,708	128,625	125,726
11	20,000	271,136	40,670	230,466	65,855	118,125	112,341
12	20,000	243,035	36,455	206,580	60,001	107,625	98,955
13	20,000	215,163	32,274	182,889	55,361	99,750	83,139
14	20,000	196,597	29,490	167,107	50,991	91,875	75,232
Totals	$280,000	$7,686,232	$1,152,935	$6,533,297	$1,996,927	$2,832,375	$3,700,922

income taxes may be justified as representative of a fair rate of return on overall investment in production operations. It is probably representative of approximate future rates for the next several years. Depreciation was added back into column 23 which is net income plus depreciation before discounting. Some valuation engineers calculate present worth directly from net income after federal income taxes (column 22) and then add depreciation back in without discounting it at any time to arrive at the cumulative payout.

The accumulative present worth of the property would amount to $2,927,985 (Table 4, column 25). This amount is $204,390 or 6.5 per cent less than the asking price. It is obvious that the lower the acceptable rate of return to the purchaser, the more that might be paid for a property. In this example the purchaser would make a little less than 15 per cent on his invest-- ment if no unaccounted-for expenditures arose. Again it should be pointed out that none of the cumulative payouts shown in Table 15-4 necessarily represents a fair market value; they are merely values from which negotiations toward purchase or sale can be initiated.

Ordinarily any dollar values in all valuation estimates are rounded off to thc nearest hundreds, thousands, and tens of thousands of dollars, depending on the size of the total amount. When it comes to the profits figure it is best to scale down rather than up.

A prospective purchaser may believe that upon completion of primary recovery operations a large additional amount of residual oil could be recovered through secondary recovery operations. Thus he might be willing to pay a premium price for the property. However, any added secondary recovery prospects years in the future would have to be discounted to a low present-worth value, even if prices for crude oil doubled in the meantime. No monetary consideration was included in the present-worth study of the Rankin property for that reason.

Approximations

Approximate methods of arriving at what is called the "fair market value" of an oil property may be useful as checks during negotiations or as probable limits to bankers' loans.

The Barrel per Day Method. The barrel per day method is the oldest and simplest method of evaluating an oil property, but it has been replaced largely by more sophisticated methods. It consisted of the setting of a dollar value per daily barrel of oil production. A large number of properties were traded in California, Illinois, Indiana, and the Appalachain area before 1945 at up to $1000 per daily barrel of production. Since World War II and during the

Table 15-4

Net and Accumulative Incomes

Year	(21) Federal income tax (20) · (rate)	(22) Net after income tax (17) – (21)	(23) Net income plus depreciation (22) + (14)	(24) Present worth	(25) Accumulative present worth (24)
1	$ 333,204	$ 856,346	$ 876,346	$ 815,207	$ 815,207
2	259,370	681,860	701,860	564,970	1,380,177
3	199,894	541,302	561,302	390,979	1,771,156
4	154,158	449,754	469,754	283,147	2,054,303
5	131,146	393,324	413,324	215,583	2,269,886
6	111,968	346,299	366,299	165,326	2,435,212
7	87,701	299,010	319,010	124,593	2,559,805
8	75,210	267,102	287,102	97,031	2,656,836
9	64,503	239,753	259,753	75,966	2,732,802
10	46,848	207,503	227,503	57,574	2,790,376
11	40,424	190,042	210,042	45,997	2,836,373
12	33,998	172,582	192,582	36,494	2,872,867
13	26,407	156,482	176,482	28,939	2,901,806
14	22,611	144,496	164,496	23,341	2,925,147
			20,000*	2,838	2,927,985
Totals	$1,587,443	$4,945,854	$5,245,854	$2,927,985	$2,927,985

*Salvage value of lease equipment.

early 1950's trades at $3000 per daily barrel were common. Reports of sales during the mid-70's were as high as $16,000 per daily bbl. Many seemed to fall within the range of $6,000 to $10,000 per daily bbl. Some of those trading at the higher prices were nearly depleted primary recovery properties having attractive secondary recovery prospects. Others were primary recovery properties with large volumes of associated and/or non-associated gas. The barrel value also depended to a great extent on the quality and price of the crude oil and the condition of the wells and surface equipment. An average depletion-drive property with wells not over ten to twenty years old, depending on condition, no particular water problem at the time or in prospect, with good secondary recovery prospects, and with oil selling at $5.25 per daily bbl would be considered in a marketable range at $3,000 to $6,000 per daily bbl of production. The Rankin property evaluated in Tables 15-1 to 15-4 which was first offered at $3,132,375 would have had a daily-barrel value of $3,750.

Approximations for Bankers' Loans. Bankers' Loans is a broad term. It is usually used to include all loans made by banks, but not limited thereto, when literally secured by future production over the loan period. One of the best known once used approximations for obtaining "fair market value" for loan purposes was stated in 1949 by DeGolyer as follows:[7]

> . . . I am familiar with cases in which the seller and the buyer have arrived at a trading figure which was approximately one-half of the future net revenue discounted at four per cent. It is rather surprising more often than not the latter method, i.e., one-half of a four percent discount future net revenue is very close to the future net revenue discounted at 10.5 per cent.

Terry and Hill used the following procedure after depreciation, depletion, and federal income tax at 52 per cent for determining the approximate value of a producing property for loan purposes:[8]

> . . . the future production, the gross revenue, and the costs of doing business are estimated year by year and the resulting annual net cash revenues are discounted, in this case at 8 per cent, compounded annually, to arrive at the total present worth of future net receipts, here considered to approximate the value of the property. . . . There is nothing sacred about the 8 per cent factor . . . it appears to the authors to result in a reasonable estimate of the fair market value for this type of property, as indicated by sales of similar properties.

Loan values vary from perhaps 50 per cent to a high of 70 per cent. Terry and Hill stated that a loan ratio of 60 to 70 per cent is ordinarily good practice and a reasonable percentage of the value of the property, depending on other factors that should be considered. The use of or the stating of any loan percentage factor is misleading unless the way of figuring present worth or "fair market value" is understood.

Bankers are usually more concerned with the net cash flow from a property before depreciation and income taxes. They have no desire to be forced into operations, even at an expected profit. They too discount net cash flow, but usually at only current loan interest rates, plus perhaps a small added factor, usually not exceeding 2 per cent.

Arps called attention to three other methods of finding the value of producing properties, namely:[9]

Gruy's Formula: "Two-thirds of future net cash earnings, before depreciation and federal taxes, discounted at 5 per cent."

Fagin's "Fair Market Value Yardstick": "Which empirically relates the fair market value as a percentage of undeferred future net income (before depreciation and federal taxes) to the average discount factor at 5 per cent interest, shown by the projection in the engineer's report."

BA-CANGOC formula: "Three-fourths of the future net cash earnings before depreciation and federal taxes, discounted at 6 per cent."

Banks usually limit loans to smaller operators to 50 per cent of the discounted cash flow for proved producing properties. There are many variations depending largely upon such factors as the reliability of the operator, the period of the loan, and the general background knowledge of the area. At times a loan may be limited so that at the time of the loan at least one-half of all the calculated future net income will remain when the loan becomes due.

At times the net income from a property has been computed and the rate of return on the investment is desired, either before or after federal income tax. Other times the interest rate at which the present worth would just equal the capital investment is wanted. Such problems are not difficult to solve with sufficient accuracy when the periodic incomes are at a nearly constant to constant rate such as might exist in some oil fields during prolonged proration and on natural gas properties where the contract calls for constant periodic "takes." Goheen and Chen,[10] Arps,[11] and Schoemaker[12] have published graphs and approximate methods which at one time were sufficiently accurate for many preliminary purposes. They could be reconstructed to accommodate today's higher crude prices if so desired. No one method is applicable to all conditions with any degree of accuracy, and all methods, exclusive of the time-consuming "cut and try method," fall short when large irregularities occur in the periodic net income.

Problems of this kind often arise in connection with a number of properties that are for sale at more or less fixed prices and the prospective purchaser wants to know which properties would be most profitable on a percentage-of-earnings basis rather than total dollars of earnings. Ratios of calculated present worths at bank loan rates to net incomes after taxes are sometimes used for such purposes of comparison, but these too can be very misleading.

None of the foregoing methods for determining values of oil properties mentions the use of two interest rates (discount factors), a higher one to apply to the unreturned capital invested and a lesser one to apply to the returned capital as originally proposed by Hoskold[13] and modified by Morkill.[14] Others also have advocated the principle. However, many within the industry have found the procedure may prove to be misleading with respect to oil property valuations because there are too many uncertainties. One slightly high discount factor accomplishes the same overall purpose satisfactorily.

A principle often used in making investment decisions makes use of multiple *cutoff rates*. Its purpose and usefulness have been discussed by Lyon.[15] He advised that both the cost of capital and the profitability of ventures should be calculated net of income taxes for this purpose. He defines cutoff rate as used in capital budgeting as the practice of establishing a threshhold rate of return which must be reached or exceeded to be acceptable. The prospective purchaser of the Rankin property was going through the first step. Lyon stated that while the principle of multiple cut-off rates is theoretically justifiable, really practical use is limited. All in the petroleum industry expanding into coal, uranium, refining, petrochemicals, etc., must resort to similar reasoning.

GAS AND GAS-CONDENSATE PROPERTIES

The principles of evaluating reservoirs and/or properties classified as natural gas, condensates, and gas condensates are basically the same as for a crude oil reservoir. The chief differences are in the amount of preliminary engineering study and calculation necessary to arrive at the best method of operation and sales. Each calculation involves some evaluation principles in order to arrive at the most profitable method of operation. Maximum efficiency of recovery may not be the most profitable, yet rates that give maximum profitability may not be permissible under state regulatory procedures.

Whether it is a matter of operating a property most efficiently or most profitably, for selling or for purchasing a property, the steps to be taken in an evaluation would consist principally of the following:

A. Calculations for each property or reservoir:

1. The recoverable reserves of gas and liquids.
2. The allowable rates of recovery under proration practices, gas sales contracts (interstate or intrastate), and market for liquids.
3. All operating expenses including taxes.
4. Net income.
5. Present worth of each year's production at an acceptable discount rate, cumulative payouts.

B. Adoption of the most profitable procedure consistent with regulatory and other requirements.

The procedure depends to a great extent on the price to be received for the natural gas and, if it enters an interstate pipeline, the degree of control by the Federal Power Commission. The problem becomes more difficult in cases of gas-condensate reservoirs where the timing and quantities of dry gas sales become a determining factor.

Goodson has published a detailed economic analysis of a multizone gas-condensate reservoir with the gas production sold under either interstate or intrastate contracts.[16]

Goodson's procedure is essentially the same as that for a crude oil property. His analysis example was made when prices for both natural gas and condensate were much lower than in the mid-70's, but his method is still perfectly valid. At the time of his analysis much intrastate gas was selling for less than interstate gas, contrary to prices in Texas as of the mid-70's. That condition could be changed should the FPC also gain control of intrastate gas prices.

SECONDARY RECOVERY PROPERTIES

The terms primary and secondary recovery have been defined and described in Chapter 10. Definitions of each have been broadened over the years until they overlap considerably. Smith[17] simply states that secondary recovery involves the introduction of artificial forces, or energy, into the reservoir system. Thus so-called tertiary and quaternary recovery processes are included as special applications of secondary recovery.

Procedures used in evaluating secondary recovery reservoirs and properties depend much upon the time and the purpose of the evaluation. For example, consideration of secondary recovery possibilities with respect to the Rankin property might have changed the relative values of the property to either seller or purchaser; that is, a property producing primary oil may have in addition a discounted secondary recovery value. A second type of study would be with respect to the profitability of applying secondary recovery to a field already depleted under primary recovery operations, and a third type of study would be one pertaining to a secondary recovery project already in successful operation. In many respects these kinds of evaluations would correspond to evaluations of primary recovery properties under wildcat, partially to fully developed conditions, and properties producing under settled conditions, respectively.

Fully developed and producing secondary recovery properties which have been on production for sufficient time to judge the degree of success are no more difficult to evaluate than similar properties under primary recovery operations. The same general evaluation procedures are used for both kinds of operations.[18] Partially developed producing properties also can be evaluated in a similar manner with the exception that much greater caution must be taken with respect to estimates of production and income from the undeveloped areas. Undeveloped areas surrounded by production usually are considered as proven, but they are not necessarily proven with respect to rates and prospective recovery.

Banks usually do not show any preference in granting primary recovery and secondary recovery loans as long as either may be considered a safe commercial investment. The same principles applicable to the return of the loan and interest within reasonable time are applicable. It is questionable whether banks and other lending institutions will supply all the future borrowed funds needed by the industry under present regulatory policies. Simpson[19] has suggested that the rising costs of enhanced recovery operations will require governmental policies that will enable the industry to attract investment and increase its cash-flow generating capacity.

Brown has enumerated the following principles which are applicable in judging the soundness of a proposed loan:[20]

> To qualify as a sound investment, the loan, together with the interest charge, should have an amortization period commensurate with the nature of the proposed operation. For a water flood, we believe this should not exceed five years. A second principle is that at least fifty per cent of the estimated reserves should remain after the indicated payout. In the case of a waterflood project, half the ultimate recovery is usually realized within the five years immediately following development. If the size of the project requires progressive development over a number of years, we would anticipate a series of loan applications as the operator's needs for funds occur. Thus, our security would grow with our investment, and the basic principles would continue to apply.

Dissolved gas-drive reservoirs showing no indications of a natural water drive are usually the best prospects for secondary recovery, especially for waterflooding. The Rankin property would be such a prospect. Either buyer or seller may have thought of it as a secondary recovery prospect upon reaching its primary economic limit. At the time of the sale a future secondary recovery value could have been included with the primary estimates. The value of the wells for secondary recovery purposes likely would have been based on a value of $3000 to $5000 per daily barrel of production with a leaning toward the higher figure. Thus the secondary recovery value of the Rankin property 14 years hence may have amounted to only

$$20 \times 5 \times \$5{,}000 \times 0.1228 = \$122{,}800$$

or approximately 4 per cent of its primary value. Thousands of acres have been traded using this simple method of evaluation for prospective secondary recovery properties throughout the United States. Since the early 1970's purchasers have preferred to take diamond cores, analyze them for oil and water content, estimate recoveries from flood-pot tests, and trade at $3.00 to $6.00 per barrel of estimated recoverable oil. Despite the engineering approach other uncertainties do not always justify the cost or effort. The many uncertainties of reservoir engineering calculations are often overlooked in proposed secondary recovery operations because of a lack of operating experience.

WILDCAT OR EXPLORATORY PROPERTIES

Many in government as well as in industry have been much concerned over the lack of sufficient wildcat drilling in the United States. Ratios of success are well known and have been studied to predict the number of wildcat wells that must be drilled annually to maintain an adequate proven reserve to production ratio.

Colonel Drake found oil on his first trial. He is said to have been lucky, yet today we know that he could not have failed in that particular spot short of getting a well down. He drilled where he knew there was oil. Many other very productive fields have been discovered domestically and foreign on the same basis. We have the old adage that the place to look for oil is where we know there is oil. This has certainly been true with respect to drilling around and on old productive salt domes. There are millions of acres of untested leased lands neither on nor near productive salt domes nor near other productive acreages. Many papers have been written explaining the laws of chance and showing the risks of finding oil with respect to the number of chances taken (wildcat wells drilled).[21–25] All take into account the laws of chance as though wildcat drilling were purely a game of chance. Arps approached the problem by what he called "the law of gambler's ruin."[26] He explained:

If an operator intended to drill 5 wildcat tests in succession, all of which are considered to have a probability of success of $\frac{1}{5}$ or 20%, the probabilities of various combinations of events can be computed with the well-known "binomial distribution function."

Table 15-5 shows his calculated probability of such occurrence. The 1:5 ratio used in Table 15-5 is applicable to much of the Gulf Coast drilling, particularly on and about salt domes. Arps explained that the table shows how an operator intending to spend his entire available capital on these five ventures would stand a 32.8 per cent chance of going broke through "gambler's ruin." He prepared another table (Table 15-6) to show the number of times a certain risk venture needs to be repeated so that the probability of going broke through a series of failures is less than 10 per cent, 5 per cent, or 1 per cent.

On the basis of Table 15-6 Arps pointed out:

. . . if the probability of success is again one in five for each venture, it takes 11 ventures in a row before the chance that all of them will fail is less than 10 per cent. It takes 21 successive ventures to reduce this chance of going broke to less than 1 per cent.

In summarizing decision making in petroleum exploration and production according to statistical theory as it relates to risk and probability analysis, Northern stated:[27]

. . . Some exploitation problems can be so defined. I do not believe, however, that these conditions are met very often in a petroleum exploration setting. Obviously, the time at which evaluation advice would be of the greatest value in exploration operations is right at the start of the venture, before any substantial funds are committed. This also happens to be the time at which knowledge is least perfect. By the time the decision to drill or not is reached, possibly something like 60 to 70 per cent of the total finding cost dollar has already been committed. . . . Whether or not probability analysis in its many forms pays off in an exploration and producing setting cannot be answered with an outright yes or no. . . . [W]here the physical and financial alternatives are reasonably well defined, it has definite utility. I question whether the certain expenditure of time and temper will be rewarded by improved performance.

Thus it would be impractical if not impossible to assign a proper dollar

Table 15-5
Law of Probability Applied to Five Wildcat Tests in Succession

Combination	Probability of occurrence
5 dry holes $(100)(0.80)^5$	32.768%
4 dry holes and 1 hit $(100)(5)(0.20)(0.80)^4$	40.960
3 dry holes and 2 hits $(100)\left(\frac{5 \times 4}{2}\right)(0.20)^2(0.80)^3$	20.480
2 dry holes and 3 hits $(100)\left(\frac{5 \times 4 \times 3}{2 \times 3}\right)(0.20)^3(0.80)^2$	5.120
1 dry hole and 4 hits $(100)\left(\frac{5 \times 4 \times 3 \times 2}{2 \times 3 \times 4}\right)(0.20)^4(0.80)$	0.640
5 hits $(100)(0.20)^5$	0.032
Total	100.00 %

Table 15-6
Number of Times Necessary to Repeat a Venture to Reduce Risk by "Gambler's Ruin"

Probability of success, %	Probability of going broke through a run of "bad luck" 10%	5%	1%
1	229	298	461
2	114	148	228
5	45	58	90
10	22	28	44
20	11	14	21
50	4	5	7

value to an undrilled property on the basis of the laws of chance alone. Nevertheless wildcat and undrilled properties must be evaluated at times for specific purposes. Hardin[28] has described a method of appraising wildcat prospects according to a profit to risk ratio. In general, estimates are made of the prospective productive acreage, sand thickness, recovery per acre, development wells required, development costs, and future net revenue. An estimate is also made of the cost of leasing, drilling a test well, legal and brokerage costs and overhead, which total is the cost of the risk. With a break-even ratio of approximately 25:1, Hardin assumes that a net profit to risk ratio of less than 50:1 is sufficient reason for not drilling the prospect. Under some circumstances where geological and other information is especially favorable, a location with a risk ratio as low as 30:1 might be drilled.

In writing on methods of decision making under uncertainty, Grayson described ways in which decisions are made by oil and gas operators. A method used by Helmerich and Payne was described in detail.[29] The chief differences between their method and Hardin's are that they go into greater detail and refinement by allowing for one development well out of five being dry, estimate future net production rates, and also calculate a discounted future net income. Grayson also described a method used by the Beard Oil Company:[30]

1. Estimate the potential dollar profit if the well is successful.
2. Estimate the probability of a dry hole, and the probability of a producer.
3. Multiply the cost of the dry hole times the estimated number of dry holes that must be drilled, on the average, to secure a producer. Multiply the cost of a producer times one (the success). Add the two investment costs together to get the amount that, *on the average*, would have to be invested, at those odds, to secure a producer.
4. Divide the profit (1) by the potential investment (3) to get a profit-risk ratio.

The company adopted a risk ratio of 3:1 but found that few prospects ever met this requirement and used the method only as a general decision guide. The necessity of having to drill and complete undrilled locations on a property being evaluated might be handled in one of these manners. And in the case of offset wells the alternative of paying compensatory royalty may be considered.

Actually the laws of chance or of "Gamblers Ruin" have very little meaning in wildcat drilling operations. The laws are based on the fact that there is a certain probability in a certain number of chances. That is, there is definitely a chance of winning. This is a matter of past experience, which is to say that one has never failed to find spots on the cards, or on the dice, or heads and tails on the coins. Such is not the case with respect to an oil reservoir. One simply cannot pick out five locations for wildcat wells, or ten locations, or in fact any number of locations and be certain one or more overlies an oil reservoir.

Those properties with undrilled locations, semicommercial well locations, and semiwildcats as contrasted with true wildcats are not as a general rule too difficult to evaluate. It is the untested isolated property that presents the many problems. Basically there is no reason for not making a study along lines of those made by Hardin and by Helmerich and Payne. Nevertheless, what might be thought of as a "fair market value" as described in Chapter 4 is a better approach. That is the value of a property in question is more nearly the *going value* of similar properties in that or similar areas than it is a matter of mathematical computation, particularly when it comes to sales and trades and matters to be decided by the courts. For this same reason properties can be obtained cheaply at times by those who have or think they have better knowledge and have actually evaluated the property as described here.

REFERENCES

1. *History of Petroleum Engineering*, American Petroleum Institute, New York, 1961, pp. 1039-1041.
2. Dorsey Hager, "Valuation of Oil Properties," *Engr. and Mining J.*, May 27, 1916, pp. 930-932.
3. M. L. Requa, "Methods of Valuing Oil Lands," *Trans. AIME*, Vol. 59, 1918, pp. 526-556.
4. Earl Oliver, "Appraisal of Oil Properties," *Trans. AIME*, Vol. 65, 1921, pp. 353-364.
5. Carl H. Beal, "Essential Factors in Valuation of Oil Properties," *Trans. AIME*, Vol. 65, 1921, pp. 344-352.
6. Requa, *op. cit.*, p. 526.
7. E. L. DeGolyer, *Evaluation, First Annual Institute on Oil and Gas Law*, Mathew Bender and Co., New York, 1949, pp. 591-592.
8. L. F. Terry and K. E. Hill, "Valuation of Producing Properties for Loan Purposes," *J. Petr. Tech.*, July 1953, Sec. 1, p. 23.
9. J. J. Arps, "Profitability of Capital Expenditures for Development Drilling and Producing Property Appraisal," *J. Petr. Tech.*, July 1958, pp. 13-20.
10. H. C. Goheen and Henry C. Chen, "New Table and Graphs Help Solve Rate of Return Problems Rapidly," *World Oil*, January 1962, pp. 81-84.
11. Arps, *op. cit.*, pp. 15-17.
12. R. P. Schoemaker, "A Graphical Short-Cut for Rate of Return Determinations," *World Oil*, July 1, 1963, pp. 73-84; Aug. 1, 1963, pp. 69-73; and Sept. 1, 1963, pp. 64-68.
13. Arps, *op. cit.*, p. 14.
14. D. B. Morkill, "Formulas for Mine Valuation," *Mining and Sci. Press*, May 18, 1918, pp. 276-277.

15. J. R. Lyon, "Using Multiple Cutoff Rates for Capital Investment," *J. Petr. Tech.*, July 1975, pp. 822-826.
16. William C. Goodson, "Development and Evaluation of Gas-Condensate Reservoirs," Pt. 8, *Petr. Engr.*, March 1960, pp. B-38–B-50.
17. Charles Robert Smith, "Mechanics of Secondary Oil Recovery," Reinhold Publishing Corp., New York, N. Y., 1966, p. 1.
18. W. W. Wilson and A. J. Pearson, "How to Determine the Market Value of Secondary Recovery Reserves," *J. Petr. Tech.*, August 1962, pp. 829-833.
19. J. J. Simpson, "Financing Enhanced Recovery," *J. Petr. Tech.*, July 1977, pp. 771-775.
20. C. L. Brown, "Bank Financing of Secondary Recovery Projects," *J. Petr. Tech.*, March 1957, pp. 22-24.
21. Sylvain J. Pirson, "Probability Theory Applied to Oil Exploration Ventures," *Petr. Engr.*, February 1941, pp. 27-28; March 1941, pp. 177-182; April 1941, pp. 46-48; and May 1941, pp. 53-56.
22. Edgar W. Owen, "Petroleum Exploration–Gambling Game or Business Venture," *Economics of Petroleum Exploration, Development, and Property Evaluation*, Prentice-Hall, Englewood Cliffs, N. J., 1961.
23. C. Jackson Grayson, "Decisions under Uncertainty," *Symposium on Petroleum Economics and Valuation*, AIME Society of Petroleum Engineers, Dallas, Texas, 1962, pp. 97-101.
24. J. J. Arps and J. L. Arps, "Prudent Risk Taking," *J. Petr. Tech.*, July 1974, pp. 711-716.
25. P. D. Newendorp, "A Method for Treating Dependencies Between Variables in Simulation Risk-Analysis Models," *J. Petr. Tech.*, October 1976, pp. 1145-1150.
26. J. J. Arps, "The Profitability of Exploratory Ventures," *Economics of Petroleum Exploration, Development, and Property Evaluation*, Prentice-Hall, Englewood Cliffs, N. J., 1961, pp. 154-155.
27. Ian G. Northern, "Investment Decisions in Petroleum Exploration and Production." *J. Petr. Tech.*, July 1964, pp. 727-731.
28. George C. Hardin, Jr., "How to Appraise Wildcat Prospects," *World Oil*, Pt. I, March 1959, pp. 119-120; Pt. 2, April 1959, pp. 138-141.
29. C. Jackson Grayson, Jr., *Decisions under Uncertainty–Drilling Decisions by Oil and Gas Operators*, Harvard University, Division of Research, Graduate School of Business Administration, Boston, Mass., 1960, pp. 151-165.
30. *Ibid.*, pp. 165-166.

Additional References

J. J. Arps, "A Strategy for Sealed Bidding," *Oil and Gas Property Evaluation and Reserve Estimates*, Petroleum Transactions Reprint Series No. 3, AIME, 1970 ed., pp. 220-226.

F. Brons and M. W. McGarry, Jr., "Methods for Calculating Profitabilities," *Oil and Gas Property Evaluation and Reserve Estimates*, Petroleum Transactions Reprint Series No. 3, AIME, 1970 ed., pp. 129-139.

F. Brons and M. Silbergh, "The Relation of Earning Power to Other Profitability Criteria," *Oil and Gas Property Evaluation and Reserve Estimates*, Petroleum Transactions Reprint Series No. 3, AIME, 1970 ed., pp. 147-153.

J. M. Campbell, "Optimization of Capital Expenditures in Petroleum Investments," *Oil and Gas Property Evaluation and Reserve Estimates*, Petroleum Transactions Reprint Series No. 3, AIME, 1970 ed., pp. 140-146.

W. M. Campbell and F. J. Schuh, "Risk Analysis: Over-All Chance of Success Related to Number of Ventures," *Oil and Gas Property Evaluation and Reserve Estimates*, Petroleum Transactions Reprint Series No. 3, AIME, 1970 ed., pp. 168-175.

E. C. Capen, R. V. Clapp, and W. W. Phelps, "Growth Rate–A Rate-of-Return Measure of Investment Efficiency," *J. Petr. Tech.*, May 1976, pp. 531-543.

D. V. Carter, "Engineering Appraisals," *J. Petr. Tech.*, February 1967, pp. 193-196.

L. B. Davidson, "Investment Evaluation Under Conditions of Inflation," *J. Petr. Tech.*, October 1975, pp. 1183-1189.

L. B. Davidson and D. O. Cooper, "A Simple Way of Developing a Probability Distribution of Present Value," *Trans. AIME.*, Vol. 261, 1976, Pt. I, pp. 1069-1078.

Joel Dean, "Measuring the Productivity of Capital," *Harvard Business Review*, January-February 1954, pp. 120-130.

C. R. Dodson, "Application of the Petroleum Engineer's Report to Financing," *J. Petr. Tech.*, February 1967, pp. 187-192.

Forrest A. Garb and H. J. Gruy, "Practical Application of Digital Computers to Economic Analysis of Producing Properties," *J. Petr. Tech.*, February 1965, pp. 145-150.

Arthur J. Henry, "Appraisal of Income Acceleration Projects," *J. Petr. Tech.*, April 1972, pp. 393-398.

David B. Hertz, "Risk Analysis in Capital Investment," *Oil and Gas Property Evaluation and Reserve Estimates*, Petroleum Transactions Reprint Series No. 3, AIME, 1970, pp. 176-187. (Reprinted from *Harvard Business Review*, January-February 1964).

Melvin Kaitz, "Percentage Gain or Investment–An Investment Decision Yardstick," *Trans. AIME*, Vol. 240, 1967, Pt. I. pp. 679-687.

D. R. Knop and J. F. Roorda, "Economic Restraints on U. S. Energy Supply and Demand," *J. Petr. Tech.*, July 1975, pp. 803-812.

Arthur W. McCray, *Petroleum Evaluations and Economic Decisions*, Prentice-Hall, Inc., 1975, 448 pp.

Gary A. McDaniel, "How to Quickly Evaluate Exploration Projects," *World Oil*, July 1970, pp. 131-137.

John G. McLean, "How to Evaluate New Capital Investments," *Harvard Business Review*, November-December 1958, pp. 59-69.

"The Money Market: Can It Handle Industry Needs?" *World Oil*, March 1976, pp. 81-82.

Paul D. Newendorp, "Bayesian Analysis–A Method for Updating Risk Estimates," *J. Petr. Tech.*, February 1972, pp. 193-198.

Michael Silbergh and Folkert Brons, "Profitability Analysis–Where Are We Now?" *J. Petr. Tech.*, January 1972, pp. 90-100.

Ezra Soloman, "Return on Investment: The Relation of Book Yield to True Yield," *Oil and Gas Property Evaluation and Reserve Estimates*, Petroleum Transactions Reprint Series, No. 3, AIME, 1970, pp. 154-161.

Joseph S. Szyliowicz and Bard E. O'Neill, "The Energy Crisis and U. S. Foreign Policy," Praeger Publishers, New York, N. Y., 1975, 258 pp.

J. E. Walstrom, T. D. Mueller, and R. C. McFarlane, "Evaluating Uncertainty in Engineering Calculations, *Trans. AIME.*, Vol. 240, 1967, Pt. I, pp. 1595-1603.

Wallace W. Wilson, "Oil and Gas Property Acquisition," *J. Petr. Tech.*, June 1961, pp. 527-530.

L. D. Woody, Jr., and T. D. Capshaw, "Investment Evaluation by Present-Value Profile," *Oil and Gas Property Evaluation and Reserve Estimates*, Petroleum Transactions Reprint Series, No. 3, AIME, 1960 ed., pp. 171-174.

16 THE EVALUATION REPORT

The following five major purposes for making oil property evaluation studies and reports were listed in Chapter 2:

1. Determining security values for loans.
2. Consummating sales, trades, and mergers of properties.
3. Aiding management in formulating policies and operations.
4. Guiding the investor.
5. Complying with government rules and regulations.

Evaluation reports for any of these purposes will vary widely in scope and detail, but basically all those involving producing properties have to start with estimates of future production, gross income, expenses, production and operating taxes, and net income before state and federal income taxes. Some reports may be no more than a summary of less than one page irrespective of the dollar value; others concerning properties of only a few thousand of dollars in value may be voluminous. With that kind of variation there can be no generalizations as to how a report should be prepared; it is all a matter of the purpose of the individual report, that is, for whom and what it is intended, which may be a little different from, yet closely related to, the real purpose of the evaluation.

REPORTS FOR DETERMINING SECURITY VALUES FOR LOANS

Evaluation reports prepared for purposes of justifying or supporting applications for loans from financial institutions often are referred to as applications for bankers' loans. Most of them go to banks. Banks usually cater to short-term loans and insurance companies to long-term loans. The size of the loan is not too important because two or more banks may split a large loan request whereas an insurance company might handle it alone.

At times banks also split large loans to reduce individual risk. And quite often one company with some surplus cash will lend money to another under either a short- or a long-term basis and under many different plans of repayment. At times intercompany loans may call for a choice of repayments in oil, cash, or an option on corporate shares at some fixed price.

The majority of loans made by banks are to small independent producers which very often involve personal negotiations as contrasted to company negotiations. For this reason many in industry at some time in the past swore they would rather go broke than humiliate themselves by disclosing all the bankers wanted to know about their financial affairs. Fortunately times have changed much for the better; those who handle the bank loans know that production operations have gone "scientific" since the early 1930's. Almost all of the larger city banks in and close to the oil-producing states and in New York City now have oil and gas loan departments headed by and staffed with technically trained engineers with first-hand knowledge of the oil production business. All banks are in business for profit. Nearly all the income of a bank comes from interest on loans of surplus monies. For that reason every bank must make every effort to keep the risk factor to a minimum. If banks appear to be miserly and ultraconservative in granting a loan it is because they are serving as custodians of the public's monies. Their business operations are closely watched and regulated by state and federal laws. Thus the loan must be granted on the basis of the integrity of the borrower and his ability to meet the interest payments and to repay the loan when due.

Most bank loans are made on producing properties. They are usually one of two kinds, namely, a direct loan or a production payment loan. In the first case the bank will hold what amounts to a first mortgage on the entire property. The bank will not have a direct mortgage on the property in the second case, hence its investigation and security requirements may be more stringent. A new borrower may be asked to furnish financial data that may be analyzed by the bank to determine his business ability and financial responsibility. He also may have to agree to open a business checking account and to maintain a balance of 10 to 20 per cent of the principal amount of the loan.

In discussing the financing of oil properties Rice[1] has prepared a relative suitability rating for loan collateral which is given in Table 16-1. Only producing proved leaseholds and producing proved royalties are rated as excellent collateral. All unproven property interests are listed as unsuitable collateral.

Some prospective borrowers for the first time may have a property evaluated and a report prepared before talking with a prospective banker. It is generally best to discuss the matter carefully with the banker first and to find just what information will be required in the evaluation report. The

Table 16-1

Relative Suitability of Oil Properties as Loan Collateral[1]

Oil collateral	General rating as sole collateral	Repayment calculated upon and determined by	Instruments of borrowing	Repayment term*
Proved leaseholds, producing	Excellent	Projected monthly net revenues	Deed of trust or chattel mortgage, assignment of runs, term loan agreement	3 to 7 years
Proved leaseholds, nonproducing	Fair, depending on drilling commitments and capabilities	Value obtainable through liquidation	Deed of trust or chattel mortgage, demand note, assignment of runs	On demand
Unproved leaseholds	Unsuitable	...	...	...
Proved royalty, producing	Excellent	Projected monthly income	Deed of trust, assignment of runs, term loan agreement	3 to 7 years
Proved royalty, nonproducing	Fair	Value obtainable through liquidation	Deed of trust, demand note, assignment of runs, term loan agreement	On demand
Proved overriding royalty, producing	Good	Value obtainable through liquidation	Deed of trust, assignment of runs, term loan agreement	2 to 5 years
Proved overriding royalty, nonproducing	Poor, depending on lease commitments	Value obtainable through liquidation	Deed of trust, demand note	On demand
Unproved royalty	Unsuitable	...	...	...
Unproved overriding royalty	Unsuitable	...	...	...
Unproved fee minerals	Fair, if located in good oil provinces	Estimated annual lease rentals	Deed of trust, demand note	On demand
Oil payments	Good, depending on character and responsibility of operator	Projected monthly income	Deed of trust, assignment of payments, term loan agreement	2 to 5 years

* Depends entirely on the nature and quality of collateral and internal policy of lender.

kinds of information as well as the detail vary depending chiefly on the amount and kind of loan desired, payout time, acquaintance with operator, and how well the banker knows the field or perhaps nearby or adjoining operations. As a general rule every evaluation should be made by an independent evaluation engineer who is recommended by or acceptable to the bank. The report had better err in the direction of too much information as long as it is reliable rather than take the request for a brief report too literally.

The Evaluation Report

Most banks will not lend money on any kind of unproven properties. Money needed for wildcatting or proving-up a property may be secured by a mortgage on other proven properties or other acceptable collateral. The banker is usually quite curious as to how the money is to be used. Also the fewer the number of producing wells on a property the smaller the loan value and the more difficult it is to get without other security. Dodson has prepared the following checklist of questions to which the banker must have adequate and convincing answers before approving a loan application:[2]

1. Is the operator or borrower known to the bank and will he stand up under checking as to experience, competence and business ability? What are his history and reputation?
2. Have there been previous loans on these properties, and what has been the experience?
3. Are the properties to support the loan in areas acceptable for bank loans, and do they conform to bank policy in regard to such things as the number of wells, production, sales contracts and legal requirements?
4. Are the financial statements, balance sheets, income and expense, and source and use of funds set up by certified and acceptable auditors? Are they representative and comparable to companies with top credit ratings?
5. What factors are influenced or controlled by outside parties?
6. Is the engineering report prepared by a consultant or engineering firm known and acceptable to the bank?
7. Does the bank have permission to check the properties and findings independently? Were the properties checked by the consultants or engineers preparing the report?
8. Could a new operator for the properties be readily found if necessary?
9. Are there purchase contracts for the oil and gas production?
10. Have the reserve estimates been compared with others, and do they seem reasonable?
11. What are the lease conditions that could affect the loan? Are there any special clauses, requirements or exploration conditions?
12. What are the royalties, overrides, carried interests or net profit interests in the properties? Are there any reversionary rights or options?

13. Is this a joint or unit operation, or is one contemplated?

14. Will the loan pay off if certain properties are failures? Is there a good diversity?

15. Do an analysis and a review of the engineering report include the following or answer the following questions?

a. The reserves of oil, gas, gasoline, and LPG.

b. The estimated fair-market value of the properties and reserves.

c. A realistic forecast by years of production, sales, costs and taxes and net revenue to service debt.

d. Are there arithmetic or other errors?

e. Are the properties on the downgrade (past their prime), or are they recently developed and not seasoned? Are they protected against drainage?

f. Is production prorated or not? Are the wells producing on discovery allowables or special exemptions?

g. Were the wells tested?

h. Are the wells or equipment in proper condition and adequate?

i. What are the characteristics of the reserves? Are they readily marketable? Are there sulphur, salt water or other problems? Can the unfavorable conditions be corrected?

j. How are the properties located in regard to structural position? Are they edge properties, top properties, or what? Is the structural position poor or advantageous?

k. Are there any pluses such as tested and non-producing zones behind pipe or deeper zones? Do the properties have waterflood prospects or other fluid injection programs to increase production rate or reserves?

l. What are the safety factors in the proposed loan relative to reserves and ultimate net revenue? Are they adequate to protect the loan if reserves are found to be substantially below estimates or if future prices drop, proration is more severe, markets are lost and wells are less productive due to corrosion, sand or wax trouble or low capacity?

m. Is the estimated economic limit realistic, and how were the operating costs, taxes and remedial costs estimated? Are they acceptable?

n. What are the producing mechanisms? That is, is it a depletion-type reservoir, is there gravity drainage or gas-cap drive, water drive, or a combination of these drives?

o. What are the well completion techniques? Have they been fractured or acidized? Were the results satisfactory and is there a good history of production to support the forecast?

This list of questions that must be answered to the satisfaction of the bank indicates the scope of a complete evaluation report. The amount of the loan granted depends on the answers as well as a number of factors including the overall economics of the industry, state oil and gas regulations, quality of the production, the bank's policy on so-called loan ratios, etc. The maximum loan usually is limited to that amount which can be repaid within two to five years with reasonable certainty and still have remaining 50 per cent of the estimated original future net revenue.

Most banks with oil property loan departments have checklists for use in the preparation of an evaluation report for loan purposes. Some go into considerable detail, but the subject headings of all such reports are essentially as follows:

1. Summary (one page, preferably).
2. Introduction.
3. Producing property interests.
   a. Location, legal description, kind of ownership, special lease conditions, deferred charges and obligations.
   b. Development.
      (1) Discovery wells, tests, reservoirs, reservoir depths, development wells, spacing or pattern, tests, completion methods, need for recompletions, drill-deeper, plug-backs, etc., other wells and/or dry holes, tests.
4. Description of reservoirs.
   a. General geologic structure and characteristics.
   b. Geology of reservoirs, thickness, extent, characteristics, outstanding features.
   c. Characteristics and properties of all reservoirs fluids.
   d. Kind of drive, indication of any change.
   e. Oil-water and/or oil-gas contacts, and change in.
   f. Undrilled locations, reservoirs, prospects of production.
   g. Probable future development and/or recompletions and costs.
   h. Secondary recovery prospects, if under primary operation.
5. Production history, oil, gas, water, and other fluids.
   a. By wells, reservoirs, leases, and properties.
   b. Reasons for any abandonments.
   c. Reasons for any large past production expenses.
   d. Any anticipated major production costs.
   e. Present and future operating costs and taxes, including transportation or pipelines, treating, etc.
6. Production forecasts.
   a. Estimates of gross and net future recoverable reserves of oil, gas, LPG, natural gasoline, and plant products.
   b. Estimated gross and net yearly production of all oil, gas, LPG, natural gasoline, and plant products.
   c. Effects of present and/or future proration practices.
   d. Recoveries per acre, per acre-foot, per reservoir, and estimated per cent depletion.
   e. Methods used in making estimates.
   f. Probable reliability of estimates.

7. Economics.
   a. Estimated gross income.
   b. Estimated production, development, testing, recompletion, and other property expenses.
   c. Estimated tax obligations.
   d. Estimated net income from property by years and total.
   e. Estimated salvage value upon abandonment.
   f. Estimated secondary recovery value, if any.
8. Miscellaneous information of significance.
9. Evaluation summary in more detail than in item 1 above.
10. Well, structural, isopach, isobaric, and other maps, cross sections, and other pertinent exhibits, preferably in back pocket.

Obviously, any one report may contain far more or far less information than just indicated, and the order of presentation may be different, all as indicated by any suggestions or preferences the lending institution may make.

One suggested form of an evaluation summary (item 9, above) may be prepared as follows:[3]

Evaluation Summary.

A. Evaluation Table.
   1. By years, for minimum of 5 years, set out in first column estimated gross future reserve, lump remainder, and show total–total should be identical with gross future reserve presented in Estimation of Reserve section of report.
   2. Multiply gross future reserve numbers by appropriate ownership interests, as set out in Introduction of report, to arrive at net future reserve.
   3. Multiply column of future net reserve numbers by product price to obtain future net revenues.
   4. Subtract net operating expenses and taxes from net revenues to yield future net operating income.
   5. Subtract out net investment and add in net salvage to obtain future net cash profit from future net operating income.
   6. Provide detailed footnotes explaining arithmetic factors and procedures used in construction of table.

B. Summary Statement. Prepare summary statement of evaluation tabulation pointing out significance and directing attention to any limitations contained therein, particularly those associated with the reserve estimate.

The report will need a one-page letter of transmittal with dates, the purpose of the request, the date the loan is required, its proposed duration, method of repayment, and extent of security being offered. All these matters usually are handled by the consultant who prepared the report.

REPORTS FOR CONSUMMATING SALES, TRADES, AND MERGERS

Reports intended for purposes of making sales and trades in oil properties also may be very short and simple to very long and detailed. Much depends on whether a property is being put on the market (peddled) in hopes of finding a buyer or if there has been some "meeting of the minds" with a prospective purchaser before commencing any detailed study. At times with respect to a sale, but more often in the case of a trade, the negotiations are between companies that are somewhat familiar with each other's property. The transaction may be intended for convenience, that is, what may be unfavorable about one property to its owner may be much more favorable to the other. Very often the idea originates in an effort to eliminate reservoir drainage problems, bypassing of returning gas, and to consolidate properties in a field or an area in an effort to reduce operating and maintenance costs. Thus the evaluations made for these purposes probably would stop with estimates of reserves and net operating income and an inventory and depreciated value of all production facilities. The net reserves involved are the primary interest and, although considerable gathering of data might be involved, nothing more than a memorandum to management with brief supporting information may be necessary instead of a report. Often the effort required to complete these sales and trades is inversely related to the values of the properties. Each party usually makes up his mind ahead of time as to what he would give or take in return and a close evaluation rarely changes matters. The transaction is more a matter of "horse-trading."

Mergers cannot be handled as conveniently. They normally involve volumes of evaluation data and paper work, but the evaluation engineer usually will find himself merely furnishing data to the accounting and legal departments for compilation and presentation. The legal department normally plays the major role.[4] The larger the merger the greater the care in compiling all data and information in order to answer all questions that stockholders and the Department of Justice may ask. Quite often a good deal of the work is done in a preliminary way and the detailed and accurate studies are not started until after the intent of merger has been announced.

REPORTS FOR AIDING MANAGEMENT

Reports intended for management as an aid in formulating policies and operational procedures are to be distinguished from those intended for use in making sales, trades, and mergers. Reports for aiding managment are more in the form of economic studies. The studies may be thought of as engineering evaluations as contrasted with property evaluations, but the basic principles are the same whether they involve one well or a whole property. These studies should carry an investment to its discounted present worth so that it may be compared with alternative procedures.[5] No general rules can be stated for preparing reports of these kinds except that they should be short and pertinent, showing present costs, investment needed and for what, and savings or payout over a specified period of time.

REPORTS FOR GUIDING THE INVESTOR

The evaluation engineer will only contribute data to reports aimed at guiding the investor. These and other data usually go to an investment house that prepares a brochure to be used in raising capital for particular purposes for the company. A good deal of information may have been compiled for management's use and much of this may have been in the form of engineering and operating estimates. The evaluation engineer may be asked to provide estimates of future net incomes from properties, but he seldom prepares any formal report.

REPORTS FOR COMPLYING WITH GOVERNMENTAL RULES AND REGULATIONS

Reports intended for judicial and governmental bodies are best prepared by the legal department or consultants experienced in handling such matters. As in the case of investment brochures, the evaluation engineer will have little to do with compilation of the report other than to contribute the material requested. He must be prepared to testify if necessary as to how it was compiled and its accuracy. The majority of reports in this category have been intended for presentation before the Federal Power Commission in support of reserves of natural gas, availability, and costs of production. Others may be in support of income tax deductions, dissolution of estates and partnerships, and sales of stocks and securities.

REFERENCES

1. C. L. Rice, Jr., "Private Financing of Oil Producing Properties," *J. Petr. Tech.*, January 1949, Sec. I, pp. 24-28.
2. Charles R. Dodson, "The Petroleum Engineer's Function in Oil and Gas Financing," *J. Petr. Tech.*, April 1960, pp. 19-22.

3. John H. Ferry, "Bank Use of Petroleum Appraisal Reports," unpublished thesis, The Stonier Graduate School of Banking, Rutgers University, New Brunswick, N. J., 1964, Appendix 3, p. 169.
4. Richard J. Archer, "Mergers in the Oil Industry," *J. Petr. Tech.*, May 1965, pp. 531-536.
5. C. E. Reistle, Jr., "Applications of Engineering Valuations in Management," *J. Petr. Tech.*, June 1956, pp. 11-12.

Additional References

R. I. Bradley, "How to Present Property Appraisal Data," *World Oil*, January 1953, pp. 68-72.

C. R. Dodson, "Facts the Banker Needs in Making Oil Loans," *World Oil*, Feb. 1, 1958, pp. 32-34.

Greg Ireton, "How to Finance Oil and Gas Production," *Oil and Gas J.*, May 9, 1960, pp. 165-168.

Jerol M. Sonosky, "Engineering the Oil Loan, " *J. Petr. Tech.*, January 1961, pp. 19-23.

J. Ed. Warren, "Considerations Concerning Bank Financing of Oil and Gas Properties," *J. Petr. Tech.*, May 1956, pp. 11-14.

APPENDIX

Table A-1

Present Worth of $1.00 Due t Years Hence at the Following Rates of Interest r Compounded Semiannually $2t - 1$ Times

Years	5%	6%	7%	8%	9%	10%	11%
1	$0.975610	$0.970874	$0.966184	$0.961539	$0.956938	$0.952381	$0.947867
2	.928600	.915142	.901943	.888996	.876297	.863838	.851614
3	.883854	.862609	.841973	.821927	.802451	.783526	.765134
4	.841265	.813092	.785991	.759918	.734828	.710681	.687437
5	.800728	.766417	.733731	.702587	.672904	.644609	.617629
6	.762145	.722421	.684946	.649581	.616199	.584679	.554910
7	.725421	.680951	.639404	.600574	.564272	.530321	.498560
8	.690466	.641862	.596891	.555265	.516720	.481017	.447933
9	.657195	.605016	.557204	.513373	.473177	.436296	.402446
10	.625528	.570286	.520156	.474643	.433302	.395734	.361579
11	.595386	.537549	.485571	.438834	.396788	.358942	.324861
12	.566607	.506692	.453286	.405726	.363350	.325571	.291872
13	.539391	.477606	.423147	.375117	.332731	.295303	.262234
14	.513400	.450189	.395013	.346817	.304691	.267848	.235604
15	.488661	.424346	.368748	.320652	.279015	.242946	.211679
16	.465115	.399987	.344231	.296460	.255502	.220359	.190184
17	.442703	.377026	.321343	.274094	.233971	.199872	.170871
18	.421371	.355383	.299977	.253416	.214254	.181290	.153519
19	.401067	.334983	.280032	.234297	.196199	.164435	.137929
20	.381742	.315754	.261413	.216621	.179665	.149148	.123923

Years	12%	13%	14%	15%	16%	18%	20%	25%
1	$0.943396	$0.938967	$0.934579	$0.930234	$0.925926	$0.917431	$0.909091	$0.888889
2	.839619	.827849	.816298	.804961	.793832	.772184	.751315	.702332
3	.747258	.729881	.712986	.696558	.680583	.649931	.620921	.554929
4	.665057	.643506	.622750	.602755	.583490	.547034	.513158	.438462
5	.591898	.567353	.543933	.521583	.500249	.460428	.424098	.346439
6	.526788	.500212	.475093	.451343	.428883	.387533	.350494	.273730
7	.468839	.441017	.414964	.390562	.367698	.326179	.289664	.216280
8	.417265	.388827	.362446	.337966	.315242	.274538	.239392	.170888
9	.371364	.342813	.316574	.292453	.270269	.231073	.197845	.135023
10	.330513	.302244	.276508	.253069	.231712	.194490	.163508	.106685
11	.294155	.266476	.241513	.218989	.198656	.163698	.135131	.084294
12	.261797	.234941	.210947	.189498	.170315	.137781	.111678	.066603
13	.232999	.207138	.184249	.163979	.146018	.115968	.092296	.052662
14	.207368	.182625	.160930	.141896	.125187	.097608	.076278	.041580
15	.184557	.161013	.140563	.122787	.107327	.082155	.063039	.032853
16	.164255	.141959	.122773	.106252	.092016	.069148	.052099	.025958
17	.146186	.125159	.107235	.091943	.078889	.058200	.043057	.020510
18	.130105	.110348	.093663	.079562	.067634	.048986	.035584	.016205
19	.115793	.097289	.081809	.068847	.057986	.041231	.029408	.012804
20	.103056	.085776	.071455	.059576	.049713	.034703	.024304	.010117

Table A-2

Present Value of $1.00 Due t Years Hence at the Following Rates of Interest r Compounded Semiannually

Years	5 %	6 %	7 %	8 %	9 %	10 %	11 %
1	$0.951814	$0.942596	$0.933511	$0.924556	$0.915730	$0.907029	$0.898452
2	.905951	.888487	.871442	.854804	.838561	.822702	.807217
3	.862297	.837484	.813501	.790315	.767896	.746215	.725246
4	.820747	.789409	.759412	.730690	.703185	.676839	.651599
5	.781199	.744094	.708919	.675564	.643928	.613913	.585430
6	.743556	.701380	.661783	.624597	.589664	.556837	.525981
7	.707727	.661118	.617782	.577475	.539973	.505068	.472569
8	.673625	.623167	.576706	.533908	.494469	.458111	.424581
9	.641166	.587395	.538361	.493628	.452800	.415520	.381466
10	.610271	.553676	.502566	.456387	.414643	.376889	.342729
11	.580865	.521893	.469151	.421955	.379701	.341850	.307925
12	.552876	.491934	.437957	.390122	.347703	.310068	.276656
13	.526235	.463695	.408838	.360689	.318402	.281240	.248563
14	.500878	.437077	.381655	.333478	.291571	.255093	.223322
15	.476743	.411987	.356279	.308319	.267000	.231377	.200644
16	.453771	.388337	.332590	.285058	.244500	.209866	.180269
17	.431906	.366045	.310476	.263552	.223896	.190355	.161963
18	.411094	.345032	.289833	.243669	.205028	.172657	.145516
19	.391285	.325226	.270562	.225286	.187750	.156605	.130739
20	.372431	.306557	.252573	.208289	.171929	.142045	.117463

Years	12 %	13 %	14 %	15 %	16 %	18 %	20 %	25 %
1	$0.889997	$0.881659	$0.873439	$0.865333	$0.857339	$0.841680	$0.826446	$0.790123
2	.792094	.777323	.762895	.748800	.735030	.708425	.683013	.624295
3	.704961	.685334	.666342	.647961	.630169	.596267	.564474	.493270
4	.627412	.604231	.582009	.560702	.540269	.501866	.466507	.389744
5	.558395	.532726	.508349	.485194	.463193	.422411	.385543	.307946
6	.496969	.469683	.444012	.419854	.397114	.355535	.318631	.243315
7	.442301	.414100	.387817	.363313	.340461	.299246	.263331	.192249
8	.393646	.365095	.338734	.314387	.291890	.251870	.217629	.151901
9	.350344	.321890	.295864	.272049	.250249	.211994	.179859	.120020
10	.311805	.283797	.258419	.235413	.214548	.178431	.148644	.094831
11	.277505	.250212	.225713	.203711	.183940	.150182	.122846	.074928
12	.246979	.220602	.197146	.176277	.157699	.126405	.101526	.059202
13	.219810	.194496	.172195	.152539	.135202	.106393	.083905	.046777
14	.195630	.171479	.150402	.131997	.115914	.089548	.069343	.036960
15	.174110	.151186	.131367	.114221	.099377	.075371	.057309	.029203
16	.154957	.133295	.114741	.098839	.085200	.063438	.047362	.023074
17	.137912	.117520	.100219	.085529	.073045	.053395	.039142	.018231
18	.122741	.103613	.087535	.074011	.062644	.044941	.032349	.014405
19	.109239	.091351	.076457	.064044	.053690	.037826	.026735	.011382
20	.097222	.080541	.066780	.055419	.046031	.031838	.022095	.008993

Table A-3

Present Value of $1.00 Due t Years Hence at the Following Rates of Interest r Compounded Quarterly

Years	5%	6%	7%	8%	9%	10%	11%
1	$0.951524	$0.942184	$0.932958	$0.923846	$0.914843	$0.905951	$0.897166
2	.905399	.887711	.870411	.853491	.836938	.820747	.804906
3	.861509	.836388	.812057	.788493	.765668	.743556	.722134
4	.819747	.788031	.757616	.728446	.700466	.673625	.647874
5	.780009	.742471	.706824	.672972	.640817	.610271	.581250
6	.742197	.699544	.659437	.621722	.586247	.552876	.521478
7	.706219	.659100	.615227	.574375	.536324	.500878	.467852
8	.671985	.620993	.573981	.530634	.490653	.453771	.419741
9	.639410	.585090	.535501	.490224	.448870	.411094	.376577
10	.608414	.551263	.499600	.452891	.410646	.372431	.337852
11	.578921	.519391	.466106	.418401	.375677	.337404	.303109
12	.550857	.489362	.434857	.386538	.343685	.305671	.271939
13	.524154	.461069	.405704	.357102	.314418	.276923	.243974
14	.498745	.434412	.378505	.329907	.287643	.250879	.218885
15	.474568	.409296	.353129	.304783	.263149	.227284	.196376
16	.451563	.385633	.329455	.281572	.240740	.205908	.176182
17	.429673	.363337	.307367	.260129	.220239	.186542	.158065
18	.408845	.342330	.286761	.240319	.201484	.168998	.141810
19	.389026	.322538	.267536	.222018	.184327	.153104	.127227
20	.370167	.303890	.249600	.205110	.168630	.138705	.114144

Years	12%	13%	14%	15%	16%	18%	20%	25%
1	$0.888487	$0.879913	$0.871442	$0.863073	$0.854804	$0.838561	$0.822702	$0.784665
2	.789409	.774247	.759412	.744895	.730690	.703185	.676839	.615699
3	.701380	.681270	.661783	.642899	.624597	.589664	.556837	.483118
4	.623167	.599458	.576706	.554869	.533908	.494469	.458111	.379085
5	.553676	.527471	.502566	.478892	.456387	.414643	.376889	.297455
6	.491934	.464128	.437957	.413319	.390122	.347703	.310068	.233403
7	.437077	.408392	.381655	.356724	.333478	.291571	.255093	.183143
8	.388337	.359350	.332590	.307879	.285058	.244500	.209866	.143706
9	.345032	.316196	.289833	.265722	.243669	.205028	.172657	.112761
10	.306557	.278225	.252573	.229338	.208289	.171929	.142045	.088480
11	.272372	.244814	.220103	.197935	.178046	.144173	.116861	.069427
12	.241999	.215415	.191807	.170832	.152195	.120898	.096142	.054477
13	.215013	.189546	.167148	.147441	.130097	.101380	.079096	.042746
14	.191036	.166784	.145660	.127252	.111207	.085013	.065073	.033541
15	.169733	.146756	.126935	.109828	.095060	.071289	.053535	.026319
16	.150806	.129132	.110616	.094790	.081258	.059780	.044044	.020651
17	.133989	.113625	.096396	.081810	.069460	.050129	.036235	.016204
18	.119047	.099980	.084003	.070608	.059375	.042037	.029811	.012715
19	.105772	.087974	.073204	.060940	.050754	.035250	.024525	.009977
20	.093977	.077409	.063793	.052596	.043384	.029559	.020177	.007829

Table A-4

Present Value of $1.00 Due t Years Hence at the Following Rates of Interest r Compounded Monthly

Years	5 %	6 %	7 %	8 %	9 %	10 %	11 %
1	$0.951328	$0.941905	$0.932583	$0.923362	$0.914238	$0.905213	$0.896203
2	.905026	.887184	.869712	.852597	.835832	.819410	.803324
3	.860977	.835643	.811079	.787256	.764149	.741740	.720006
4	.819071	.787096	.756398	.726922	.698615	.671432	.645329
5	.779206	.741370	.705404	.671212	.638700	.607789	.578397
6	.741281	.698299	.657848	.619722	.583924	.550178	.518408
7	.705201	.657732	.613498	.572274	.533846	.498028	.464640
8	.670878	.619520	.572138	.528416	.488062	.450821	.416449
9	.638225	.583529	.533567	.487919	.446205	.408089	.373257
10	.607162	.549629	.497595	.450526	.407938	.369407	.334544
11	.577610	.517698	.464049	.415998	.372953	.334392	.299816
12	.549497	.487622	.432765	.384117	.340968	.302696	.268747
13	.522752	.459294	.403589	.354679	.311726	.274004	.240873
14	.497309	.432611	.376380	.327497	.284991	.248032	.215891
15	.473104	.407478	.351006	.302398	.260550	.224521	.193499
16	.450077	.383806	.327342	.279223	.238205	.203240	.173430
17	.428171	.361508	.305274	.257824	.217776	.183975	.155442
18	.407331	.340506	.284693	.238065	.199099	.166537	.139320
19	.387506	.320724	.265500	.219820	.182024	.150751	.124871
20	.368645	.302092	.247601	.202973	.166413	.136462	.111919

Years	12 %	13 %	14 %	15 %	16 %	18 %	20 %	25 %
1	$0.887450	$0.878709	$0.870063	$0.861509	$0.853046	$0.836388	$0.820083	$0.780804
2	.787567	.772130	.757010	.742197	.727687	.699544	.672535	.609654
3	.698926	.678478	.658647	.639410	.620750	.585090	.551534	.476020
4	.620261	.596185	.573065	.550857	.529528	.489362	.452304	.371678
5	.550451	.523873	.498603	.474568	.451711	.409296	.370926	.290208
6	.488497	.460332	.433816	.408845	.385330	.342330	.304190	.226595
7	.433517	.404498	.377447	.352223	.328704	.286321	.249461	.176926
8	.384724	.355436	.328403	.303443	.280400	.239475	.204579	.138145
9	.341423	.312325	.285731	.261419	.239194	.200294	.167771	.107864
10	.302996	.274443	.248605	.225215	.204043	.167523	.137586	.084221
11	.268894	.241156	.216302	.194025	.174058	.140115	.112832	.065760
12	.238630	.211906	.188196	.167154	.148479	.117190	.092532	.051345
13	.211772	.186203	.163743	.144005	.126660	.098016	.075884	.040091
14	.187937	.163619	.142466	.124061	.108047	.081980	.062231	.031303
15	.166784	.143773	.123955	.106880	.092169	.068567	.051034	.024441
16	.148013	.126335	.107849	.092078	.078624	.057348	.041852	.019084
17	.131354	.111012	.093835	.079326	.067070	.047965	.034322	.014901
18	.116570	.097547	.081642	.068340	.057214	.040112	.028147	.011635
19	.103450	.085715	.071034	.058876	.048806	.033554	.023083	.009084
20	.091807	.075319	.061804	.050722	.041634	.028064	.018930	.007093

Table A-5

Present Worth of $1.00 Due t Years Hence at the Following Rates of Interest r Compounded Continuously (∞ Times per Year)

Years	5%	6%	7%	8%	9%	10%	11%	12%
1	$0.951229	$0.941764	$0.932394	$0.923116	$0.913931	$0.904837	$0.895834	$0.886920
2	.904837	.886920	.869358	.852144	.835270	.818731	.802519	.786628
3	.860708	.835270	.810584	.786628	.763379	.740818	.718924	.697676
4	.818731	.786628	.755784	.726149	.697676	.670320	.644036	.618783
5	.778801	.740818	.704688	.670320	.637628	.606531	.576950	.548812
6	.740818	.697676	.657047	.618783	.582748	.548812	.516851	.486752
7	.704688	.657047	.612624	.571209	.532592	.496585	.463013	.431711
8	.670320	.618783	.571209	.527292	.486752	.449329	.414783	.382893
9	.637628	.582748	.532592	.486752	.444858	.406570	.371577	.339586
10	.606531	.548812	.496585	.449329	.406570	.367879	.332871	.301194
11	.576950	.516851	.463013	.414783	.371577	.332871	.298197	.267135
12	.548812	.486752	.431711	.382893	.339596	.301194	.267135	.236928
13	.522046	.458406	.402524	.353455	.310367	.272532	.239309	.210136
14	.496585	.431711	.375311	.326280	.283654	.246597	.214381	.186374
15	.472367	.406570	.349938	.301194	.259240	.223130	.192050	.165299
16	.449329	.382893	.326280	.278037	.236928	.201897	.172045	.146607
17	.427415	.360595	.304221	.256661	.216536	.182684	.154124	.130029
18	.406570	.339596	.283654	.236928	.197899	.165299	.138069	.115325
19	.386741	.319819	.264477	.218712	.180866	.149569	.123687	.102284
20	.367879	.301194	.246597	.201897	.165299	.135335	.110803	.090718

Years	13%	14%	15%	16%	18%	20%	25%
1	$0.878095	$0.869358	$0.860708	$0.852144	$0.835270	$0.818731	$0.778801
2	.771052	.755784	.740818	.726149	.697676	.670320	.606531
3	.677057	.657047	.637628	.618783	.582748	.548812	.472367
4	.594521	.571209	.548812	.527292	.486752	.449329	.367879
5	.522046	.496585	.472367	.449329	.406570	.367879	.286505
6	.458406	.431711	.406570	.382893	.339596	.301194	.223130
7	.402524	.375311	.349938	.326280	.283654	.246597	.173774
8	.353455	.326280	.301194	.278037	.236928	.201897	.135335
9	.310367	.283654	.259240	.236928	.197899	.165299	.105399
10	.272532	.246597	.223130	.201897	.165299	.135335	.090718
11	.239309	.214381	.192050	.172045	.138069	.110803	.063928
12	.210136	.186374	.165299	.146607	.115325	.090718	.049787
13	.184520	.162026	.142274	.124930	.096328	.074274	.038774
14	.162026	.140858	.124456	.106458	.080460	.060810	.030197
15	.142274	.122456	.105399	.090718	.067206	.049787	.023518
16	.124930	.106458	.090718	.077305	.056135	.040762	.018316
17	.109701	.092551	.078082	.065875	.046888	.033373	.014264
18	.096328	.080460	.067206	.056135	.039164	.027324	.011109
19	.084585	.069948	.057844	.047835	.032712	.022371	.008652
20	.074274	.060810	.049787	.040762	.027324	.018316	.006738

SUBJECT INDEX